Chemistry and Light

Chemistry and Light

Paul Suppan
Professor of Physical Chemistry, University of Fribourg, Switzerland

A catalogue record for this book is available from the British Library.

ISBN 0-85186-814-2

Published by The Royal Society of Chemistry,
Thomas Graham House, The Science Park, Cambridge CB4 4WF

Typeset by Keytec Typesetting Ltd.
Printed and bound Hartnolls Ltd., Bodmin

To Professor Lord George Porter, OM, FRS with gratitude for his teaching and scientific collaboration.

Foreword

It is a pleasure to introduce and commend this volume about photochemistry by my one-time student, my colleague, and friend, Paul Suppan. We shared for many years, first at the University of Sheffield and then at the Royal Institution in London, our fascination with chemical reactions induced by light—electronic flash lamps in the early days, lasers later. This account of photochemistry continues the enthusiasm and talent for clear exposition that Dr Suppan has always shown. It is a notable addition to the literature of photochemistry and will be welcomed by students and research workers alike.

George Porter

Preface

The aim of this book is to provide a concise introductory overview of the various light-induced processes in physics, chemistry, biology, as well as, in medicine and industry. It is largely based on the course of photophysics and photochemistry given at the University of Fribourg, and every effort has been made to keep it up to date with the latest developments in this field, in particular the probing of the fastest light-induced reactions on picosecond and femtosecond time scales.

The general level of the book is suitable for final year undergraduate students of chemistry as well as for post-graduate students. In addition, it should be a useful source of information for professional scientists and teachers who may find it valuable for the preparation of courses and seminars devoted to 'photochemistry' in the broadest sense. The mathematics have been kept deliberately simple, without any lengthy derivation of important equations for which a qualitative interpretation is sufficient for our purpose. The suggestions for 'Further Reading' provide references to source books and original publications for readers who wish to go into more details in these fields.

Paul Suppan

Contents

CHAPTER 1

Introduction

The chemical effects of light play a most important role in our life, although we may often not be aware of it. The photosynthesis of green plants is the basis of our whole food chain, through the combination of water and carbon dioxide to form organic matter; in this way the heat released in the combustion of wood for example is of photochemical origin. The same is true of the fossil fuels coal, oil and natural gas; these were formed long ago by the decay of organic matter and therefore represent a storage of the energy of sunlight through a photochemical process.

There are other energies which depend indirectly on sunlight, but these do not involve photo*chemical* reactions: hydroelectric power, wind power, and the thermal energy of oceans depend on the degradation of the energy of light into heat. This is a photo*physical* process, as it implies no chemical change of the light-absorbing matter. The distinction between photophysical and photochemical processes will be discussed further on, as it is more than just a matter of semantics.

So far we have considered the importance of light-driven processes for the supply of energy in more or less indirect ways. The direct conversion of the energy of light into electricity or hydrogen gas does not exist in nature, but it is the aim of current research which will be described in section 6.6.

1.1 LIGHT-INDUCED PROCESSES IN EVERYDAY LIFE

In biology, *vision* is probably the most important photochemical process after photosynthesis. The basic photochemistry of vision is now well understood and is considered in section 5.3 but there are other biological effects of sunlight which remain rather mysterious: *phototropism* is the orientation of plants towards the direction of sunlight, best known in the case of sunflowers; *photomorphogenesis* is the control of the growth of plants by the intensity of light, to give only a few examples.

In these processes light is used to produce a chemical change which acts as a trigger for some complex enzyme-catalysed reaction. The primary photochemical process may be relatively simple, but the following 'dark' reactions are often quite complex. These secondary (dark) reactions are not considered in this book, since they are part of biochemistry rather than photochemistry.

In the processes of vision, phototropism, photomorphogenesis, *etc.*, light acts only as a trigger and there is no permanent change either in the energy or in the composition of the chemical system. There is however one photobiological reaction in which the natural synthesis of an essential compound relies on a key photochemical step, and this is the biosynthesis of vitamin D considered in section 5.6. This leads to an important question concerning the use of photoinduced reactions in industrial synthesis. There are indeed some synthetic applications of photochemistry on an industrial scale, but these are quite small compared with the many dark reactions used in large-scale industrial synthesis. The major problem is the high cost of light as a source of energy (see section 6.3). This cost problem restricts the industrial applications of photochemical synthesis to chain reactions (*e.g.* the chlorination of polymers like PVC) and the production of high-price chemicals such as pharmaceutical and cosmetic products; here the industrial syntheses of vitamin D and of 'rose oxide' provide examples of useful synthetic applications on small industrial scales.

1.1.1 Photodegradation Processes: The 'Negative' Actions of Light

So far we have been concerned with what could be called the 'positive' or useful aspects of photoinduced processes: light as a source of energy, light as the energy needed to build organic matter, light as information in visual and other biological processes, and light as a reactant in chemical synthesis. All these can be described as 'positive' within the needs and wishes of human beings. The point of view of an insect about to be killed by an ultraviolet lamp may of course be quite different, but since this book is aimed at a human readership the meanings of 'positive' (beneficial) or 'negative' (detrimental) effects of light must be seen in this context.

In photobiology these negative actions start with the processes which modify or destroy the essential molecules of life, specifically the nucleic acids and the proteins. The former carry the genetic code in a sequence of nucleotides and even a minor modification of the nucleotide sequence can have profound consequences. By the laws of chance alone these modifications and their consequences are almost always detrimental and result usually in the death of the organism (*e.g.* microbes) exposed to short-wavelength ultraviolet radiation.

In industrial applications the photodegradation of synthetic polymers as well as of dyes and pigments, of agrochemicals, *etc.*, are major causes for concern. The mechanisms of these photochemical reactions are now relatively well understood and various protective measures can be used to delay (but never to stop totally) the complex processes of photodegradation (see section 6.2). Still, even here some photodegradation reactions can be put to good use. The design of photodegradable polymers provides one example, for plastic bags and containers made of such materials will decay to powder under the action of sunlight, instead of building up virtually stable refuse in the environment.

Many of these degradation processes involve photo-oxidations and these cannot be avoided since molecular oxygen is an essential part of our surroundings. Here also the definitions of 'positive' and 'negative' actions of light must depend on one's point of view: in the phototherapy of cancer the malignant cells are destroyed by photo-oxidation, so from the point of view of these cells the action of light is surely highly negative, but from the point of view of the organism which will be cured of the cancer the effect is altogether positive.

1.1.2 Imaging Processes

Some of the major industrial applications of photochemistry are found in the various 'imaging' processes which include *photography, photopolymerization/photodepolymerization, photochromism* and *electrophotography*—the process used in photocopying machines.

Photopolymerization is not restricted to imaging processes. However, such processes form the most spectacular applications of photochemical reactions, as they are used to make printed circuits and integrated circuits for the electronics industry.

1.2 GENERAL FEATURES OF PHOTOCHEMICAL AND PHOTOPHYSICAL PROCESSES

The distinction between 'photophysical' and 'photochemical' processes depends on the definition adopted for 'chemical species' and 'chemical change'. It is often held that a chemical change must involve the breaking or making of chemical bonds; in that case it may be stated that the addition or the removal of an electron from a molecule would not be a 'chemical' change, so that the positive or negative ions would not be distinct chemical species but only 'states' of the neutral molecule. According to such a definition, the stereoisomers of a molecule would not be different 'chemical' species, but simply 'states' of the same molecule, and every chemist would recognize that this would be an absurd definition.

The concept of chemical species in terms of bonding patterns presents many problems, for even excited states of molecules differ in this respect from the ground state. To take one example, consider the valence bond structures of butadiene in its ground state (S_0) and first (singlet) excited state (S_1; Figure 1.1). The double bonds between carbon atoms C_1 and C_2,

Figure 1.1 *Valence bond structures of butadiene in the ground state S_0 and the first excited state S_1*

and C_3 and C_4, are reduced to single bonds, while the single bond between C_2 and C_3 becomes a double bond. According to such a definition, electronic excitation should be considered as a 'chemical' change.

In order to avoid such ambiguities, the definition of 'chemical species' will depend on the simple concept of stability. In the absence of chemical reactions, a chemical species will last indefinitely. Thus an ion is a distinct chemical species, and an electron transfer reaction must be seen as a chemical change. However, an electronic excited state of an atom or molecule must inevitably decay back to the ground state, so the processes of excitation, emission and non-radiative deactivation are photophysical processes.

1.2.1 The Pathways of 'Dark' Reactions and Photochemical Reactions

In the Arrhenius–Eyring model of a chemical reaction which takes place without the intervention of light, the reactant(s) R go over to the product(s) P through a transition state (X) which determines the activation barrier E_a in the rate constant equation

$$k = A \exp(E_a/RT) \tag{1.1}$$

When the molecule is excited by light to reach one of its electronically excited states (M* in Figure 1.2) it may undergo some other chemical reaction, leading to some high-energy product P′ through an activation barrier E_a^*. Both in the ground state (dark) and excited state (photoinduced) reactions the activation barriers (E_a and E_a^* respectively) must be overcome

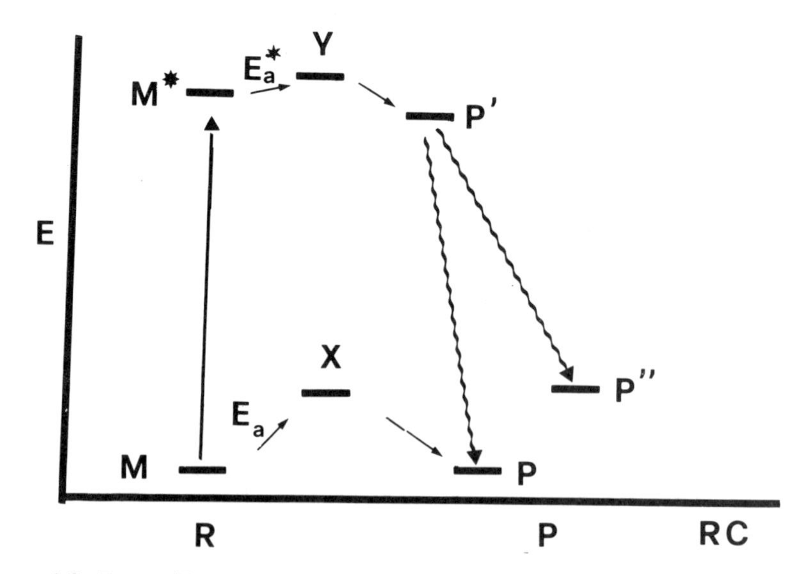

Figure 1.2 *Energy (E)–reaction coordinate (RC) diagram of ground state (dark) and light-induced chemical reactions*

from the thermal energy of the chemical system. For this reason, it may be somewhat misleading to define ground state reactions as 'thermal' and excited state reactions as 'light-induced'. The photochemical reaction is in fact the *thermal* reaction of the electronic excited state M* of the molecule M, while the 'dark' reaction of M is the thermal reaction of the ground state.

1.2.2 The Mistaken Concept of 'Catalysis' by Light

In a photochemical reaction light always acts as a reactant, never as a catalyst. By definition, a catalyst in a chemical reaction must be recovered unchanged, and can in principle be re-used again and again. In a photochemical reaction, light is absorbed and its energy is used to make electronically excited molecules; it will not come out and cannot be used after the reaction, so by definition light is not a catalyst and the concept of 'light catalysed' reaction is fundamentally incorrect.

The idea that light can act as a catalyst is however still rather widely held, and it comes from a consideration of the energy–reaction coordinate diagram shown in Figure 1.2. In a ground state (dark) reaction the role of a true catalyst is to lower the activation barrier E_a; it cannot change the energy levels of the reactants or products, or the free energy change of the overall reaction. Now, when light is absorbed by the reactant, the energy of the excited state (M*) is almost always largely in excess of the activation barrier E_a of the ground state reaction; so it may seem quite likely that light would have supplied the energy required to reach the transition state of the ground state reaction (the decay from M* to X providing the thermal energy). Actually this is practically never the case, as we shall see through many examples in this book. The photochemical reaction takes place on the high-energy excited state potential surface, *e.g.* $M^* \rightarrow P'$ and leads in many cases to some high-energy products such as free radicals or radical ions. These may eventually react to form the final, stable products through dark reactions. In some cases these may resemble the products of the ground state (dark) reaction, but this similarity is purely coincidental.

The non-radiative decay of the excited state (M*) to the *transition state* (X) of the ground state reaction is so unlikely that it can be altogether forgotten. Even if there was a single example of such a process, it could not be described as 'catalysis' by light. The expressions 'photoinduced' or 'photoactivated' reactions are accurate; they do not imply that light acts as a catalyst, but rather as a reactant which is consumed in the chemical process.

1.2.3 The Range of Photochemical Reactions: Vibrational Photochemistry and Radiation Chemistry

The definition of a 'photochemical' reaction depends on the definition of 'light'. Indeed, a photochemical reaction is a chemical reaction induced by light, a reaction in which the energy of light is used to promote molecules from their ground state to excited states. The question then arises as to the nature of these excited states, because electromagnetic radiation covers a virtually continuous spectrum of wavelengths extending from infinity to zero. The details of these excitation processes will be considered in chapter 3. For the moment, a summary of the properties of energy states of molecules will be sufficient. Figure 1.3 gives a simple picture of these various energy states, related to the wavelengths of electromagnetic radiation which can be used to bring molecules to different energy states.

The translational motions of molecules represent the lowest levels of 'excitation'. The energy of a molecule of mass m is then

$$E = (1/2)\, mv^2 \qquad (1.2)$$

and these energy levels form a continuum; that is to say that this energy varies nearly continuously, unlike the other forms of energy which vary in discrete steps.

Thermal energy, the energy contained in the heat of a chemical substance, is essentially the translational energy of molecules. There are two forms of kinetic energy which play an important role in photophysics: the rotational and vibrational energies of molecules.

The rotational energies represent the spinning motions of a molecule, when the entire molecule rotates around one of its inertial axes. This should not be confused with *internal rotation* which is the rotational motion of one part of a molecule with respect to some other part of the same molecule.

In most molecules of interest in photochemistry, these 'external' rotational levels are quite closely spaced and many of these states are populated even at room temperature. From Figure 1.3 it is clear that the energies of rotational states are usually below the level of the average thermal energy kT, and they can be reached by the absorption of 'light' (or rather, electromagnetic radiation) of very long wavelength, corresponding to the microwave region of the spectrum.

At higher energies the molecules acquire vibrational motion, the bonds between the atoms behaving like springs connecting mass centres. The spacing of vibrational energy levels is usually quite large compared with the average thermal energy kT and only the lowest ones are populated at room temperature. These vibrational levels play, however, a most important role in the photophysical processes of large molecules, and these will be discussed in Chapter 3.

At much higher energies the *electronic* excited states of molecules are reached; these correspond to 'light' in the usual sense, in the visible (VIS), near infrared (NIR) and near ultraviolet (NUV) regions of the spectrum of electromagnetic radiation. In these electronic excited states the atomic structure of the molecule remains unchanged but one or several electrons are promoted to 'orbitals' of higher energy. These electronic excited states are at the basis of photochemical reactions.

Going further up the energy scale the molecule will reach its ionization limit where the impact of electromagnetic radiation is so strong that an

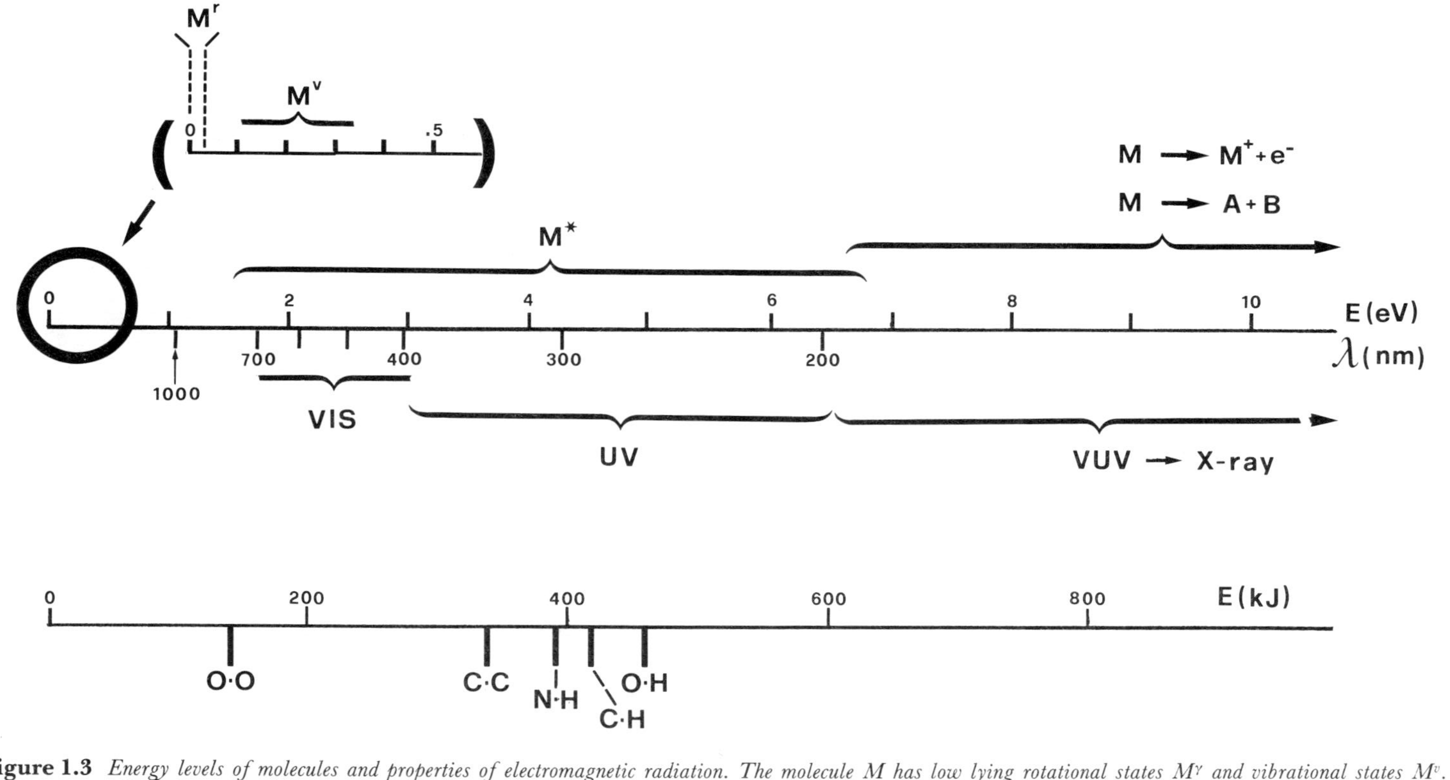

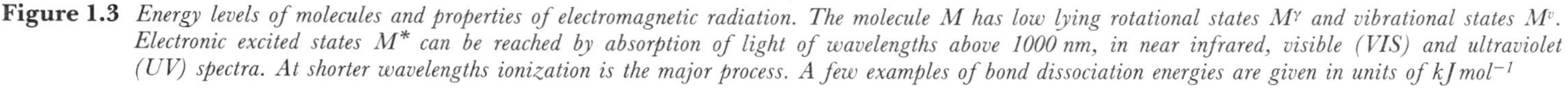
Figure 1.3 *Energy levels of molecules and properties of electromagnetic radiation. The molecule M has low lying rotational states M^r and vibrational states M^v. Electronic excited states M^* can be reached by absorption of light of wavelengths above 1000 nm, in near infrared, visible (VIS) and ultraviolet (UV) spectra. At shorter wavelengths ionization is the major process. A few examples of bond dissociation energies are given in units of kJ mol^{-1}*

electron is ejected and the neutral molecule ceases to exist; it becomes a positive ion separated from its departing electron

$$M \xrightarrow{h\nu} M^+ + e^- \tag{1.3}$$

Electromagnetic radiation of such high energy falls within the 'vacuum ultra-violet' (VUV), *X*-ray and γ-ray regions of the spectrum. They are called 'ionizing' radiations and their effects on matter are part of the science of *radiation chemistry*.

It has often been stated that photochemistry is the chemistry of electronically excited molecules. According to this definition 'light' is the NIR/VIS/NUV part of the spectrum of electromagnetic radiation, in the range of wavelengths covering about 100 to 1000 nm. Within this range molecules are promoted to electronic excited states. There remains however an area which can be considered as part of photochemistry, although no 'electronically' excited states are involved: this is the process of infrared multiphoton absorption which can result in 'vibrational' photochemistry. This will be discussed in section 8.6.

With this one exception of vibrational photochemistry through multiphoton infrared light absorption, photochemistry is restricted to the chemical reactions of electronic excited states of molecules. Radiation chemistry is outside the scope of this book, so a very short section is devoted to it to conclude this introduction.

1.2.4 Chemical Effects of Ionizing Radiations

Ionizing radiation includes fast electrons, protons and neutrons as well as photons in the frequency range of hard *X*-rays and γ-rays. These particles often have kinetic energies of the order of millions of electron-volts (MeV = 10^6 eV), while the energies of electronically excited states of molecules span the range of 2 to 4 eV. When a particle of ionizing radiation passes through matter, its energy is therefore dispersed through a wide region along its path (Figure 1.4). Electrons ejected from molecules can still

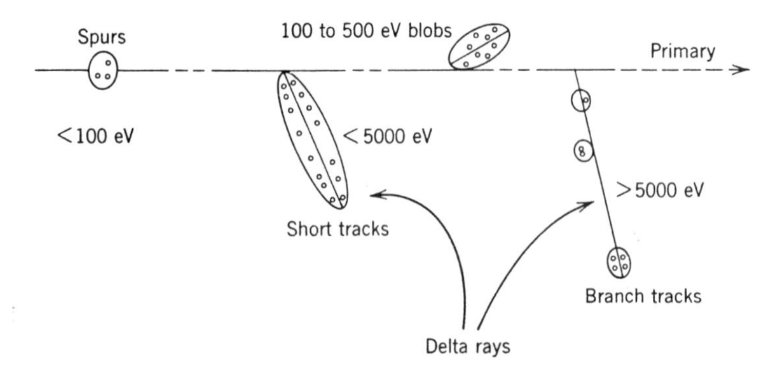

Figure 1.4 *Energy decay of a high energy particle of ionizing radiation. Ionization along the path of travel produces secondary electrons that lead to further ionization*

have sufficient energy to produce secondary ionizations through further collisions with neighbouring molecules. Both positive and negative ions are formed in the track of the particle as well as in the 'spurs', but the negative ions are solvated electrons, not free electrons as they would be in a gas.

The structure of a solvated electron depends on the solvent, and it is often difficult to describe accurately. The existence of the solvated electron as a distinct chemical species is however ascertained by its absorption spectrum; this is a broad, structureless spectrum which covers the far VIS and NIR regions (Figure 1.5).

The major chemical processes in radiation chemistry are reduction and oxidation reactions, according to the following examples. In the gas phase, ionization predominates

$$H_2 \rightsquigarrow H_2^+ + e^- \tag{1.4}$$

often followed by recombination leading to dissociation into neutral radicals and further reactions

$$H_2^+ + e^- \rightarrow 2H; \qquad H + O_2 \rightarrow HO_2^{\bullet} \tag{1.5}$$

In water and aqueous solutions, dissociation and ionization take place

$$H_2O \xrightarrow{h\nu} H + OH \qquad \text{and} \qquad H_2O \rightsquigarrow H_2O^+ + e^- \tag{1.6}$$

Both VUV light and ionizing radiation can lead to the formation of dissociative excited states, *e.g.*

$$H_2O \rightsquigarrow H_2O^* \rightarrow H + OH \tag{1.7}$$

Organic molecules such as hydrocarbons can be excited to high energy states which lead to ionization or dissociation

$$RH_2 \rightsquigarrow RH^{**} \rightarrow RH_2^+ + e^- \tag{1.8}$$

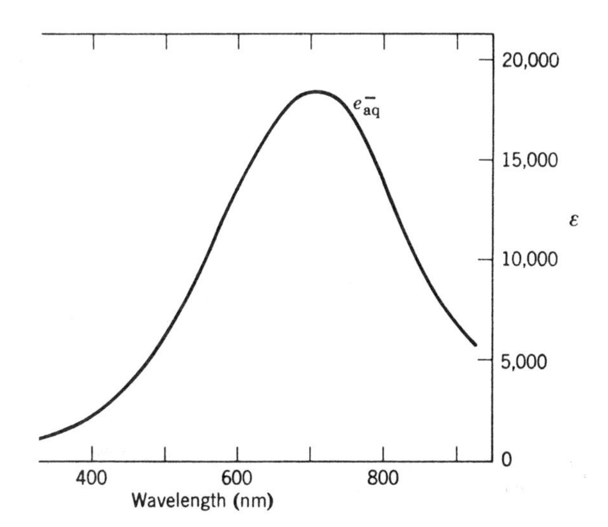

Figure 1.5 *Absorption spectrum of the hydrated electron; the ordinate is the molar decadic extinction coefficient* (M^{-1} cm^{-1})

Another important process is ion recombination, when a solvated electron meets a positive ion. One of the neutral products can be formed in an electronically excited state and its chemical reactions are then similar to photochemistry

$$M^+ + e^- \rightarrow M^* \rightarrow \text{products} \qquad (1.9)$$

Reactions of this type are also observed in photoinduced electron transfer processes and these are discussed in section 4.2.

CHAPTER 2

Light and Matter

Photophysical and photochemical processes result from the interaction of light and matter. When a beam of light passes through matter, several processes can take place as shown in Figure 2.1.

The intensity of the incident beam B_i is reduced to that of the transmitted beam B_t through scattering (S), luminescence (L) and absorption (A). Assuming that the incident beam follows a well defined path, the scattered light and the luminescence light are emitted in all directions.

The absorbed light (A) disappears altogether and becomes excitation energy of the sample. Photochemistry originates only from this type of excitation, in which atoms or molecules are promoted to states of higher energy. In many cases this absorbed energy is eventually degraded into heat, but this does not result in chemical effects (except possibly for the small rise in the temperature of the sample which could cause a chemical reaction).

2.1 ELECTROMAGNETIC RADIATION

Visible light is a small part of the spectrum of electromagnetic radiation which also includes radiowaves, microwaves, infrared light on the long-wavelength side, and X-rays, γ-rays and components of cosmic rays at the short-wavelength side. These are all fundamentally similar, and are represented in classical physics as the combination of oscillating electric and magnetic fields in perpendicular planes, these planes being themselves perpendicular to the direction of propagation of the beam (Figure 2.2). This classical picture of light waves explains most optical properties of light, such

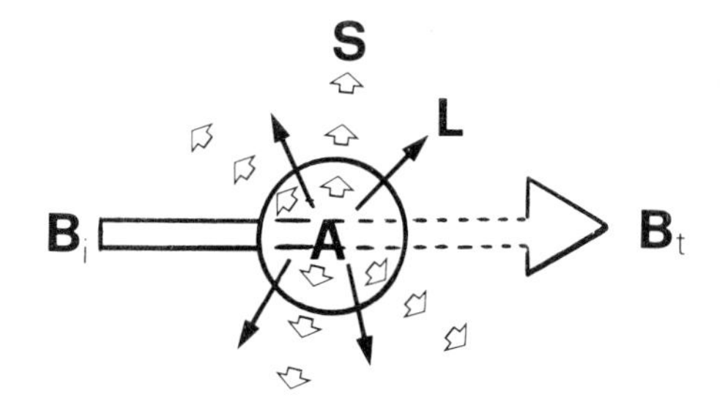

Figure 2.1 *Photophysical processes of a beam of light interacting with matter: absorption (A), scattering (S), luminescence (L); the rest of the beam is transmitted*

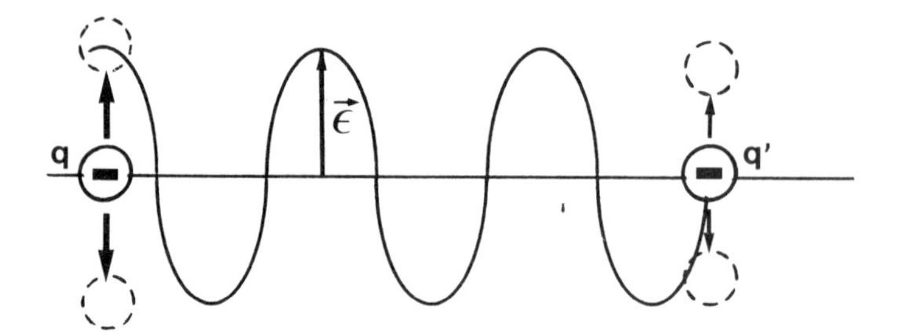

Figure 2.2 *Electromagnetic radiation couples two distant electrical charges q, q' through the oscillating electric field $\vec{\epsilon}$*

as reflection, refraction, interference, polarization, *etc*. The term 'electromagnetic' radiation comes from the fact that this radiation is emitted by electrically charged particles such as electrons, and it will induce a mechanical force on other electrically charged particles at a distance. From the point of view of photophysics and photochemistry this wave picture of light does not give an accurate description, and it is necessary to consider that electromagnetic radiation consists, in fact, of particles called photons. Experimental evidence shows that these particles must have dimensions similar to those of electrons or atomic nuclei, as discussed in the next sections.

This dual nature of light appears puzzling to most students of this field, and cannot be resolved by any simple picture. From our point of view it is sufficient to consider that light is a stream of photons which travels in a straight line at a constant velocity c ($c = 3 \times 10^8\ \mathrm{m\,s^{-1}}$). Each photon has an electric vector $\vec{E}$ and a magnetic vector $\vec{H}$ that allow interactions with electrons and nuclei through electric and magnetic forces.

In the classical picture of light waves, the frequency of light is simply the frequency of oscillation of the electrical charge which emits the light. In the photon picture of light this concept of frequency is replaced by that of the energy, E, of individual photons, this energy being proportional to the classical frequency according to

$$E = h\nu \qquad \text{(the first Planck equation)} \tag{2.1}$$

A beam of light is monochromatic if all photons have the same energy (the same frequency or wavelength in the wave picture). A beam of light is completely polarized if all the photons have parallel electric and magnetic vectors ($\vec{E}$ and $\vec{H}$).

2.1.1 Energy, Frequency, Wavelength and Velocity of Electromagnetic Radiation

In the classical picture of a wave which travels at a velocity v, the wavelength λ and the frequency ν are related by

$$\lambda = v/\nu \tag{2.2}$$

The kinetic energy of a material particle of mass m is $E = (1/2)mv^2$. According to the theory of relativity the energy of a photon is then $E = mc^2$, missing the factor (1/2) for the 'immaterial' particle (the photon can have only the velocity c, it does not exist 'at rest'). Combining the equation which relates energy and frequency with that of energy and mass, we obtain the fundamental equation of De Broglie

$$\lambda = h/p \tag{2.3}$$

This relationship provides the bridge between corpuscular physics and wave physics, since the momentum $p = mv$ (in this case $p = mc$) is then related to the wavelength λ. This equation holds much deeper implications, since each particle of mass m and velocity v is associated with a wavelength λ which in effect defines the distribution of the particle in space; when the wavelength is short, the particle is more localized.

2.1.2 The Photoelectric Effect

This is the emission of electrons from metals (or metal alloys) irradiated by light of suitable wavelength.

A photoelectric cell consists of a photosensitive surface (the metal or alloy) deposited inside an evacuated glass bulb fitted with a second metal electrode kept at a positive potential (Figure 2.3). Electrons emitted by the photosensitive surface called the photocathode will then be captured by the positive anode and a current will flow in an external circuit which connects the two electrodes.

When the photocathode is illuminated by light of wavelength below a

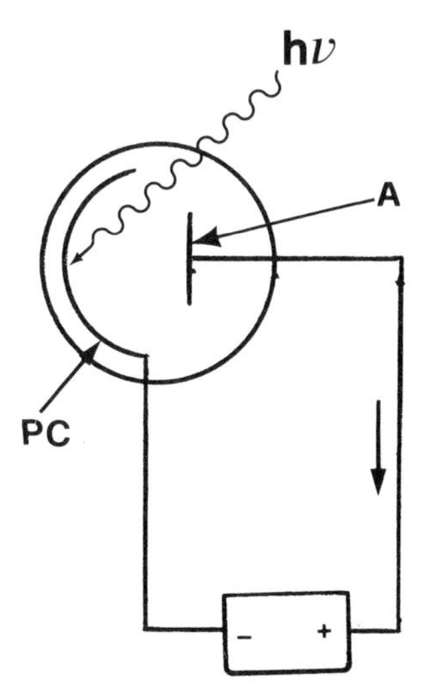

Figure 2.3 *A photoelectric cell consists of an evacuated glass bulb which contains a photocathode PC, and an anode A, which is connected to an external DC generator*

critical level which depends on its material, a flow of electrons starts immediately. If the intensity of light is low, the number of electrons will be small, but there is no delay at all between the start of illumination and the flow of current.

The photoelectric effect shows that the energy of light is not distributed evenly over a uniform wavefront, but is concentrated in space within the dimensions of atomic particles, in particular electrons.

An important aspect of the photoelectric effect is that it requires a minimum frequency of light which depends on the material of the photocathode. This means that each particle of light or 'photon' carries an energy E proportional to the frequency ν; $E = h\nu$. The factor h is known as 'Planck's constant', one of the most fundamental physical constants in all of nature.

The kinetic energy of the photoelectron E_{kin} of mass m_e and velocity v is then

$$E_{kin} = (1/2)m_e v = h\nu - W \tag{2.4}$$

W being the 'work function' of the photosensitive material. The threshold of the photoelectric effect corresponds to the condition $h\nu = W$, for which the velocity of emitted electrons is 0.

The photoelectric effect is not only important as experimental evidence for the corpuscular nature of light but also finds applications in light detectors, and it is a fundamental process in photoelectrochemistry.

2.1.3 Compton Scattering

When a beam of X-rays irradiates a light element such as carbon (in the form of graphite for instance), some of it is scattered at various angles θ as shown in Figure 2.4. The remarkable phenomenon is however that the scattered radiation has a different wavelength from the incoming beam. The detailed law of this effect known as Compton scattering can be established following the classical picture of the collision of two billiard balls, one being

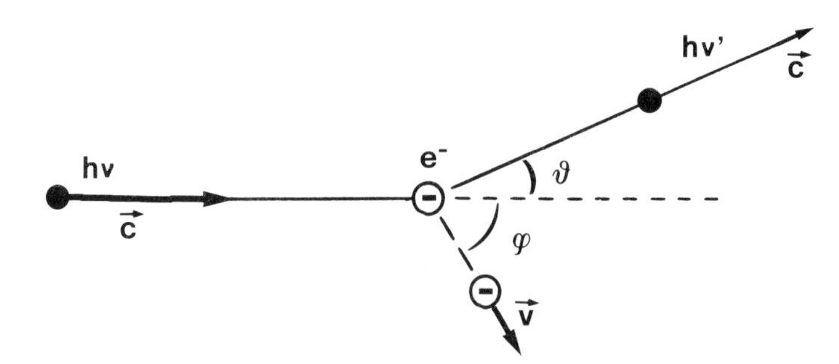

Figure 2.4 *In Compton scattering a photon of energy $h\nu$ and velocity c hits a stationary electron e^-; the outgoing (scattered) photon at an angle θ keeps this velocity but has a different energy $h\nu'$ while the electron acquires a velocity v*

an electron of mass m_e and the other a photon of X-rays of initial frequency ν and final frequency ν' for a scatter angle of θ. Such a collision of two bodies is described in classical physics following two fundamental laws of conservation. These play such an important role in photophysics that it may be worth our while to consider them briefly here.
(1) The law of *conservation of energy* states that the total energy of a 'closed' physical system must remain constant. In the present case the electron and the X-ray photon constitute a 'closed' system inasmuch as they do not give or receive any energy from the rest of the Universe during their impact.
(2) The law of *conservation of momentum* states that the total momentum of a closed system must also remain constant. The momentum of a body of mass m and velocity v is the vector $\vec{p} = m\vec{v}$, so that the total momentum of a physical system is the vector sum of the individual momenta. Collision of a particle of momentum MV with a stationary particle (of zero momentum) will scatter the particles in different directions in such a way that the sum of their momenta remains equal to the initial momentum MV

$$MV' + mv = MV \tag{2.5}$$

In the case of a photon of light the velocity cannot change, so it is the frequency which changes so as to conserve both the energy and the momentum. Such behaviour cannot be explained by a wave picture of light. It is clear that the photon of an X-ray must be a kind of particle of dimensions of the same order as the electron.

2.1.4 Photon Mass, Photon Spin and Momenta

Although the photon has no mass at rest, its inertial mass $m = h\nu/c^2$ gives a linear momentum of $m\vec{c}$. Experimental evidence for this linear momentum is found in the observation of recoil electrons in Compton scattering, or on a much larger scale in the 'solar wind' which blows the tails of comets away from the sun. (It should not be forgotten that material particles also contribute to the solar wind.)

The photon must also be given an intrinsic angular momentum which is known as photon spin; experimental evidence concerning this property of the photon is found in selection rules of atomic transitions and in one direct observation at least which is considered in Appendix 2. The absolute value of the photon spin is unity, so it can take the values $+1$ or -1 but the value 0 does not exist. Photon spin is a microscopic property of each individual particle, but it is related to the macroscopic properties of polarization of light. Thus circularly polarized light consists of photons of spin $+1$ (*e.g.* right circular polarization), or -1 (*e.g.* left circular polarization) only. Linearly polarized light contains equal numbers of photons of spins $+1$ and -1 all following the axis of polarization. Natural, unpolarized light is made up of photons of randomly oriented spins, and elliptically polarized light contains unequal numbers of photons of spins $+1$ and -1.

No such thing as photon–photon collision has ever been observed, and to all practical purposes photons must be considered as non-interacting particles; the collisional cross-section of the photon has been estimated from theory to be less than 10^{-70} cm^2. This means that electromagnetic radiation (even cavity radiation) cannot be compared with a gas of molecules that can reach thermodynamic equilibrium through collisions which result in exchange of momentum and energy.

2.2 MATTER: MOLECULES AND ATOMS, NUCLEI AND ELECTRONS

It has been one of the great unifying discoveries of natural science that all matter, however complex, is made up of atoms and molecules. The molecules are combinations of atoms, there being about 100 different kinds of atoms in nature. Some molecules are very small and contain only two atoms. Others can be extremely large, such as the molecules of proteins and other components of living matter.

For the purposes of photophysics and photochemistry it is therefore sufficient to keep in mind the simple picture of an atom as a heavy, positively charged nucleus around which move light, negatively charged electrons. In the smallest atom, that of hydrogen, there is a single electron, whereas in the uranium atom, which is the heaviest natural element known on Earth, there are 92 electrons. It is the motion of these electrons which determines the chemical properties of an atom or a molecule so that it is now necessary to consider in a qualitative way the structure of these elementary particles of matter.

2.2.1 The Planetary Model of the Atom: Orbits and Orbitals

There is a very instructive analogy between the motion of the Earth around the Sun and the motion of an electron around the nucleus of an atom, as shown in Figure 2.5. In both cases there is an 'orbital' motion (such as the

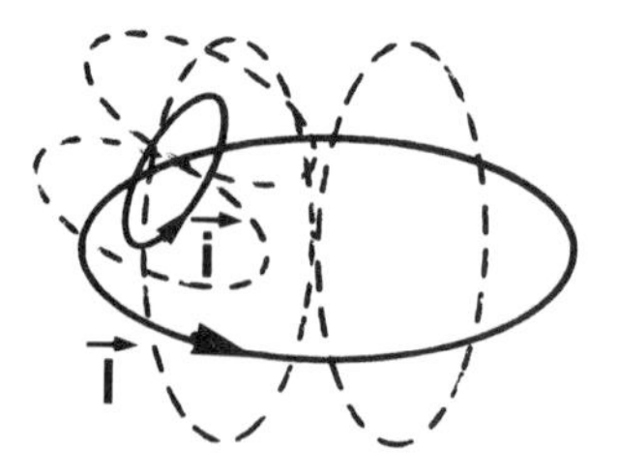

Figure 2.5 *The orbital and spin motions of an electron around a nucleus are equivalent to circular electric currents I and i; these produce magnetic fields shown by their force lines (broken curves)*

annual revolution of the Earth around the sun) and a 'spinning' motion (such as the daily rotation of the Earth on its own axis). There is however one major difference: the orbital motion of the Earth (and of the other planets) indeed follows a planar orbit which is an ellipse, but the 'orbit' of an electron does not keep to a single plane, and its distance from the atomic nucleus varies. The electron in a hydrogen atom for instance therefore follows a complex path in three-dimensional space, this path being represented by a sphere rather than a (plane) circle. This sphere is called an 'orbital', by analogy with the orbit of a planet around the Sun. The reason for this is that the gravitational force between the Sun and a planet is strictly a central force, the force vectors passing through the mass centres. According to Newton's laws the motion is then constrained to a plane, since there is no out-of-plane force component. However, the electron is an electrically charged particle so its orbital and spin motions are equivalent to small circular electric currents which create magnetic fields. The interaction between these two currents $\vec{I}_o$ (orbital) and $\vec{i}_s$ (spin) and the respective fields $\vec{H}_o$ and $\vec{H}_s$ produce out-of-plane forces

$$\vec{F} = \vec{i} \times \vec{H} \tag{2.6}$$

These forces are responsible for spin-orbit coupling, which plays a very important role in photophysics.

2.2.2 Wavefunctions and Operators: The Schrödinger Equation

It would be hopeless to try to describe the complex motions of electrons and nuclei as a function of coordinates in space and time. The only useful concept is that of distribution or probability of presence. This distribution is represented by the *wavefunction* ψ, $\psi(x, y, z)$ such that ψ^2 gives the probability of presence of any one particle in the volume element $x \pm \Delta x$, $y \pm \Delta y$, $z \pm \Delta z$ at any time. This is similar to a map of population distribution, which does not depend on time. We cannot track the travels or the instantaneous positions of the people in a large city, or even in a small village, but we can draw a map of population density which tells us where we are most likely to find any one person. The square of the wavefunction is such a map of population density of electrons and nuclei. These distributions are called 'wavefunctions' because their properties are similar to standing waves in classical physics.

In general the energies of motion are continuous in classical mechanics; a car can move at any speed, not just in increments of say 10 m.p.h. The absorption and emission spectra of atoms show that this is not true at the microscopic level, where the energy varies in discrete steps. The properties of standing waves are best illustrated by the vibrations of a string held within a limited space, something like a violin string. This gives a picture of standing waves in one dimension, but wavefunctions extend of course in three dimensions.

Figure 2.6 provides a simple one-dimensional illustration of the properties

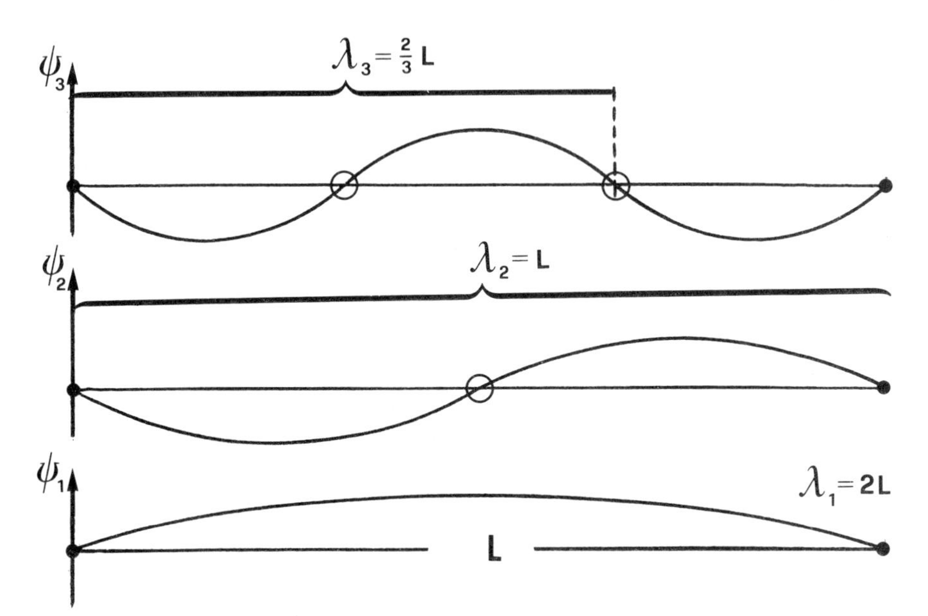

Figure 2.6 *Nodal properties of standing waves. A one-dimensional oscillation (wave) constrained within a space of length L can have amplitudes (wavefunctions) of discrete wavelengths only. The open circles are the nodes where the amplitude is always zero*

of standing waves. There are two boundaries which define the overall length within which waves can exist. At these points the wave amplitude must necessarily go to zero. At various frequencies there can be other points where the amplitude is always zero, and these are called the 'nodes' of the wavefunction.

The lowest frequency standing wave has no node (ψ_1) and is called the 'fundamental' (frequency). Higher frequency modes are 'harmonics' which have increasing numbers of nodes.

These concepts of the classical physics of standing waves have important implications in photophysics, in particular for the understanding of orbital symmetries and laser light emission. In the case of a standing wave the propagation velocity does not exist, and the important relationship defines the wavelengths as a function of the distance between the boundaries and the number of nodes in the wavefunction

$$\lambda = 2nL \tag{2.7}$$

In eqn. 2.7 the number n is a 'quantum number'. It is in fact related to the number of nodes in the wavefunction and must in this case be a positive integer ($n = 1$, 2, *etc.*). This would apply to a wave which follows one direction only. Since real space is three-dimensional a standing wave must be defined by three quantum numbers. The motions of electrons around nuclei are essentially circular, so that the use of polar coordinates is preferable and the three quantum numbers are:

n, the principal quantum number, which is related to the average distance between the electron and the nucleus as well as to the number of nodes in the wavefunction. The energy of the electron is determined essentially by the value of this principal quantum number n.

l, the secondary quantum number, defines the spatial symmetry of the orbital. An 's' orbital follows a spherical symmetry, a 'p' orbital has one symmetry axis, and a 'd' orbital has a more complex shape.

m, usually called the 'magnetic quantum number' describes the orientation of the symmetry axis of an orbital in space. This makes sense only if there is some reference direction, otherwise all orientations have the same energy. Experimentally such a reference can be provided by an external magnetic field.

s, the spin quantum number which defines the direction of the spinning motion of the electron.

While it is obvious that the principal quantum number n must be a positive integer (it is impossible to have less than zero nodes in a wavefunction), the values of the other quantum numbers can be negative when they are defined in polar coordinates. The allowed values of the various quantum numbers are

$$
\begin{aligned}
h &= 1, 2, 3 \ldots \\
l &= 0, 1 \ldots, n-1 \\
m &= 0, \pm 1, \pm 2, \ldots \pm l \\
s &= +\tfrac{1}{2} \text{ or } -\tfrac{1}{2}
\end{aligned}
\tag{2.8}
$$

2.2.2.1 Operators, Eigenfunctions and Eigenvalues. Inasmuch as all matter is (from a chemist's point of view) a combination of atomic nuclei and electrons, it is clear that the motions, or rather the distributions, of these particles must explain all observable properties, and there are many of these: for example the energy of a molecule, its dipole moment, and its shape. In the end the observable property must be a number (with the relevant units) and these values are called 'eigenvalues'. How can one obtain them?

The distributions of nuclei and electrons are described quantitatively by their wavefunctions ψ. The instantaneous positions of nuclei and electrons change all the time, so the instantaneous value of the property in question changes as well; *i.e.* this instantaneous value is a function of ψ which can be described as

$$V = O(\psi) \tag{2.9}$$

Here O is a mathematical 'operator' which changes a function ψ into a different function V. There must be such an operator for each property V, and if the form of this operator is known the instantaneous value of V can be obtained. What we really want, however, is a measurable property which is the weighted average of the instantaneous V values. This weighting is essential because not all instantaneous positions of electrons or nuclei are equally probable, but depend on the value of the wavefunction, so that this instantaneous weighted value must be

$$V = \psi \cdot O(\psi) \tag{2.10}$$

The observable value V is finally the integration of all these weighted instantaneous values V, so we come to the final equation

$$V = \iiint_{-\infty}^{+\infty} \psi \cdot O(\psi)\, \mathrm{d}x\, \mathrm{d}y\, \mathrm{d}z = \langle \psi | O | \psi \rangle \tag{2.11}$$

(The symbol of broken brackets denotes integration over all space.)

The most important property of any atom or molecule is its total energy, and this is given by the Schrödinger equation in which the operator is called the 'Hamiltonian' with the symbol H

$$E = \langle \Psi | H | \Psi \rangle \tag{2.12}$$

The wavefunction which fits the equation and leads to discrete values of V is the eigenfunction. The search for such eigenfunctions and eigenvalues can be a most demanding mathematical excercise, and need not be considered here. Let us note however that the solutions of the Schrödinger equation lead to the definitions of the orbital quantum numbers n, l and m. The quantum numbers of rotational and vibrational levels are also derived from the Schrödinger equation.

A wavefunction ψ and its eigenvalue E define an orbital. The orbital is therefore an energy level available for electrons and it implies the relevant electron distribution. In mathematical models, these distributions extend to infinity, but in a pictorial representation it is sufficient to draw the volume in which the probability of presence of the electron is rather arbitrarily around 90%. The spatial distribution of atomic and molecular orbitals have implications for processes of electron tunneling (section 4.2.1).

The number of orbitals in an atom or molecule is not infinite, even though mathematically the principal quantum number n has no limit. In reality, the energy range is restricted to the ionization level, and this defines the only absolute energy scale of electrons. When ionization takes place the electron goes to 'infinite' distance from the nucleus, in practice to a distance so great that their electrostatic interaction is negligible. The coulombic force and the potential energy are

$$F = qq'/r^2; \qquad E = qq'/r \tag{2.13}$$

so that they both go to 0 when r goes to infinity (Figure 2.7).

2.2.3 The Principle of Exclusion and The Uncertainty Principle

An orbital defines an energy level available to electrons; it can of course remain vacant, but it can hold only one or two electrons and no more. This results from the principle of exclusion as stated by Pauli.

There is a rather obvious principle of exclusion in classical physics which states that no two material bodies can be found at the same time at the same

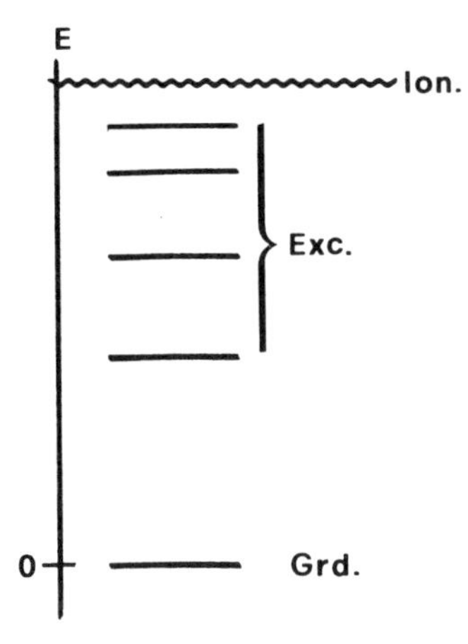

Figure 2.7 *The absolute energy scale of orbitals must be referred to the energy of an isolated electron in a vacuum*

place. The equivalent statement in quantum mechanics is that within a finite space (atom or molecule for example) no two electrons can have the same wavefunction. Remember that the wavefunction ψ defines the probability of the presence of a particle at space coordinates x, y, z through its square $\psi^2(x, y, z)$. If two particles have identical wavefunctions, this implies that there is a finite probability of finding them at the same time at the same place, which is ruled out. As the wavefunction of an orbital is defined by the four quantum numbers n, l, m and s the principle of exclusion implies that no two electrons can have the same quantum numbers. The smallest possible difference resides in the spin quantum number s which can take only two values $+\frac{1}{2}$ or $-\frac{1}{2}$ in units of angular momentum $h/2\pi$.

Any one orbital can therefore contain two electrons at most, and these must have opposite spin motions (Figure 2.8).

2.2.3.1 The Uncertainty Principle. This principle of quantum mechanics is of significance in photophysics because it establishes a relationship between the

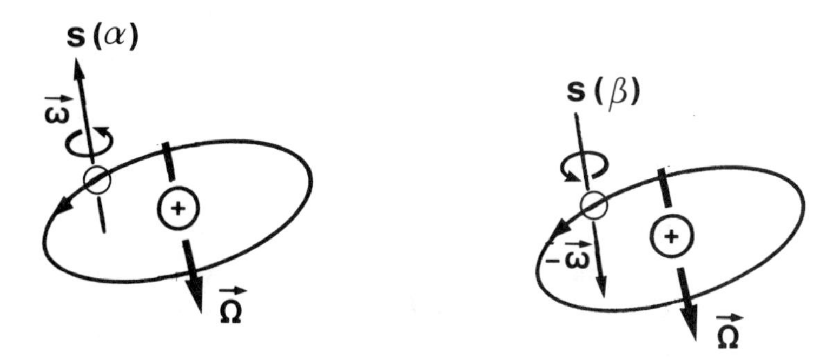

Figure 2.8 *The two possible arrangements of spin and orbital motions of an electron; for a given orbital angular momentum the spin can be* $\alpha(+\frac{1}{2})$ *or* $\beta(-\frac{1}{2})$

energy levels of an excited state and its lifetime. It is based on the concept that any observation of a physical system implies a change in that system so that its state is not the same before and after observation. Arguably the smallest possible change is the emission of a photon of very low energy, then the change in the 'action' of the system (*e.g.* an atom) is of the order of Planck's constant h.

This constant h has the physical dimensions of an 'action', ML^2T^{-1}. Any two quantities, the product of which has the physical dimensions of an action, are 'conjugate' quantities, for instance

$$\text{energy and time, } E \cdot T = (ML^2T^{-2}) \cdot T \tag{2.14}$$

$$\text{momentum and distance, } (MLT^{-1}) \cdot L \tag{2.15}$$

Although we know that the action of the system has changed by approximately h, we cannot tell to what extent this change is due to E and to T, where E can be the energy of an excited molecule of lifetime T. It follows that if E is defined perfectly ($\Delta E = 0$), the lifetime is totally uncertain, *i.e.* $\Delta T = \infty$. The usual form of the uncertainty principle is given as

$$\Delta E \cdot \Delta T \simeq h \tag{2.16}$$

In the case of a ground state the energy can be determined to any degree of accuracy since its lifetime is infinite. This is not so when excited states are considered, for in this case the lifetime is finite. This implies that the energy of an excited state can be defined only to a limited accuracy, the shorter-lived states having the greater energy spread.

2.3 THE INTERACTION OF LIGHT WITH MATTER

A simple 'thought experiment' due to Einstein gives a basic idea of the interaction of electromagnetic radiation with matter. Consider a space surrounded on all sides by perfectly reflecting mirrors (Figure 2.9). Inside this cavity there is a hot material body which is in thermal equilibrium with the radiation which fills the cavity. This radiation is then isotropic, as it fills, in a random manner, all the space of the cavity; its intensity can be defined as an energy per unit volume.

Now for the sake of simplicity it will be assumed that the atoms in the

Figure 2.9 *In the thought-experiment of the basic interaction of radiation with matter, a material body at vanishing density is in thermal equilibrium with monochromatic electromagnetic radiation held within a perfectly reflecting enclosure*

material body have only two energy levels, a ground state M and an excited state M*, so that the radiation is perfectly monochromatic. Excited states are formed by the absorption of radiation by ground states $M + h\nu \rightarrow M^*$ and excited states can then decay back to the ground state through spontaneous emission $M^* \rightarrow M + h\nu$ or through stimulated emission $M^* + h\nu \rightarrow M + 2h\nu$. This latter process must be included in order to obtain equilibrium conditions between radiation and matter. It is also a necessity according to the principle of microscopic reversibility, for when a photon $h\nu$ strikes an excited molecule M* the upward transition $M^* \xrightarrow{h\nu} M^{**}$ and the downward transition $M^* \xrightarrow{h\nu} M + 2h\nu$ must be equally probable. Within our assumption of a two-level atom, there is no excited state M** at twice the energy of M* so that only the downward transitions can exist (stimulated emission). Let the concentrations of ground and excited states of M be [M] and [M*] respectively, and the isotropic photon density of the radiation be I, then the kinetic equations are

absorption $$\mathrm{d}[M^*]/\mathrm{d}t = BI[M] \tag{2.17}$$

spontaneous emission $$\mathrm{d}[M^*]/\mathrm{d}t = A[M^*] \tag{2.18}$$

stimulated emission $$\mathrm{d}[M^*]/\mathrm{d}t = -BI[M^*] \tag{2.19}$$

These equations are similar to those of first- and second-order chemical reactions, I being a photon concentration. This applies only to isotropic radiation. The coefficients A and B are known as the Einstein coefficients for spontaneous emission and for absorption and stimulated emission, respectively. These coefficients play the roles of rate constants in the similar equations of chemical kinetics and they give the transition probabilities.

2.3.1 Transition Moments

The Einstein coefficients are related to the most fundamental quantity which describes the transition probability, known as the transition moment. During an electronic transition for instance, an electron jumps from one orbital to another. Its distance from the nucleus changes, so there is a change in the instantaneous dipole moment. The greater this change, the more probable the transition because it is the interaction between this transition dipole and the electric vector of light.

In a general picture, the motions of all particles should be included and the transition moment is

$$\vec{M} = \sum_i v_1 | q_i \vec{r}_i | v_2 \tag{2.20}$$

where v_1 and v_2 are the wavefunctions of the initial and final states, respectively, q is the electrical charge of the ith particle and r_i is its distance from the origin of coordinates.

The dipole moment operator is $\sum q_i \vec{r}_i$, just as in classical physics. If the same wavefunctions are introduced in eqn. 2.14, then the permanent dipole moment of the relevant state is obtained. It is however essential to distinguish between the transition moment between two states and the difference between their permanent dipole moments. In an atom of spherical symmetry or a molecule which has an inversion centre there is no permanent dipole moment. They can have large transition moments, as these describe in effect the motion of an electron from one orbital to another. The overall symmetry is restored after the electron jump.

2.3.1.1 The Planck Distribution of Black-body Radiation. The Planck relationship between the energy of the photon and the frequency of monochromatic light leads to the equation of distribution of the intensity of light as a function of frequency (or wavelength)

$$I(w)\,\mathrm{d}w = \frac{\hbar w^3\,\mathrm{d}w}{\pi^2 c^2 (\mathrm{e}^{-\hbar w/kT} - 1)} \qquad \text{with } w = 2\pi\nu \tag{2.21}$$

This equation applies to 'black-body' radiation, that is the isotropic radiation in equilibrium with a perfectly absorbing material body. For one particular temperature T this distribution goes through a maximum, shifting to higher frequencies with higher temperatures. This equation is based on the assumption that the energy of the atoms or molecules (oscillators in the classical sense) is of strictly thermal origin; there should be no contributions of 'high grade' energies such as electrical energy which could modify the population distribution. The Planck distribution does not take into account discrete absorption or emission frequencies specific to certain types of atoms or molecules, so that the distribution of black-body radiation is shown as a continuous function (Figure 2.10).

2.3.2 Electro-optical Phenomena

Polarized light is obtained when a beam of 'natural' (unpolarized) light passes through some types of anisotropic matter. In optical instruments this is usually a birefringent crystal which splits the incident unpolarized beam into two beams of perpendicular linear polarization, known as the 'ordinary' and 'extraordinary' beams. Anisotropy can also be created by the effect of an electric field, this being known as the Kerr effect.

Randomly distributed dipolar molecules form an isotropic medium characterized by a single refractive index n. When these molecules are more or less aligned in an electric field the medium becomes anisotropic and its optical properties must be described by two refractive indices. The Kerr cell consists of a glass vessel fitted with two parallel optical windows and two metal electrodes (Figure 2.11). The cell contains a liquid or a gas (solids can also be used); when an electric field exists between the electrodes the phase difference between the electric vectors is proportional to the square of the field and to the value of the electrical Kerr constant of the dielectric

$$\frac{\delta}{\lambda} = \frac{l(n_e - n_0)}{\lambda} = C_e l E^2 \tag{2.22}$$

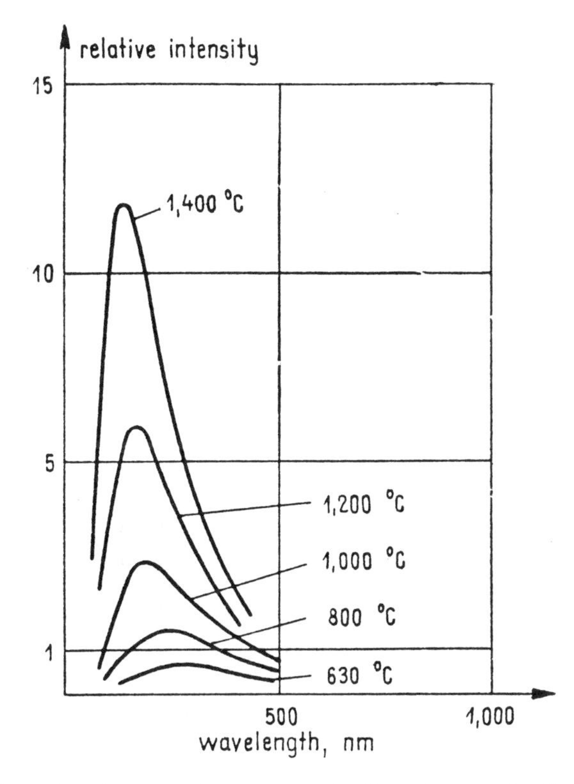

Figure 2.10 *Examples of the intensity versus wavelength or frequency distribution of black-body radiation according to the Planck equation. Each distribution corresponds to a temperature T of the radiating body*

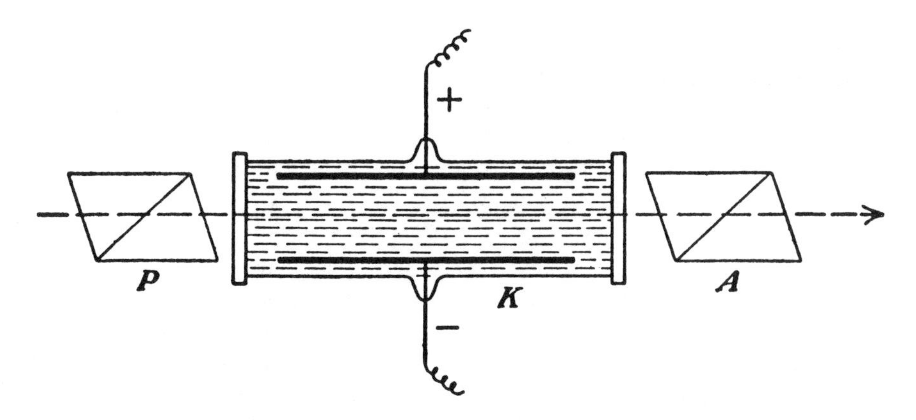

Figure 2.11 *Kerr cell. P = Polarizer; K = Kerr Cell; A = Analyser*

In eqn. (2.22) δ is defined as the retardation, l is the pathlength of the cell, and n_e and n_0 are the refractive indices of the extraordinary and ordinary rays, respectively, for a wavelength λ. C_e is the electrical Kerr constant.

Since the angle of polarization of the beams changes when the dielectric is electrically polarized, the intensity of the light varies when the Kerr cell is fitted between polarizer crystals. The Kerr cell can therefore act as an electro-optical shutter, its action time depending on the nature of the dielectric. This is very slow with solids (several seconds) but reaches the ns time-scale with low viscosity liquids.

CHAPTER 3

The Energy of Light: Excited Molecules

In order of increasing energy, molecules have translational, rotational, vibrational and electronic excited states (Figure 1.3). The electronic excited states only lead to photochemistry, since their energies correspond to the frequencies of near IR, visible and near UV light. An electronic transition brings an atom or molecule from the (electronic) ground state to one of its (electronic) excited states, or *vice versa*. The upward transitions require the absorption of light, but the downward transitions can take place with emission of light (radiative deactivation) or with conversion of the excess energy into heat (non-radiative deactivation).

3.1 ORBITALS AND STATES

An orbital is an energy level available to an electron in an atom or molecule. Since atoms and molecules usually contain many electrons, there are many orbitals, and they form the rungs of an energy ladder. Each rung can accept two electrons, but it can also take just one electron or remain altogether empty.

Orbitals are however only mathematical concepts and the observable properties of atoms and molecules are their energy states. Such a 'state' must be defined by the distribution—the wavefunction—of all the nuclei and all the electrons of the molecule. This would be a formidable problem and in practice the wavefunctions of the heavy nuclei are separated from those of the light electrons.

$$\psi_{\text{mol}} = \varphi_{\text{nucl}} \cdot \varphi_{\text{electron}} \tag{3.1}$$

This separation is known as the Born–Oppenheimer approximation. This assumes that the nuclei (being thousands of times heavier than electrons) remain motionless while electrons move around them.

3.1.1 Excited States of Atoms and Atomic Spectra

The energy states of atoms are purely electronic (atoms have of course no rotational or vibrational states) and these are obtained in a simple picture by fitting the available electrons in the available orbitals. In the ground state all these orbitals are filled in the order of increasing energy, subject to the exclusion principle which states that any one orbital can contain no more than two electrons. A special case does arise when two or more orbitals happen to have the same energy; these are then called 'degenerate' orbitals and several electron configurations can be drawn (Figure 3.1). Here we consider two orbitals which receive two electrons. It is possible to fit the two electrons in the same orbital, provided they have opposite spins (the other orbital then remains empty), but it is also possible to fit two electrons of parallel spins in the two separate orbitals. In this simple picture the energy of the two states are identical and these would be called degenerate states. In fact a state of higher total spin quantum number is lower in energy than one of lower total spin quantum number because of the concept of electron correlation. When two electrons are in the same orbital they are on average close together so that their electrostatic repulsion is large. However, electrons that must reside in different orbitals because of their parallel spins stay on average at a greater distance and the repulsion of their negative electric charges is lower. Electron correlation is the basis of 'Hund's rule' which states that for any specified scheme of orbitals the energy of the states increases in the order of decreasing multiplicity (total spin quantum number).

Figure 3.1 shows how two isoenergetic orbitals φ, φ' give rise to two states S and T of different energies, although their electron distribution differs only by the total spin quantum number.

3.1.2 Atomic Absorption and Emission Spectra

The ground state of an atom is obtained by placing all available electrons in the orbitals in the order of increasing energy, following Hund's rule, when there are isoenergetic orbitals. An excited state is formed when one or several of these electrons are displaced to other orbitals. These states then correspond to 'one-electron' or 'two-electron', *etc.*, excitations, as illustrated in Figure 3.2.

The jump of an electron from one orbital to another is a *transition* between two energy states, such that an upward transition requires an input of energy, in the form of a photon of light for example. This is the absorption of light by the atom. A downward transition is accompanied by the release of energy, for instance in the shape of a photon of light. This is the process of emission or *luminescence*.

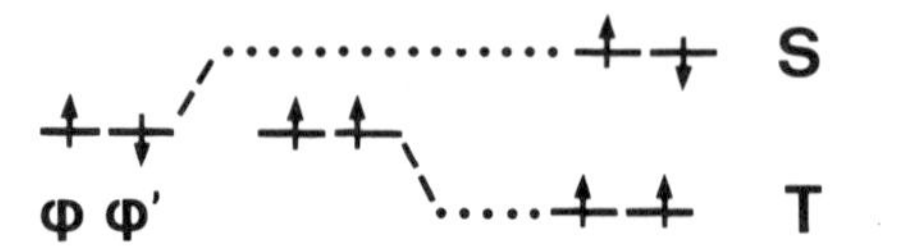

Figure 3.1 *Two electrons in two orbitals (φ, φ') can form either a singlet state of 0 total spin (S) or a triplet state of total spin 1 (T); these states have different energies as a result of electron correlation*

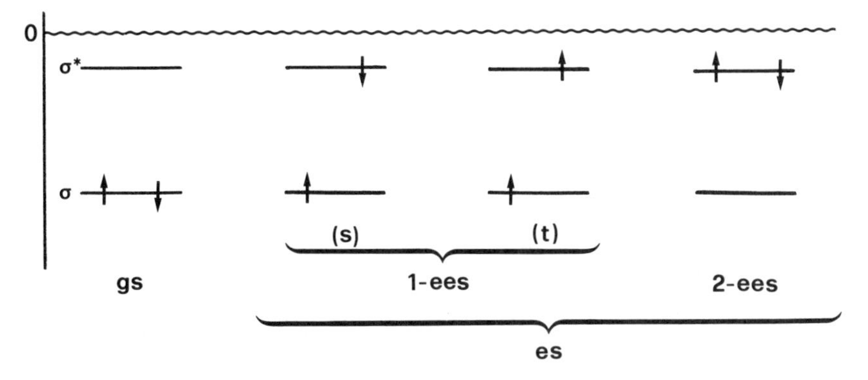

Figure 3.2 *Ground state (gs) and excited electronic states. The displacement of one electron is a one-electron excitation (1-ees), that of two electrons is 2-ees*

In principle such upward or downward transitions can take place between any two energy states. The absorption spectrum of an atom consists of very sharp lines, the frequencies of which correspond to the difference of energies between the two states, $E_2 - E_1 = h\nu$. Similarly the luminescence spectrum of an atom consists of sharp emission lines of the same frequency. Figure 3.3 gives a simple picture of the energy states of an atom and of the transitions which can be observed in the absorption and emission spectra. The

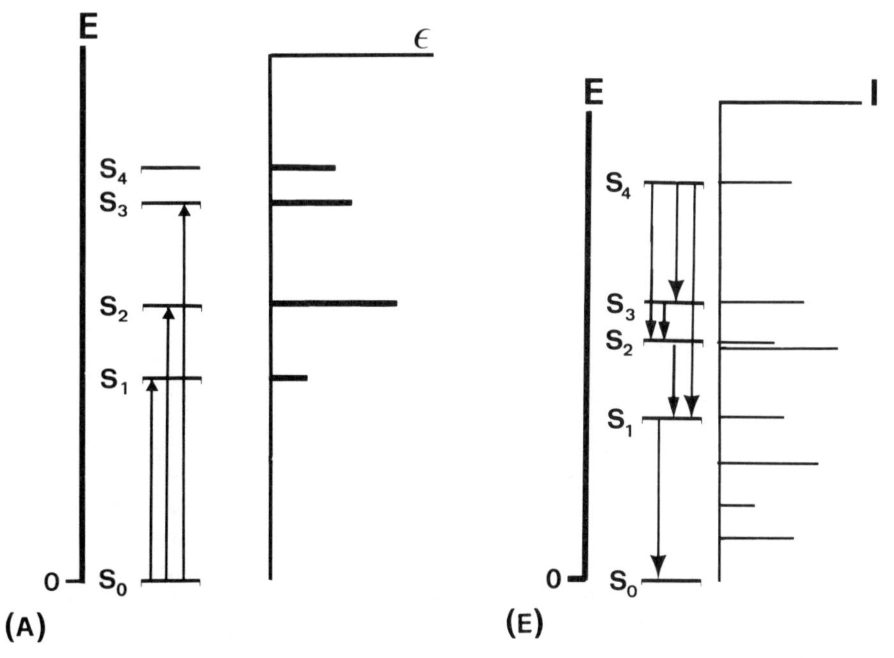

Figure 3.3 *Outline of atomic absorption (A) and emission (E) spectra. These arise from pure electronic transitions and therefore consist of very sharp lines*

luminescence of atoms is a type of 'resonance luminescence', so called because the emission frequencies match exactly the absorption frequencies.

3.1.3 Real Atomic Spectra: Broadening of Absorption and Emission Lines

The properties of absorption and luminescence emissions of atoms are important in analytical techniques as well as in spectroscopy in general. The absorption and emission spectra of atoms are line spectra which provide the unmistakable 'fingerprint' of each element, and this is used in the analytical technique known as *atomic absorption spectroscopy* for example. Although the energy levels of atoms are shown as simple lines in a qualitative picture such as that of Figure 3.3, the absorption and emission lines which correspond to transitions between these levels are not infinitely narrow (that is, absolutely monochromatic) because of several effects.

(1) The *natural broadening* which results from the finite lifetime of excited states. The energy of a state and its lifetime are related by the principle of uncertainty (section 2.2) which implies a minimal spread of the actual energy of any excited state of finite lifetime; this gives an absolute limit to the width of atomic spectral lines.

(2) In practice, the effect known as *Doppler broadening* is a much more severe limitation. This results from the motions of the atoms with respect to the observer, and produces a shift in the frequency ν of a line according to the relative velocity v.

(3) When the excited atoms can be deactivated by collisions with other atoms in competition with the emission of light, there is the additional process of *collisional broadening* (also known as pressure broadening, since collisions become more frequent as the pressure of a gas increases). The shorter excited state lifetime leads to a wider spread of excited state energies.

The last two mechanisms of the broadening of atomic spectral lines are in most cases the real experimental limitations in atomic spectroscopy. The half-widths of such lines are usually of the order of 10^{-3} nm.

3.2 EXCITED STATES OF MOLECULES

3.2.1 Diatomic Molecules

In a very simple molecule such as hydrogen, H_2, two H atoms come together to form a chemical bond, and this bond is shown as a line in the structural formula, *i.e.* H–H. In a more detailed picture, the two atomic orbitals interact when they are very close together, and this interaction can lead to two opposite situations, as shown in Figure 3.4 for the general case of two different atoms A and B. A *bonding* molecular orbital is formed if the two electrons are found mostly in the space between the nuclei; they provide a shield between the two positively charged protons so that their electrostatic repulsion is lowered. An antibonding molecular orbital is formed if the two

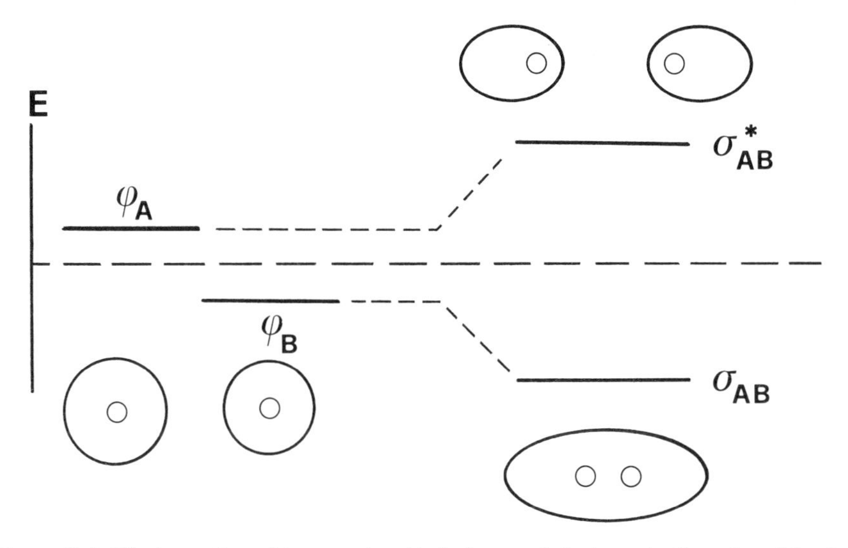

Figure 3.4 *The interaction of two atomic orbitals (φ_A, φ_B) leads to the formation of bonding (σ_{AB}) and antibonding (σ^*_{AB}) molecular orbitals of different energies*

electrons are localized essentially outside the internuclear region, so that the repulsion of the nuclei is actually greater than in the case of two separate atoms at the same distance. The energies of the bonding and antibonding orbitals are therefore lower and higher, respectively, than that of the constituent atomic orbitals.

Here it must be remembered that orbitals define the distribution of an electron in space (through the wavefunction) and the energy which corresponds to this particular distribution. An orbital is defined independently of its occupation, and the actual energy state depends on the way the electrons are placed in the available orbitals.

3.2.1.1 The Various Types of Atomic and Molecular Orbitals. When two atomic orbitals of s type interact to form molecular orbitals as shown in Figure 3.4, these new orbitals are labelled σ (bonding) and σ^* (antibonding), from the greek letter sigma (σ). A doubly occupied σ-orbital is in fact a 'single' bond in usual structural formulae. A double bond, however, results from the more complex interaction of four atomic orbitals, as shown in Figure 3.5. The σ and σ^* orbitals are formed as before, following the axis of the two nuclei, and these molecular orbitals combine atomic orbitals which are either spherical (s) or which follow this internuclear axis (p_x). The atomic orbitals p_y or p_z which are perpendicular to that axis will form new molecular orbitals labelled π (bonding) and π^* (antibonding).

The screening of the nuclei by electrons which are contained in the off-axis π orbital is of course less effective than that provided by electrons in the σ orbital, hence the energy of the π orbital is higher than the energy of the σ orbital, and conversely the energy of the π^* orbital is lower than that of the σ^* orbital.

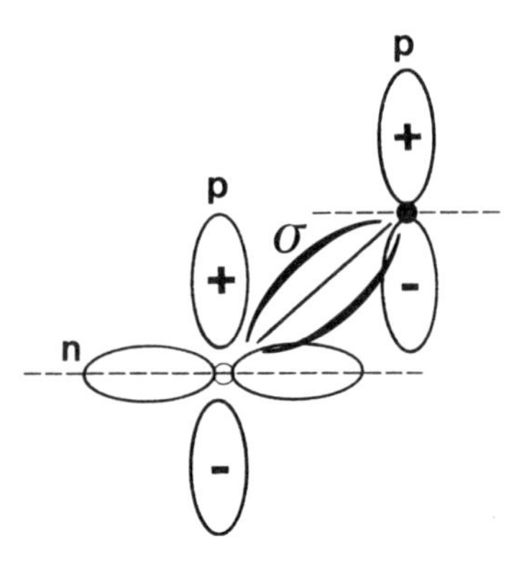

Figure 3.5 *The symmetry properties of σ, π and n orbitals in a carbonyl group. π and π* orbitals are formed from the p atomic orbitals and have the same symmetry*

In a triple bond (*e.g.* in acetylene, or ethyne) there is one σ bond and two π bonds, that is to say one set of σ and σ^* molecular orbitals and two sets of perpendicular π and π^* molecular orbitals.

There is however one more factor which must be considered for the complete description of a molecule: some atoms retain 'non-bonding' electrons which are shown as lone pairs in chemical structures. A trivalent nitrogen atom thus retains one lone pair to make up an octet of surrounding electrons, a divalent oxygen atom retains two, the halogens three such lone pairs. In the picture of orbitals, these lone pairs are assigned to non-bonding orbitals which are localized on the atom. Here a distinction must be made between lone pair orbitals which have parallel or orthogonal symmetry with respect to the p (or π) orbitals of unsaturated molecules. In a carbonyl group the two lone pairs of the O atom occupy orbitals which are orthogonal to the axis of the C atoms' p orbitals which form the CO π bond; they do not take part in the conjugated network and are therefore properly labelled as non-bonding (n) orbitals. However, the lone pair formally localized on the N atom of aniline follows closely the symmetry of the ring π orbitals so that the conjugated system extends over the six C atoms and the N atom. Resonance structures can be drawn with C=N double bonds, so that these lone pairs are not strictly speaking non-bonding; they will be labelled '*l*' orbitals to distinguish them from the n orbitals which do not take part in the conjugation of extended π systems.

3.2.1.2 Hybrid Orbitals. Orbitals, as one-electron energy levels, and corresponding wavefunctions are mathematical concepts; only 'states' are physically observable. Nevertheless, the simple picture of orbitals as the rungs of an energy ladder is very helpful, and is in many cases sufficient to account for the photophysical and photochemical properties of molecules. In more accurate pictures of orbitals it is necessary to consider their interactions, as they are not really totally independent. In this respect the concept of 'hydrid' orbitals is important; such hybrid orbitals are formed from a combination of elementary orbitals defined by their quantum numbers n, l, and m. The best known example is the hybridization of the carbon atom's 2s and 2p orbitals which are very close in energy. Depending on the bonding pattern of the C atom, the four elementary orbitals 2s, $2p_x$, $2p_y$ and $2p_z$ can combine to form one or more hybrids, sp, sp^2 or sp^3 (the suffix on p showing the relative contribution). The geometry of the bonds of the C atom depends on this hybridization; in methane, CH_4, there are four equivalent bonds which must be described as the interaction of the H 1s orbitals with the sp^3 C orbitals, but in a carbonyl group one p orbital is involved in the CO π bond. The methane molecule is tetrahedral whereas formaldehyde is planar.

The concept of hybrid orbitals in relation to molecular geometry can be important for the description of excited molecules. Figure 3.6 shows the spatial structure of these orbitals.

3.2.1.3 Rydberg Orbitals, Rydberg States. The volumes of atomic and molecular orbitals define the molecular volume in the ground state. The electrons stay quite close to the nuclei, at average distances of a few angstroms. The volumes of electronically excited molecules are larger since one or several electrons are promoted to antibonding orbitals of larger volume, and at the ionization level the molecule would have a theoretically infinite volume.

In some molecules there are orbitals of very large volume known as Rydberg orbitals which can be populated below the ionization limit. In the example of the nitrogen molecule, N_2, the average distance from the nuclei is so large that the $(N{-}N)^+$ ion looks like a single positive point charge for the excited electron.

Electronic states which result from the promotion of an electron from a localized molecular orbital to a Rydberg orbital are 'Rydberg states'. Although such states can be observed only in the gas phase, it is likely that they are involved in some photoionization processes.

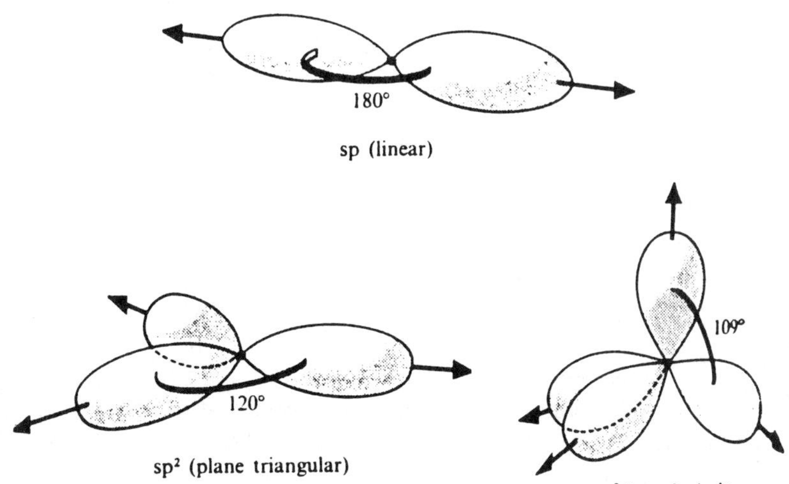

Figure 3.6 *Hybrid orbitals of the carbon atom*

3.2.1.4 Electronic States of a Diatomic Molecule. In the picture of orbitals a one-electron excited state is described by the initial and final orbitals and by the multiplicity of the excited state. In an unsaturated molecule like ethylene (ethene) there will therefore be excited states labelled $^3(\pi-\pi^*)$, $^1(\pi-\pi^*)$, $^3(\pi-\sigma^*)$, *etc.* in order of increasing energy. Two- or more electron excitations would require two or more such labels, but in practice the number of accessible excited states is quite small. For the purposes of photophysics and photochemistry it is almost always sufficient to consider one-electron excitations.

Two orbitals play a special role and are known as the Highest Occupied Molecular Orbital (HOMO) and the Lowest Unoccupied Molecular Orbital (LUMO), these labels referring of course to the situation in the ground state. The transition of an electron between these orbitals corresponds to the lowest excited state(s), *e.g.* triplet T_1 and singlet S_1, which are of special importance in the photophysics and photochemistry of polyatomic molecules.

Figure 3.7 shows the four molecular orbitals of ethylene which involve the C atoms. (The CH σ orbitals are not shown here.) The π orbital is the HOMO, the π^* orbital the LUMO.

3.2.1.5 Jablonski Diagrams. The various energy states of an atom or molecule are usually shown in Jablonski diagrams such as that in Figure 3.8. The electronic states are grouped in multiplicity manifolds. States of the same total spin quantum number follow a vertical progression. The number given to each state is simply that of its order in the energy scale and does not imply any other property (such as orbital symmetry).

More complex Jablonski diagrams can include vibrational sub-levels of the electronic states. Rotational sub-levels are not shown because they are so closely spaced as to form a near continuum. One important feature of the Jablonski diagram is that spin-allowed transitions only are shown by vertical lines; any transition which has a horizontal component is spin forbidden.

Figure 3.8(b) shows an example of a crossing of singlet and triplet states of different electron distributions resulting from different singlet–triplet splittings.

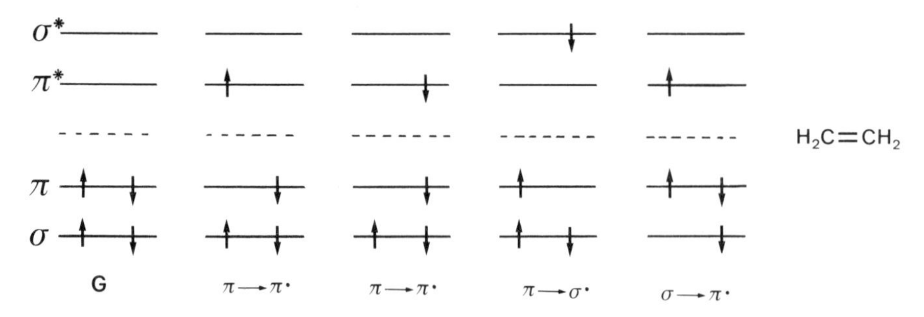

Figure 3.7 *Orbital occupation in various electronic states of an organic molecule. The states are labelled according to the orbitals involved in the excitation from the ground state (G)*

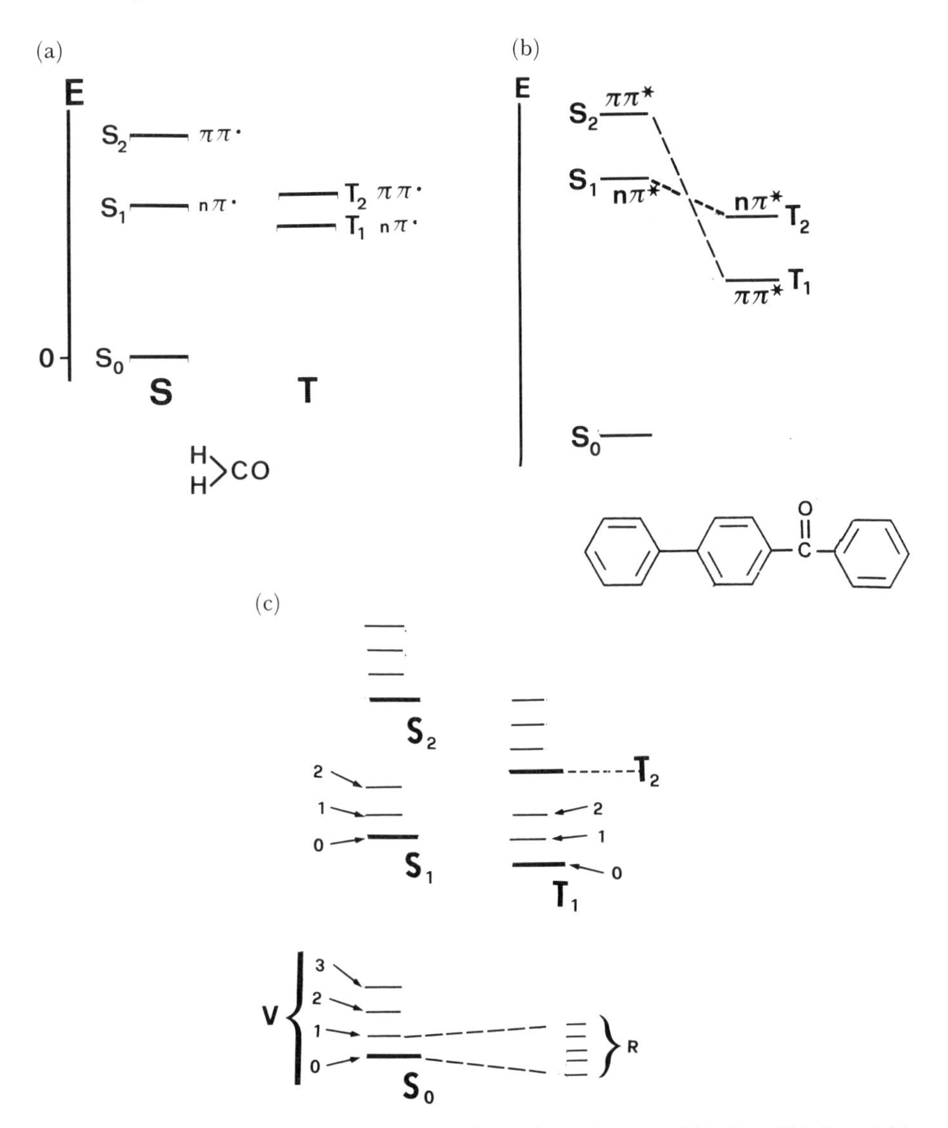

Figure 3.8 *Examples of Jablonski diagrams of pure electronic states of (a) formaldehyde and (b) 4-phenylbenzophenone, with (c) showing vibrational (V) and rotational (R) sub-levels*

3.2.1.6 Singlet–Triplet Splitting; the Theoretical Basis of Hund's Rule. In the Jablonski diagram of an organic molecule the triplet state of a specific orbital configuration is always shown at lower energy than the corresponding singlet state (*e.g.* n–π^*, π–π^*). This is one application of Hund's rule which is based on a consideration of the spatial overlap of orbitals; not to be confused with the overlap integral.

An excited state is formed by the promotion of an electron from an orbital

ψ_1 to an orbital ψ_2, which occupy different zones in space. They have however a zone of overlap, otherwise any transition between them would be forbidden. Each orbital has one electron, so that the two electrons can have parallel or antiparallel spins (triplet and singlet states respectively). According to the principle of exclusion, two electrons of parallel spins cannot occupy the same region of space at the same time, *i.e.* they cannot be found together in the overlap zone (Figure 3.9). There is no such restriction on the distribution of two electrons with antiparallel spins, so that these can be found on average closer together. The average electrostatic repulsion is therefore greater in the singlet state than in the triplet state so the former must have a higher energy.

This qualitative argument shows that the magnitude of the singlet–triplet splitting depends on the spatial overlap of the orbitals involved in the electronic transition which leads to the excited state. If the orbitals had no spatial overlap at all this splitting would be zero – this would be the case of a 100% charge transfer state. While such a situation is of course never reached, the general rule is that charge transfer states and n–π* states show small singlet–triplet splittings, while π–π* states often have large ones (with exceptions).

This argument of the spatial overlap of orbitals can be extended to any number of electrons and orbitals and is the basis of Hund's rule.

3.2.1.7 Absorption and Emission Spectra of Small Molecules. In diatomic molecules the number of vibrational and rotational levels is small, so that their energy spacing remains relatively large. Their absorption spectra are therefore line spectra which correspond to transitions to 'stable', associative excited states, but if a dissociative excited state is reached then the absorption spectrum becomes a continuum since such states have no vibrational levels.

Small molecules show resonance fluorescence from associative excited states, and these states can also take part in chemical reactions (R in Figure 3.10). Triatomic molecules are at the borderline between 'small' molecules

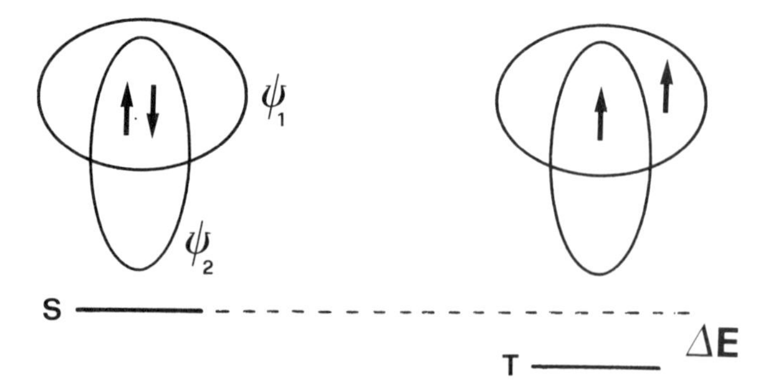

Figure 3.9 *Spatial overlap of two orbitals ψ_1 and ψ_2. In the singlet state S two electrons of opposite spin can share the overlap space*

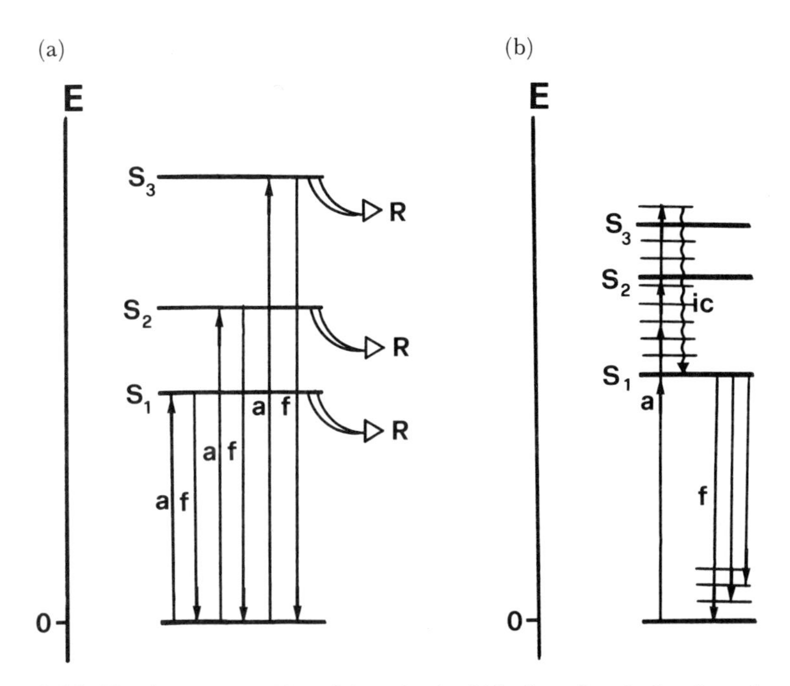

Figure 3.10 *The electronic transitions [absorption in (a)] of small molecules show vibrational and rotational lines in addition to the purely electronic spectrum. (b) Luminescence emission is resonance fluorescence (f), and chemical reactions (R) can originate from several excited states*

and 'polyatomic' molecules. As the number of atoms in the molecule increases, so the spacing of the vibrational and rotational levels becomes smaller, and transitions between these levels become so fast that resonance fluorescence (as well as chemical reactions of upper excited states) becomes generally negligible.

3.2.1.8 Potential Energy Diagram of a Diatomic Molecule. In such a diagram, shown in Figure 3.11, the energy of the molecule is plotted as a function of the internuclear distance. An 'associative' state must have a minimum which represents the most stable geometry, and this must apply of course to any stable gound state molecule such as H_2, Cl_2, *etc.* The shape of the potential energy well approximates an harmonic oscillator at low energies, but shows two important deviations which lead to nuclear repulsion at short distance and bond dissociation at large distance. Here the energy approaches a limit beyond which there are no vibrational or rotational states.

Excited states can be either associative or dissociative, *e.g.* the first triplet excited state in Figure 3.11 is dissociative whereas the excited singlet state is associative; its potential energy well resembles that of the ground state, but it is somewhat shallower and is displaced towards larger internuclear separations since the bond is weaker than in the ground state.

Two further points should be mentioned about these potential energy

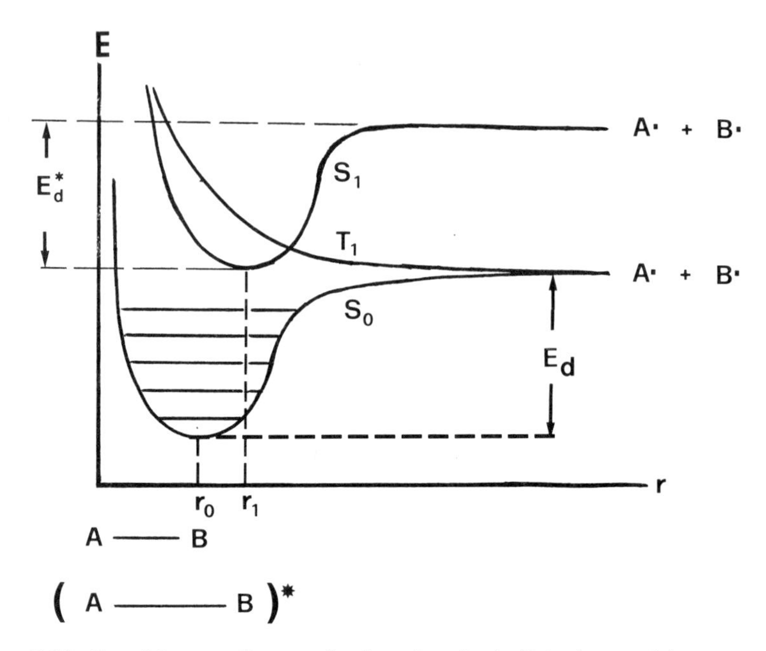

Figure 3.11 *Potential energy diagram of a diatomic molecule.* E *is the potential energy and* r *the internuclear distance. The theoretical bond dissociation energy is* E_d *in the ground state and* E_d^* *in the associative excited state* S_1*, while* T_1 *is a dissociative excited state*

diagrams. In the first place, a dissociative state shows no minimum, the energy rising continuously as the internuclear distance decreases. Secondly, the bottom of the well does not show the actual minimum of the potential energy of the molecule, because the zeroth vibrational level must retain the 'zero point' energy according to the principle of uncertainty. Even at zero degrees Kelvin there must be some vibrational motion, otherwise the internuclear distance would be defined perfectly, and then the energy would be totally unknown. For this reason there is a difference between the 'theoretical' dissociation energy, E_d in Figure 3.11, and the observable dissociation energy, which does not start from the bottom of the potential energy well.

In the potential energy diagram a radiative transition is shown by a vertical arrow since the motion of the nuclei can be neglected within the time of an electronic transition. The reorganization of nuclear geometry is shown by horizontal arrows, as seen in Figure 3.12.

Each vibrational level corresponds to a nuclear wavefunction, following the simple rule that the number of nodes increases with the vibrational quantum number. The shapes of these wavefunctions change with higher quantum numbers, the amplitudes being concentrated increasingly at the edges of nuclear motion. This situation is that of a classical pendulum that swings around an equilibrium point r_0; its velocity is zero at the edges where the

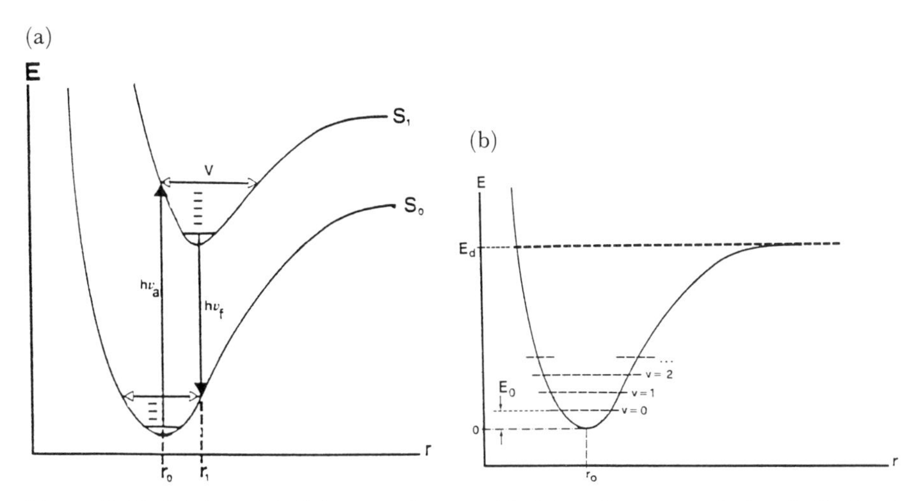

Figure 3.12 *(a) In a potential energy diagram optical transitions are vertical and vibrational motions are horizontal. (b) The zero-point energy E_0 is the energy of the $v = 0$ level*

motion is reversed, and greatest at the equilibrium point. The probability of the presence of the pendulum is therefore maximum near the edges and becomes very small near the centre of motion, this difference increasing at higher vibrational quantum numbers.

Figure 3.13 gives an illustration of a few vibrational wavefunctions of a diatomic molecule. These general shapes apply to all bonds of a polyatomic molecule, but the spacings of the vibrational levels are different and the complete nuclear wavefunctions could be shown only in multi-dimensional space. A diagram such as that of Figure 3.11 is a section through the multi-dimensional potential energy well.

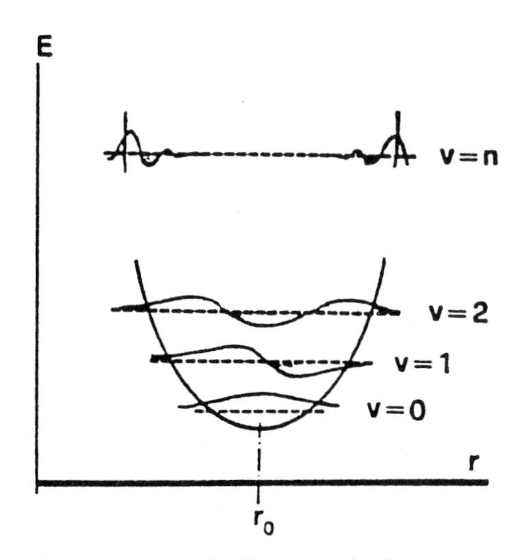

Figure 3.13 *Vibrational wavefunctions of a diatomic molecule*

3.2.1.9 Crossing of States and Avoided Crossing. In classical physics the molecule can undergo a non-radiative crossing from a higher to a lower state only at the point of intersection, since this is the only point of common energy and common geometry. In quantum mechanics crossing can occur between any two isoenergetic levels, depending on the overlap of the vibrational wavefunctions. Another quantum mechanical concept of interest to us is that of 'avoided crossing', illustrated in Figure 3.14. There is a splitting in energy V at the crossing point so that the states change over smoothly with increasing internuclear separation.

3.2.2 Polyatomic Molecules

If diatomic molecules have been considered in some detail, this is because these concepts of orbitals and energy states apply to polyatomic molecules as well, but with one important difference. Let us look first at the similarities. Taking benzaldehyde as an example, each bond corresponds to a pair of bonding/antibonding orbitals, single bonds being in fact filled σ orbitals and the double bond consisting of a filled σ orbital and a filled π orbital of somewhat higher energy, and the lone pairs of the oxygen atom are filled non-bonding orbitals, as shown in Figure 3.15. This arrangement is the ground state of the molecule, and the one-electron excitations do not distinguish between singlet and triplet states. With this distinction, the set of energy states of Figure 3.15 is obtained. The n–π^* excited state provides an example of an important property of some excited states, that is their charge transfer (CT) character. In a very precise description, such a state should be labelled 'n_O–π^*_{CO}', so as to emphasize the fact that the orbitals are localized in different regions of space. The n–π^* excitation removes an electron from the O atom, this electron then goes into the π^* orbital localized half on the C

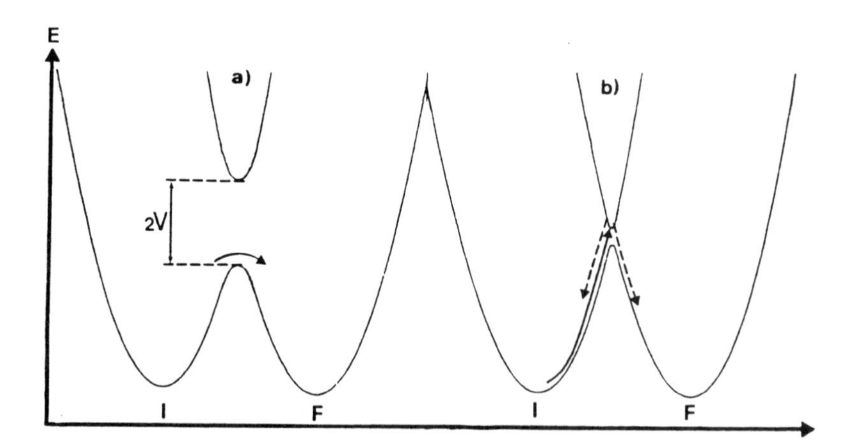

Figure 3.14 *Potential energy diagram of the crossing between a state I (initial) and F (final); (a) would be termed an 'adiabatic' process and (b) would be a 'non-adiabatic' process*

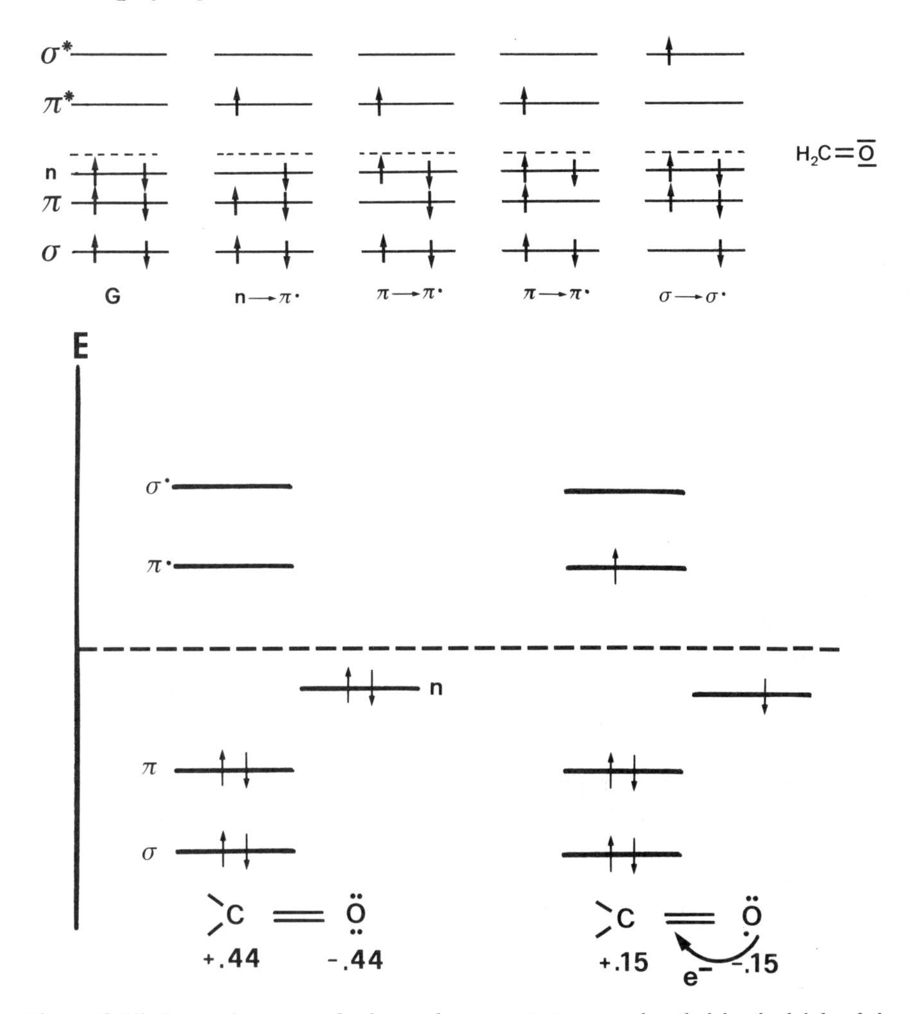

Figure 3.15 *In a polyatomic molecule one-electron excitations are described by the labels of the orbitals involved in the transition from the ground state (G). When these orbitals are localized in different zones of space the transition implies a charge transfer (CT)*

atom and half on the O atom. This is the CT of 0.5 electronic charge from O to C, with a change in the dipole moment of the molecule of some 2 Debye (D) units; the ground state dipole moment of C=O is 3 D.

In a structural formula the bonds are of course localized between pairs of atoms, so that the corresponding bonding and antibonding orbitals are by implication localized in the same way. This picture of localized orbitals is adequate in general for the σ orbitals of single bonds, and also for the π orbitals of isolated double bonds such as the one in formaldehyde. The important difference between diatomic and polyatomic molecules, which was alluded to above, arises when the molecule contains alternating single and double bonds. Such molecules are said to be 'conjugated' and the orbitals become delocalized.

3.2.3 Linear Conjugated Molecules: The Polyenes

In such molecules the carbon p orbitals which form the double bonds in the structural formula constitute a continuous array of interacting orbitals, as shown in Figure 3.16. The π orbitals which would be drawn between localized pairs of atoms now form new orbitals which are delocalized over the entire conjugated system. The wavefunctions of these delocalized orbitals are easily drawn according to their nodal properties. Remember that a standing wave has boundaries defined by the size of the molecule, and that it can have 0, or 1, 2, *etc.* nodes in between, in the order of increasing energies. Figure 3.17 shows such wavefunctions for the case of some small conjugated molecules. The molecule of ethene (ethylene) which has only one double bond can be considered as the first member of this series, although it is not conjugated. The two p orbitals of C_1 and C_2 interact to form one pair of molecular orbitals π and π^*, the first one being a bonding orbital with no node between the carbon atoms, and the second one being antibonding with one node.

Butadiene contains two double bonds, *i.e.* four carbon atoms which contribute one p orbital each; there are then four delocalized molecular orbitals which are shown in Figure 3.17 as having no node (ψ_1), one node (ψ_2), two nodes (ψ_3) and finally three nodes (ψ_4). The bonding or antibonding character of an orbital is obtained by counting the number of bonds and

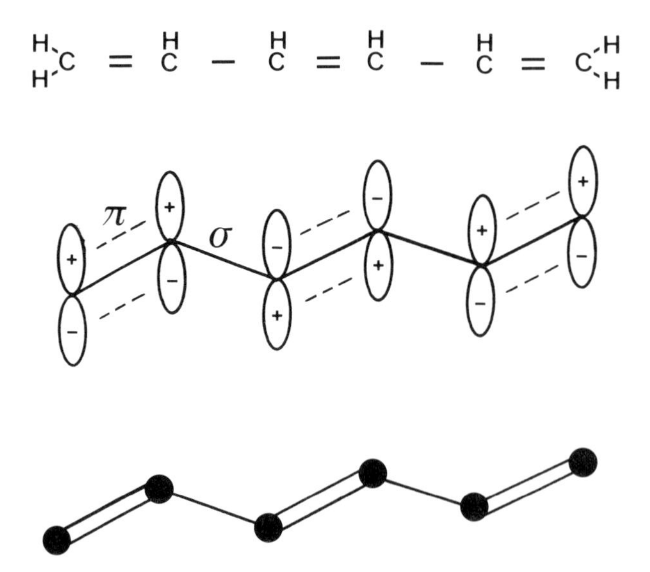

Figure 3.16 *The interaction of the carbon p orbitals in conjugated molecules leads to interactions which result in the formation of delocalized orbitals*

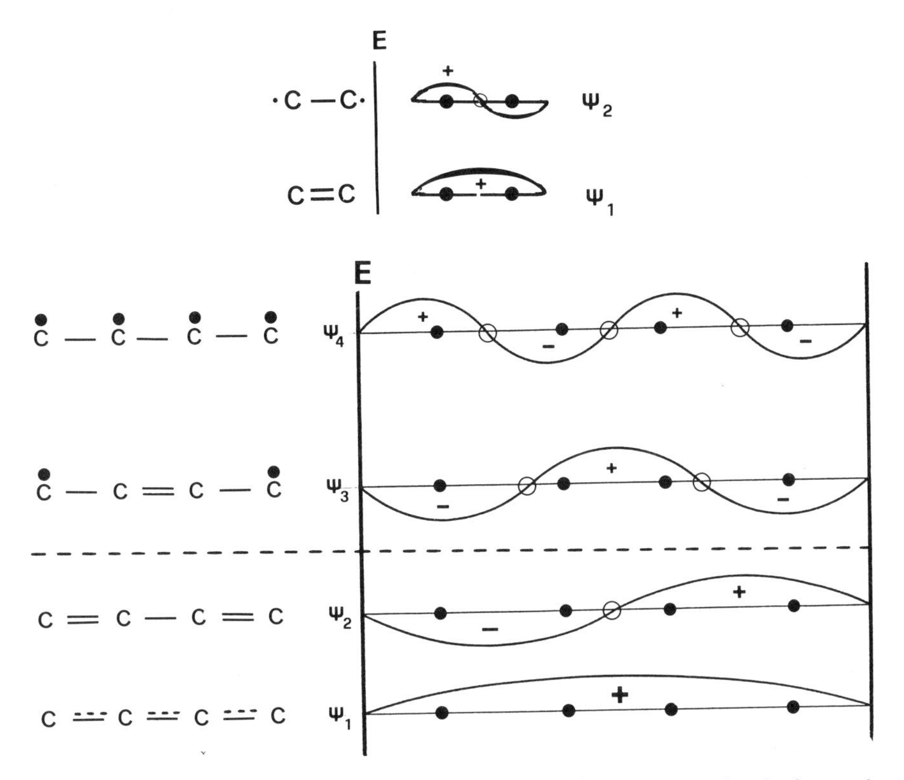

Figure 3.17 *The wavefunctions of delocalized π orbitals of linear conjugated molecules can be drawn following the number of nodes (open circles)*

antibonds between the atoms: a node is an antibond, and absence of a node being a bond.

Orbital ψ_2 is the highest occupied molecular orbital (HOMO) in the ground state. It corresponds to the structural formula of the molecule, with double bonds between C_1 and C_2, and between C_3 and C_4. Orbital ψ_3 is likewise the lowest unoccupied molecular orbital (LUMO) in the ground state and corresponds to a biradical structure of the molecule with unpaired electrons on C_1 and C_4. Such a biradical structure can be a very simple but sometimes useful representation of the excited molecule (in states S_1 or T_1).

3.2.4 Cyclic Conjugated Systems: Aromatic Molecules

The molecule of benzene is the simplest representative of a very large family of cyclic conjugated molecules which are of the greatest importance in organic chemistry. Aromatic molecules are formed essentially through the substitution, linking and fusion of the benzene ring (or rings) which consists of six carbon atoms linked in a hexagonal cycle. Since there are six p orbitals which interact to form the molecular orbitals, there will be three bonding π orbitals and three antibonding π* orbitals. The wavefunctions can be obtained also from the nodal properties of standing waves, but in the case of a cyclic system the symmetry implies that there must be nodal lines instead of just nodal points.

The orbitals φ_2 and φ_3 have one nodal line each, passing either through two atoms or through two bonds; but since these orbitals have the same number of nodes, it follows that they must have the same energy (they are degenerate). The same remark applies to the first antibonding orbitals φ_4 and φ_5, which have two nodal lines each.

The fact that these HOMOs and LUMOs have a two-fold degeneracy implies that there are four isoenergetic one-electron transitions to yield the first excited states; this complication is however resolved by the interaction of these one-electron excitations, and this is known as *configuration interaction*. The concept of configuration interaction (CI) is somewhat similar to that of the interaction of atomic orbitals to form molecular orbitals. An electron configuration defines the distribution of electrons in the available orbitals, and an actual *state* is a combination of any number of such electron configurations, the state wavefunction being

$$\Psi = \sum_i c_i \psi_i \tag{3.2}$$

where c_i is the coefficient of the ith configuration. We shall consider here only the case of benzene, which is of special importance for us. Figure 3.18 shows the calculated energies and symmetries of the first three excited singlet and triplet states. The S_1 excited state is denoted as ^1L_b, where 'b' stands for 'bond' since the wavefunction has a node through the bonds C_2–C_3 and C_5–C_6. There is also a node through the atoms (C_1, C_4) in the ^1L_a state ('a' stands for 'atom'). These symmetry properties are of importance for the understanding of the effect of substituents in the spectroscopy and photochemistry of benzene derivatives, we shall therefore consider them briefly in a qualitative way.

In benzene itself the numbering of the atoms is of course arbitrary, so we take a molecule such as toluene is which the symmetry is broken. We can now recognize two axes in the benzene ring which correspond to the 'a' and 'b' nodal planes (see Figure 3.19). In the absorption spectrum of toluene the first two bands are the S_0–^1L_b and S_0–^1L_a transitions and their transition dipole moments follow respectively the short axis 'b' and the long axis 'a' of the molecule. If a second methyl group for instance is substituted in position 4 (*para*), then the energy of the ^1L_a state is lowered, while that of the ^1L_b state remains unchanged. Conversely, substitution in the 2 (*ortho*) or 3 (*meta*) positions affects only the ^1L_b state. Thus in the photophysics of benzene derivatives the *ortho* and *meta* isomers are similar, the *para* being different, in contrast to the ground state properties where the general rule is that *ortho* and *para* substituted molecules are similar and it is the *meta* isomer which is the odd one out.

In benzene and its derivatives the transition S_0–^1L_b is symmetry forbidden, while the S_0–^1L_a is allowed; this is readily seen from the absorption

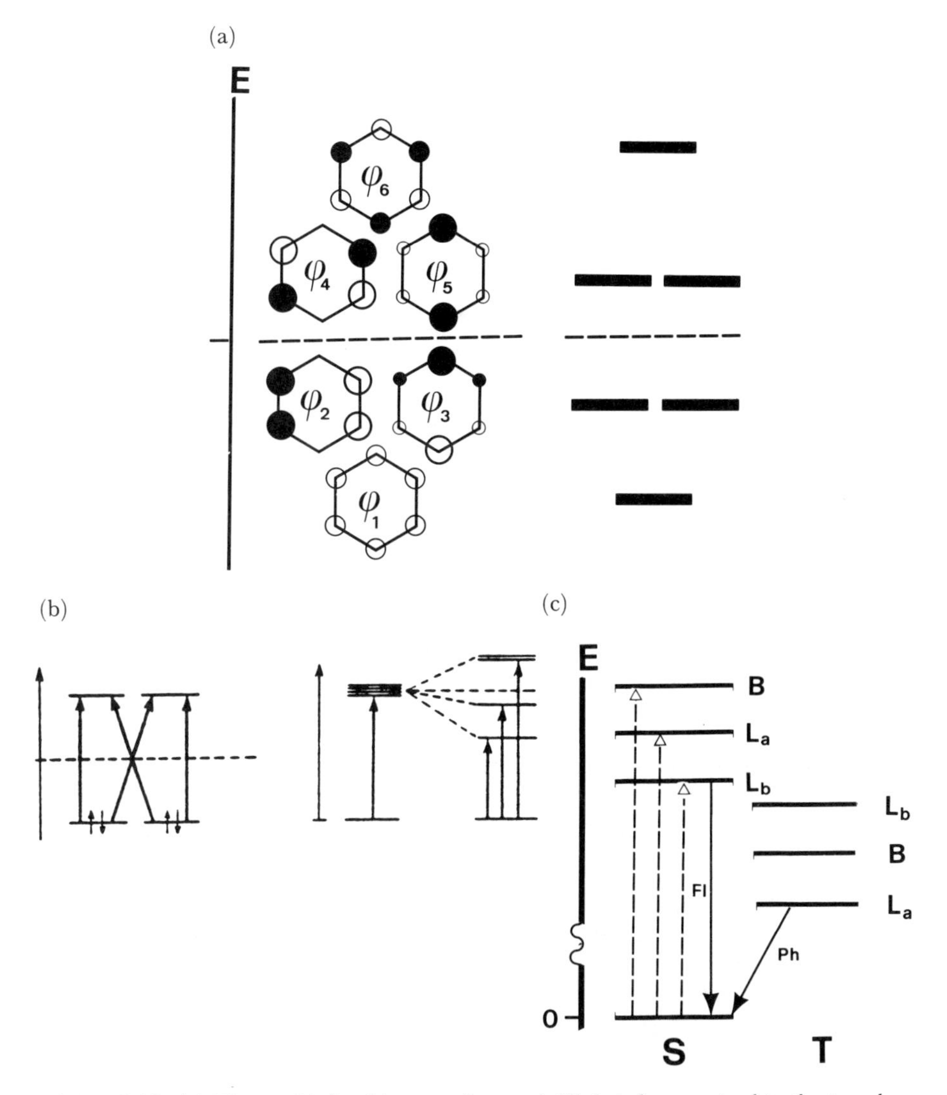

Figure 3.18 *(a) The π orbitals of benzene. Open and filled circles correspond to the + and − signs of the half-lobes, and their size represents the electron density. (b) The four HOMO–LUMO transitions and their splitting. (c) The ordering of energy states through configuration interaction*

spectra in which the second band is always more intense (extinction coefficient ε around 10^4) than the first band (ε around 10^2). In all aromatic molecules with an S_1 excited state ^1L_b the fluorescence lifetime is very long, coming close to 1 μs in some cases (the mathematical definition of ε is given in section 3.4.2).

The situation is quite different in the triplet manifold, as a result of the characteristic singlet–triplet splittings of the various states. The most important remark is that the splitting of the $^1L_a/^3L_a$ states is very large, while that

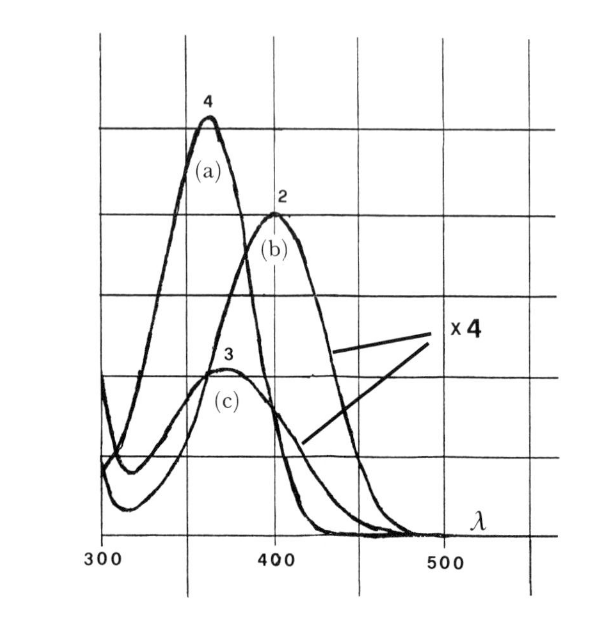

Figure 3.19 *Left: In a substituted benzene the transition moments are polarized along the short axis, b, or the long axis, a. In disubstituted benzenes the 2- and 3-isomers are similar, but differ from the 4-isomer, as shown by the absorption spectra of nitroanilines. Right: (a) 4-Nitroaniline, (b) 2-nitroaniline, (c) 3-nitroaniline. Wavelength λ in nm/100, absorbance A in arbitrary units scaled by a factor of 4 for (b) and (c)*

of the $^1L_b/^3L_b$ states is very small. The lowest triplet level T_1 is therefore ^3L_a, but intersystem crossing must go from S_1 (^1L_b) to T_3 (^3L_b) which is a strongly forbidden transition.

In the larger aromatic molecules the ordering of these L_a and L_b states can vary, and we shall mention only the cases of naphthalene and anthracene. In naphthalene the situation pertaining to benzene is kept by and large, with a lowest ^1L_b state of very long lifetime in view of the forbidden nature of its transitions to S_0 or to the triplet manifold. In anthracene however the first excited singlet state S_1 is ^1L_a, which is short-axis polarized: note that now it is the short axis which passes through atoms (a) and the long-axis which goes through bonds (b). The fluorescence lifetime of anthracene is very short, under 10 ns, and the intersystem crossing is efficient.

3.3 CHARGE TRANSFER IN EXCITED STATES; 'TICT' STATES

An electronic transition involves the motion of an electron from one orbital to another. In the simple model of a one-electron transition it is assumed that all other particles remain in their initial states. If the initial and final orbitals of the electron are separated in space there will be a change in dipole moment, so long as the molecule has no centre of symmetry. When this change is very large the excited state is described as a "charge transfer"

(CT) state. There is something arbitrary about this definition, since it depends on the magnitude of the dipole which is considered to be a charge transfer.

3.3.1 Charge Transfer States in Organic Molecules

In order to have a CT state, the molecule must have two parts which can be distinguished as 'electron donor' and 'electron acceptor'. Nitrobenzene ($ArNO_2$ in Figure 3.20a) provides a good example of this situation; the benzene ring is the electron donor, the nitro group is the electron acceptor. In a very simple picture of localized orbitals, the antibonding orbitals π^*_{Ar} and $\pi^*_{NO_2}$ are close in energy, while the bonding orbitals π_{Ar} and π_{NO_2} are well separated (Figure 3.20). In a first approximation these localized bonding orbitals do not interact, but the antibonding orbitals form two delocalized molecular orbitals. The first electronic transition promotes an electron from the orbital π_{Ar} localized on the benzene ring to the orbital $\pi^*_{ArNO_2}$ which has a coefficient c^*_{Ar} on this ring and a coefficient $c^*_{NO_2}$ on the nitro group. Assuming that the coefficient of the π_{NO_2} nitro orbital in the HOMO is zero, the CT character of the LUMO is then simply $(c^*_{NO_2})^2$. A similar situation arises in a molecule such as aniline, but here the benzene ring is the acceptor and the amino group is the donor. The HOMO orbitals of the two groups come close together, so that their interaction leads to two delocalized orbitals π_1 and π_2 with coefficients c_D (D = donor) and c_{Ar}; the localized LUMOs however are so far apart that their interaction can be neglected, so that they retain their localized character (Figure 3.20b).

From these examples it can be seen that CT characters of electronic states vary from 0 to 1, depending on the interaction of the localized orbitals. A CT character of 1 (100%) would correspond to the transfer of a whole electronic charge, and this could exist only if the donor and acceptor orbitals were totally separated in space. Such an electronic transition would be forbidden because the spatial overlap of the orbitals is zero. We shall see however that total charge separation can take place in some rather special cases, though not through a direct transition. Before discussing these twisted intramolecular CT ('TICT') states, a few words about donor–aromatic–acceptor (DArA) molecules are appropriate.

A molecule such as 4-nitroaniline provides a good example of DArA molecules. It is in a way a mixture of nitrobenzene and aniline. In the HOMO the benzene ring and the amino group interact strongly to form two delocalized orbitals, ignoring the nitro group. For the excited state the donor group is then the aniline moiety, the nitro group being the acceptor. In the first π–π^* excited state of 4-nitroaniline the dipole moment of some 14 D corresponds to a CT character of 80%. The 2- and 3-nitroaniline isomers show similar behaviour, but the CT characters of the lowest π–π^* states are not quite so high. This is linked to the polarization of the S_0–S_1 transitions, these being short-axis polarized in the 2- and 3-isomers, but long-axis polarized in 4-nitroaniline. Figure 3.19(b) shows the absorption spectra of

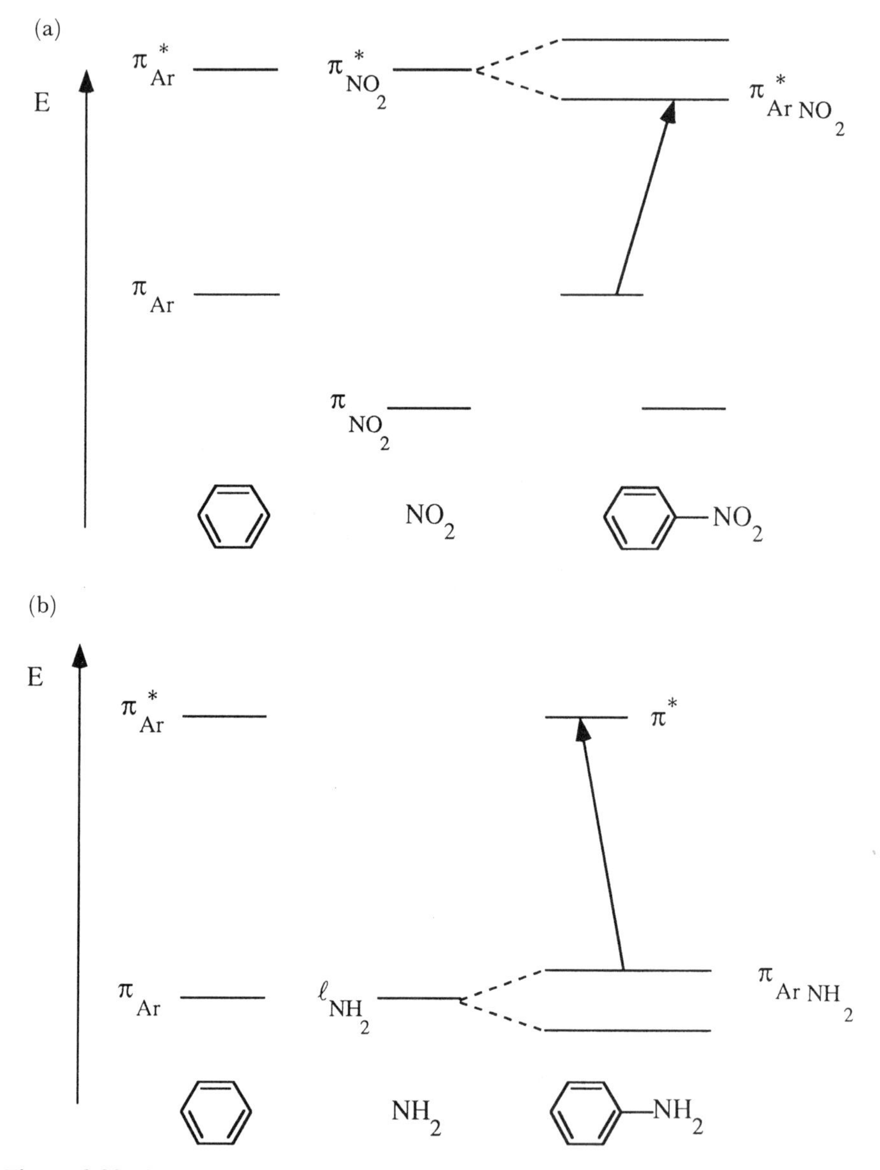

Figure 3.20 *Charge transfer in the HOMO–LUMO transitions of (a) nitrobenzene and (b) aniline*

these molecules. It is a general observation that in the benzene derivatives the excited states S_1 have similar electron distributions in the 2- and 3-isomers, the 4-isomer being different, in contrast to the ground state properties (*e.g.* substitution reactions) in which the 2- and 4-isomers differ from the 3-(*meta*)isomer.

3.3.2 Twisted Intramolecular Charge Transfer States

The molecule *N,N*-dimethylaminobenzonitrile (DMABN) is a prime example of the TICT state. In the ground state, the molecule is planar so that conjugation between the NMe_2 group and the aromatic ring is at a maximum. This geometry is retained in the excited states reached directly by the absorption of light. However, there is also a special excited state in which the NMe_2 group is twisted at a right angle so that conjugation is altogether lost; in this TICT state there is total charge separation between the NMe_2 group and the cyanobenzene moiety (Figure 3.21). The TICT state cannot be reached by direct light absorption since its geometry is totally different from that of the ground state; it is formed by the 'slow' twisting of the two parts of the molecule from the planar state S_1 structure. The solvent plays a decisive role in this process in stabilizing the highly dipolar TICT state below the level of S_1. Thus the TICT state of DMABN is formed only in polar solvents, and in such cases a dual fluorescence can be observed; the 'normal' fluorescence from S_1 and the 'anomalous' fluorescence from the TICT state (Figure 3.22).

The occurrence of TICT states was regarded for a long time as a very special property of a few unusual molecules. It seems now to be a much more general property common to many bichromophoric molecules.

3.4 TRANSITIONS BETWEEN ENERGY STATES

A detailed Jablonski diagram is useful to illustrate the various transitions which connect the energy states of a molecule (Figure 3.23). We have taken a polyatomic organic molecule as a rather general example, and we considered only the first few excited states. Transitions can be classified as 'radiative' or 'non-radiative'. The former involve a 'vertical' change in energy

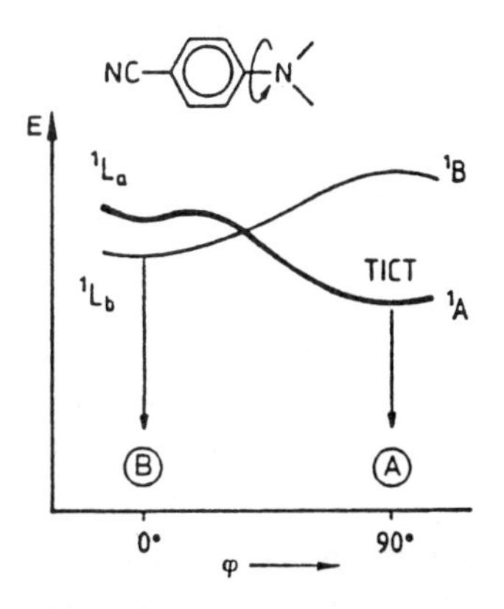

Figure 3.21 *Potential energy diagram of DMABN as a function of the twist angle of the nitrogen lone pair orbital. In the planar conformation the states ^1L_a and ^1L_b are the observable spectroscopic states; these give rise to the states A (TICT) and B through 90° internal rotation*

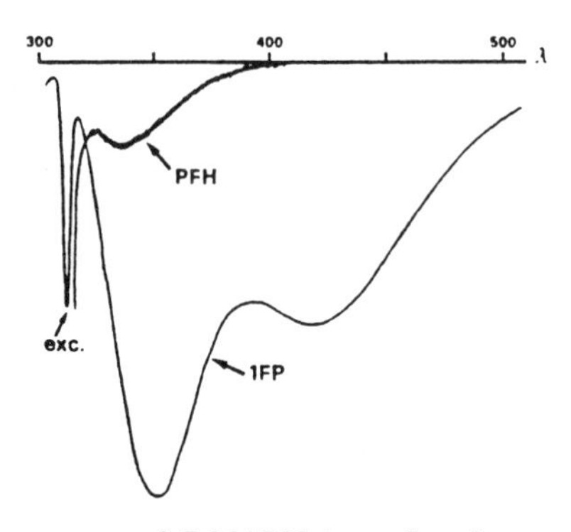

Figure 3.22 *Luminescence spectra of DMABN in perfluorohexane (PFH; non-polar solvent) and 1-fluoropentane (1FP; polar solvent); λ in nm*

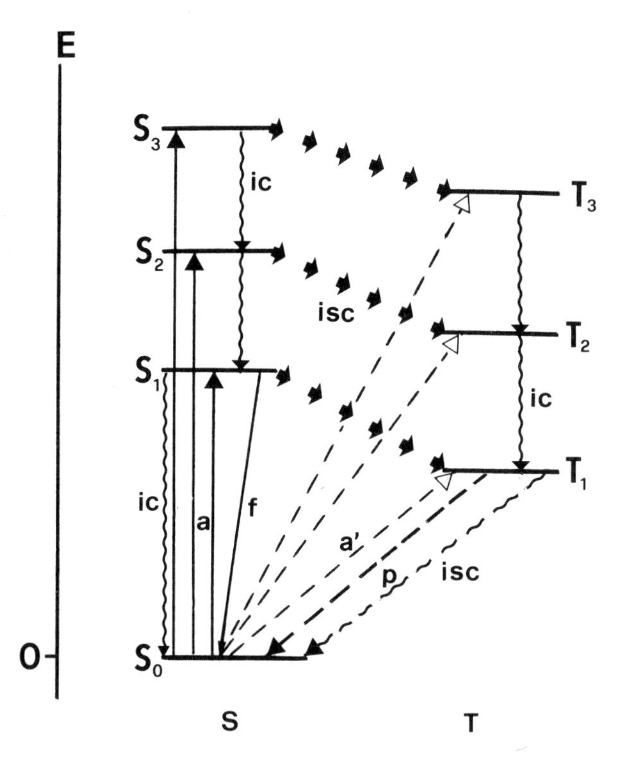

Figure 3.23 *Jablonski diagram of the transitions between electronic states of a polyatomic molecule. 'ic' stands for internal conversion, 'isc' for intersystem crossing, 'a' and 'a'' for absorption of light, 'f' for fluorescence and 'p' for phosphorescence*

which is an increase when a photon is absorbed and a decrease when a photon is emitted. Here the solid arrows correspond to such radiative transitions. Non-radiative transitions reorganize internally the energy of the molecule and eventually degrade it into heat in the surroundings; the wavy arrows which stand for these transitions always go downwards, and it may be useful to make here a few remarks about the thermal population of excited states, since this would lead to upwards non-radiative transitions. In Figure 3.23 isc (intersystem crossing) is shown by the short arrows.

3.4.1 Thermal Population of Electronically Excited States

In classical thermodynamics the Maxwell–Boltzmann distribution law gives the relative population of a state of energy E_i when the molecules have many states available, of energies E_1, E_2 .... This sum is the partition function or state sum, and it is a weighted sum of all available energy states (weighted according to their availability or probability of occupation at temperature T). This implies of course that states of higher energy are reached only by thermal activation without the intervention of external energy (*e.g.* light).

The energies of electronic states of molecules are far greater than kT, and the partition function is practically unity; this means that even if a molecule has one ground state and many excited states, only one state is actually available; the ground state. In principle it would be possible to populate electronically excited states by raising the temperature, for then kT could come close to the excited state energy. This is impossible in practice, because the temperatures required are so high that the molecules would decompose through thermal reactions well before such temperatures are reached.

3.4.2 Radiative Transitions

All upward radiative transitions in Figure 3.23 are absorptions which can promote a molecule from the ground state to an excited state, or from an excited state to a higher excited state. We have seen that the probability of these transitions is related ultimately to the transition moment between the two states and thereby to the Einstein coefficient A. In practice two other related quantities are used to define the 'intensity' of an absorption, the oscillator strength f and the molar decadic extinction coefficient ε.

The oscillator strength is used mostly in theoretical calculations. Its value is zero for a totally forbidden transition (which could not be observed!) and around 1 for the most allowed transitions which give rise to the strongest known absorption bands.

The quantity used in experimental photophysics is ε. It is defined according to Beer's law as follows: consider a beam of monochromatic light (of wavelength λ) which goes through some absorbing matter of concentration c contained in a cell of pathlength l (Figure 3.24). The intensity of the incident beam is I_0, that of the transmitted beam is I_t. If the probability of a photon being absorbed by a molecule is constant, the decrease dI in light intensity in a differential pathlength element is

$$\mathrm{d}I = -pcI\,\mathrm{d}l \qquad (3.3)$$

This leads to the practical expression of Beer's law in the form

$$I_0/I_t = 10^{-A} \qquad \text{with } A = \varepsilon cl \qquad (3.4)$$

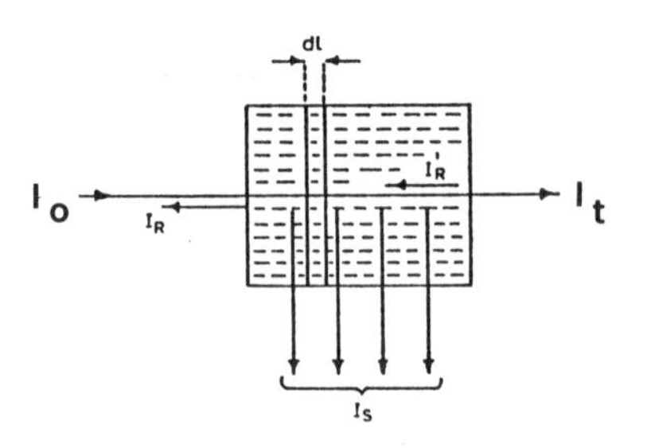

Figure 3.24 *Absorption, scattering and reflection of light as it passes through matter. I_0 is the incident beam intensity, I_t the transmitted beam intensity, I_R and I_R' are the intensities of light reflected at the optical interfaces and I_s is the total intensity of scattered light*

where A is defined as the 'absorbance' (formerly called the 'optical density'), c is the molar concentration of the absorbing species and l is the optical pathlength in cm. The molar decadic extinction coefficient does not belong to any coherent unit system (SI or cgs) but has become accepted for spectroscopic measurements. Its units are therefore $\text{M}^{-1}\,\text{cm}^{-1}$.

Values of ε span a very wide range, from near 0 to nearly 10^6. The former are characteristic of forbidden transitions such as singlet–triplet absorptions which are spin-forbidden; the latter can be found in the absorption spectra of some dyes which have very strong colour in dilute solutions. Since values of ε can vary between 0 and 10^6 for certain molecules, absorption spectra are often given on a logarithmic scale.

Figure 3.25 shows a simplified example of the absorption spectrum of an aromatic molecule such as the rigid, planar, cyclic hydrocarbons (*e.g.* benzene, naphthalene, *etc.*). The first absorption band shows a clear progression of vibrational sub-levels, but the higher absorption bands are broad and structureless; this results from the very strong vibrational coupling between S_1 and S_2, S_2 and S_3, *etc.*

It follows from Beer's law that the intensity of a beam of light decreases exponentially as it passes through an absorber (Figure 3.26). The most important limitation of Beer's law is that it applies only to strictly monochromatic light. Deviations can be traced in many cases to the non-fulfilment of this requirement, for instance in the case of highly structured spectra. It is essential that the bandwidth of the light should be small compared to the width of the absorption band or line of the spectrum.

Beer's law applies to a single absorbing species characterized by the value of the extinction coefficient at the observation wavelength. If there are several independent absorbers the total absorbance is the sum of the individual absorbances

$$A = l\sum_i \varepsilon_i c_i \qquad (3.5)$$

If the absorbers are not independent but are related through a chemical

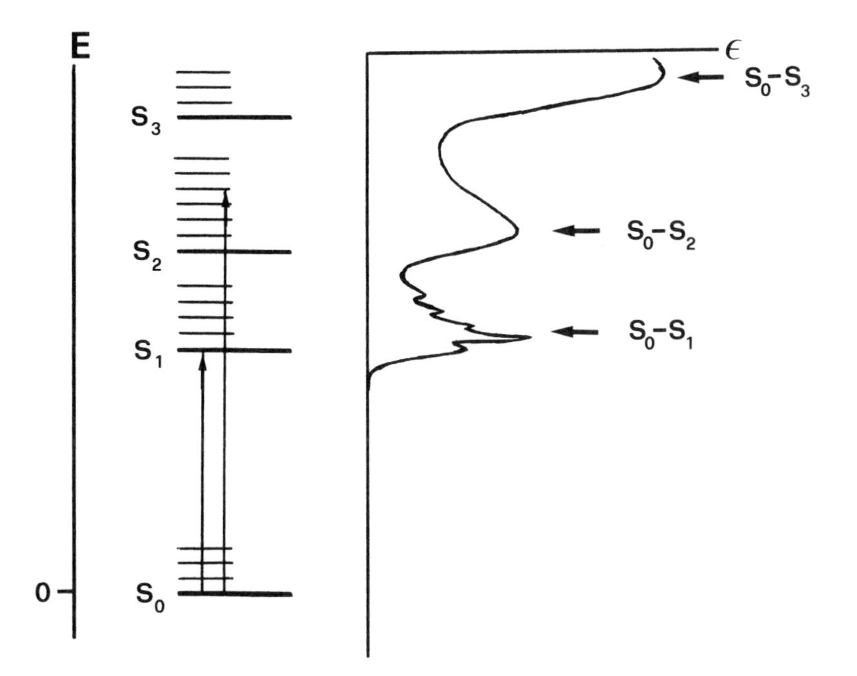

Figure 3.25 *Outline of the absorption spectrum of a rigid polyatomic molecule. The bands corresponding to electronic transitions are broad as they include vibrational and rotational transitions and they coalesce to form an absorption continuum*

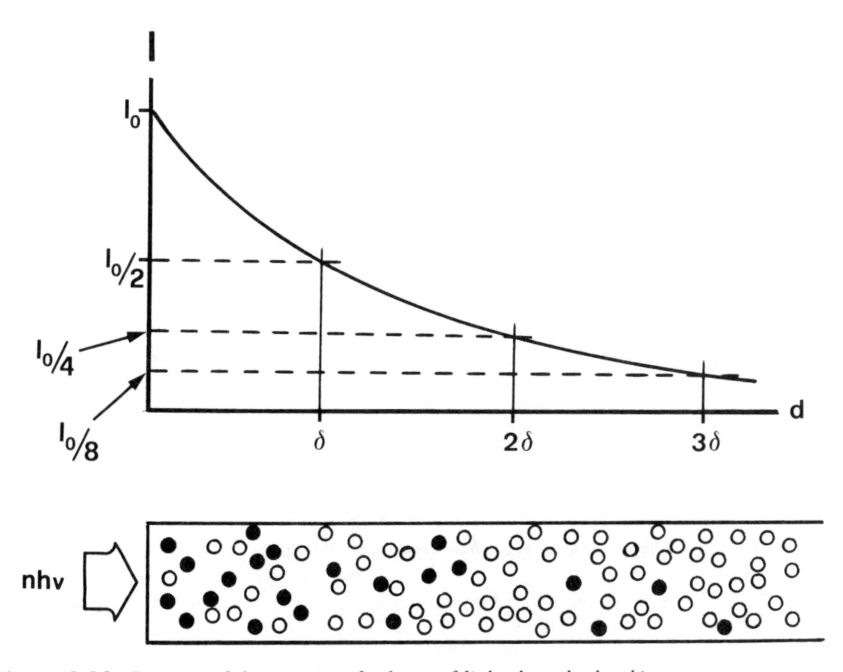

Figure 3.26 *Decrease of the intensity of a beam of light through absorbing matter*

equilibrium (*e.g.* complex formation $M + P \rightleftarrows N$) the absorbance may not be proportional to the concentration of M, if N and/or P also absorb at the same wavelength.

The only other limitation of Beer's law concerns biphotonic processes in which the apparent extinction coefficient depends on the intensity of light; this can become important in conditions of laser photolysis.

3.4.2.1 Light Scattering: Virtual States. When a beam of electromagnetic radiation passes through matter, some of it is transmitted in the original direction, some of it may be absorbed and some of it is scattered in all directions (Figure 2.1, p. 11). If the beam is monochromatic the wavelength of transmitted light must be unchanged, but the wavelengths of scattered light can either remain unchanged (Rayleigh scattering) or change in positive or negative increments of vibrational (sometimes rotational) levels of the scattering molecules (Raman scattering).

Figure 3.27 illustrates the scattering processes in the simple case of non-absorbing matter. The photon energy $h\nu$ of the incident beam is not sufficient to reach the first excited state S_1 so that light cannot be absorbed. However, the photon still promotes the molecule to an extremely short-lived 'virtual' state from which it must fall back to S_0. Most of the light is re-emitted without change of direction; this is the transmitted light. Some of it is re-emitted at different angles but still with the same wavelength, and a small part is re-emitted to other vibrational levels of the ground state. The Raman spectrum of scattered light then shows lines at the long-wavelength side of the Rayleigh scattered beam $h\nu$, the differences corresponding to the vibrational levels v_1, v_2, *etc.*

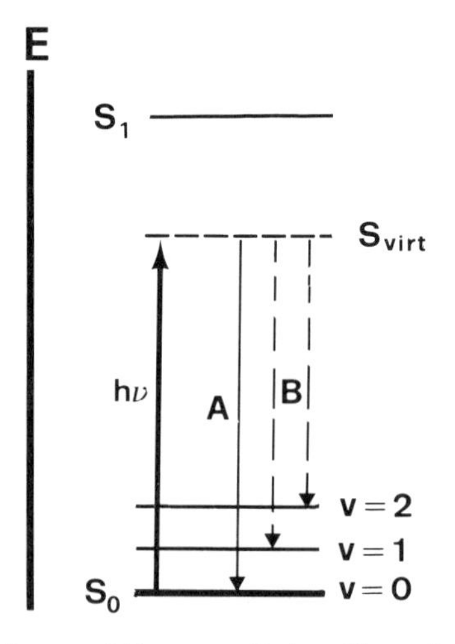

Figure 3.27 *Outline of the process of Raman scattering. The molecule promoted from the zeroth vibrational level of the ground state to a virtual state (S_{virt}) can fall back to higher vibrational levels of the ground state with the emission of scattered light of longer wavelength (B). (A) is the Rayleigh scattering without change of wavelength*

3.4.2.2 Anti-Stokes Raman Scattering. Even at ordinary temperatures a small proportion of molecules can be found in higher vibrational levels of the electronic ground state S_0. Excitation of such molecules will promote them to higher virtual states from which they can return to S_0 (v = 0); the scattered light is then of shorter wavelength than the incident light with photon energies $h\nu + \Delta E(\mathrm{v} = 0, \mathrm{v} = 1 \ldots)$, *etc.*

Luminescence emissions are the downward transitions in which an excited molecule loses energy by emission of a photon; it will then reach the ground state or some lower excited state.

A distinction is made between 'fluorescence' and 'phosphorescence' according to the change in the spin quantum number between the initial and final states. When these quantum numbers are the same $\Delta S = 0$ and the transition is spin-allowed (but there may be other factors to be taken into account as well, as we shall see); this emission is then defined as a fluorescence. If there is any change in the spin quantum number, $\Delta S \neq 0$, the transition is spin forbidden and is defined as a phosphorescence. The important difference between these two forms of luminescence resides in their kinetics: a fluorescence is shortlived, with emission lifetimes in the range 1 ns to 1 μs, while phosphorescence lifetimes go from 1 ms to many seconds or even minutes.

The luminescence spectra of rigid molecules like benzene, naphthalene, *etc.* show clear vibrational structures, especially in the gas phase. The first absorption band and the fluorescence band then show a mirror-image relationship in which the 0–0 vibrational transitions are nearly coincident (Figure 3.28). At room temperature only the lowest vibrational levels (v = 0)

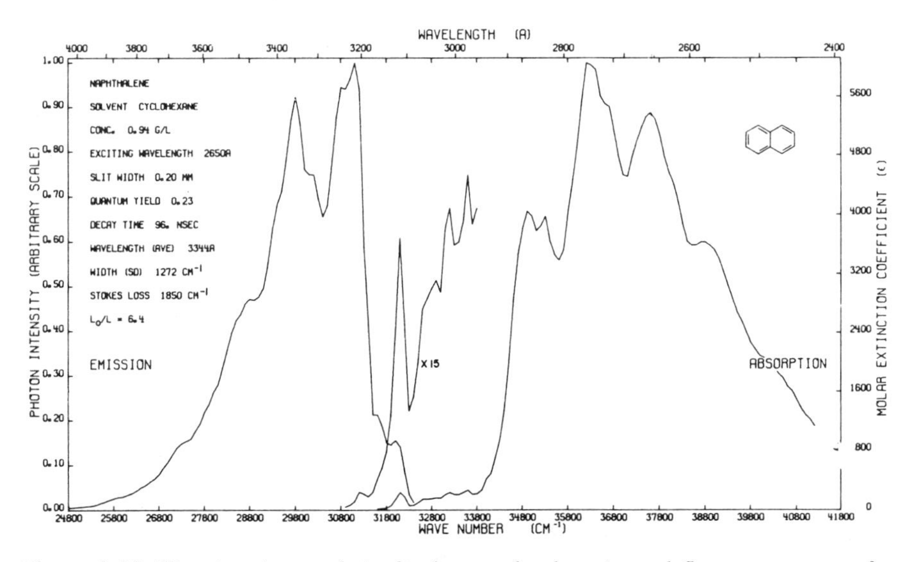

Figure 3.28 *The mirror-image relationship between the absorption and fluorescence spectra of a rigid molecule. In the gas phase the 0–0 bands would be theoretically coincident, and the spacing between the vibrational levels is not strictly identical in the ground and excited states*

of any state (S_0 or S_1) are significantly populated in conditions of thermal equilibrium, so that all absorption transitions start from the $v = 0$ level of S_0 and go to any v level of S_1. Similarly the fluorescence transitions start from the $v = 0$ of S_1 to the various v levels of S_0. Most molecular luminescence spectra are however broad and structureless because they have many vibrational and rotational levels within each electronic state. In addition, solvation in liquids leads to many different environments for the solute molecule in the solvent, and these environments yield a continuum of energy states.

The 'Stoke's shift' is defined as the difference between the maxima of the luminescence and those of the related absorption spectra. In rigid, non-polar molecules the 0–0 bands are nearly coincident, but in many cases they do not correspond to the maxima of the spectra, so that the Stoke's shift is quite large. It increases when the molecular geometry changes substantially between the ground state and the excited state, and increases with solvent polarity when the excited state has a larger dipole moment.

3.4.2.3 Luminescence Spectra of Polyatomic Molecules: Kasha's Rule. Even though polyatomic molecules have many electronically excited states, luminescence emission is observed almost exclusively from the lowest excited states of any given multiplicity, *e.g.* S_1 and T_1 in the case of closed-shell organic molecules. This is known as Kasha's rule and results from the very high rate constants of non-radiative deactivation within each multiplicity manifold (typically $10^{12}\ s^{-1}$), compared with the rate constants of radiative transitions ($10^9\ s^{-1}$ at most). This does not mean that emissions from upper excited states do not exist, simply their quantum yields are very low, usually so low as to be indetectable. Figure 3.29 shows two examples of fluorescence and phosphorescence spectra of organic molecules; the second example is an exception to Kasha's rule. The molecule azulene (an isomer of naphthalene) is characterized by an unusual pattern of excited state energy levels. The gap between S_0 and S_1 is small, and that between S_1 and S_2 is so large that the absorption spectrum is practically empty in that region. Azulene shows fluorescence from S_2 to S_0 and from S_2 to S_1, but not from S_1 to S_0. Kasha's rule is clearly related to the vibrational overlap between electronic states, *i.e.* to the energy gap law.

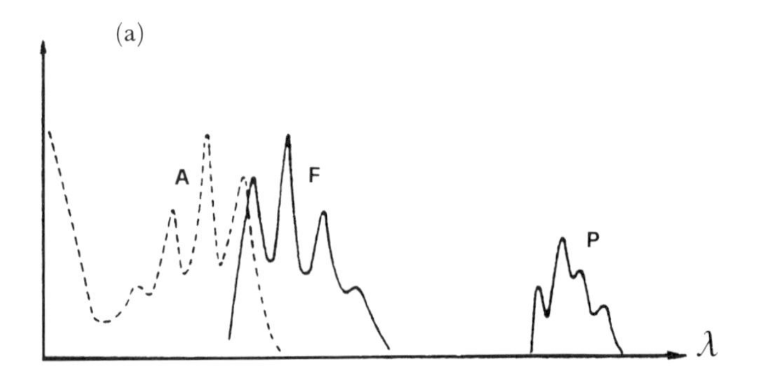

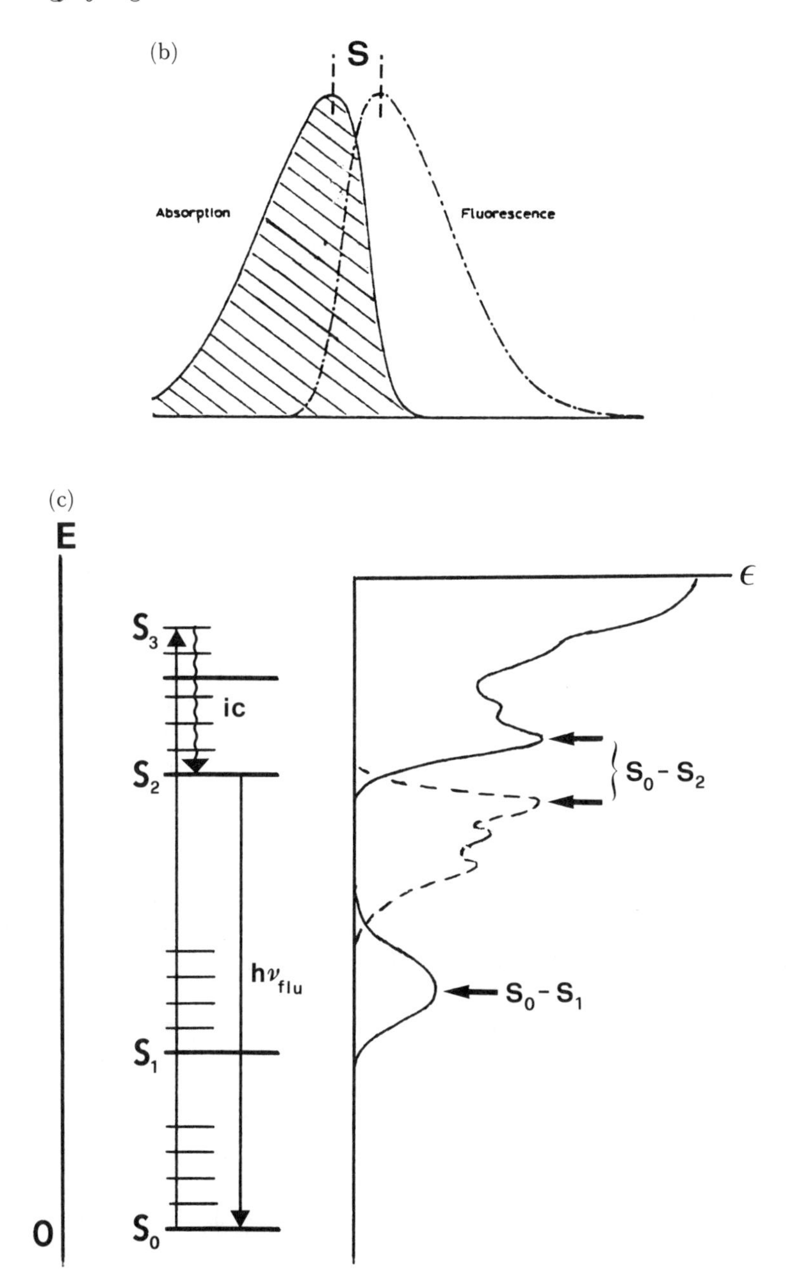

Figure 3.29 *(a) Outline of the absorption, A fluorescence, F and phosphorescence, P spectra of a rigid polyatomic molecule. λ = wavelength, vertical axis = absorbance (A) or emission intensity (F, P). (b) The Stokes shift of the absorption and fluorescence spectra is defined as the difference between their maxima. When this shift is small, there is a substantial spectral overlap between absorption and emission. (c) Jablonski diagram and outline of the absorption and fluorescence spectra of azulene, an exception to Kasha's rule. The energy gap between S_0 and S_1 is very small, that between S_1 and S_2 is very large*

3.4.2.4 Luminescence Quantum Yields of Polyatomic Molecules: Vavilov's Rule. In general the luminescence quantum yield of polyatomic organic molecules is independent of the excitation wavelength. Internal conversion between states of the same multiplicity is much faster than intersystem crossing to another multiplicity manifold (by a factor of 10^3 or 10^4). When excitation promotes the molecule to an upper excited singlet state such as S_2 or S_3, these will decay to S_1 rather than cross over to T_2 or T_3. The triplet manifold is therefore populated only by intersystem crossing from S_1, so that both the fluorescence and phosphorescence quantum yields are independent of the energy of the initial state reached in absorption.

The independence of luminescence quantum yields on excitation wavelength is known as Vavilov's rule. There are however many exceptions to this rule, in particular for molecules which contain heavy atoms such as Br and I, or metals (*e.g.* organometallic complexes). The heavy atom effect makes intersystem crossings more efficient and these can compete with internal conversions.

3.4.2.5 Polarization of Radiative Transitions. In a light-induced electronic transition an electron moves from one orbital to another. This requires a force which is provided by the electric vector of the photon. This electric vector is perpendicular to the direction of propagation of light; it defines the direction of polarization of the transition. The transition is most probable when the electric vector of the photon is exactly parallel to the transition dipole moment of the molecule; if these vectors are at right angle, then the transition cannot take place, even though the transition moment may be large. For intermediate orientations of the vectors $\vec{E}$ and $\vec{M}$, the transition probability follows a simple relationship

$$p = p_{\max} \cos \theta \tag{3.6}$$

θ being the angle between $\vec{E}$ and $\vec{M}$.

Figure 3.30 gives an illustration of this property of polarization of a

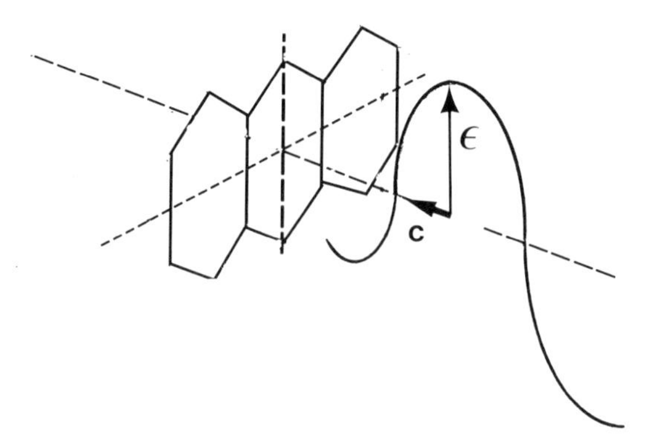

Figure 3.30 *Polarization of an electronic transition. The electric vector of light is $\vec{E}$, the direction of propagation $\vec{c}$. Interaction of this photon with the molecule of anthracene is favourable for the transition to the ^1L_a state*

radiative transition in the case of the first electronic transition of anthracene. The first absorption band (and the fluorescence band) of anthracene corresponds to the $S_0 \rightarrow S_1$ (^1L_a) transition, which is short-axis polarized; the photon should come ideally at right angles to the molecular plane, and its polarization should be such that its electric vector coincides with the short axis of the molecule for maximum probability of absorption. Different transitions in the same molecule may have different polarizations. Experimentally the direction of polarization is obtained from the measurement of the luminescence spectrum in polarized light. This is fairly straightforward when all the molecules follow a common orientation, as in certain crystals. In random samples it is still possible to measure polarized luminescence spectra, so long as the molecules do not move appreciably during the emission lifetime. This is the case in solid samples (*e.g.* frozen at low temperatures) but not in liquids in which the molecule can rotate in sub-nanosecond times. The luminescence spectrum then shows no polarization, a process known as rotational depolarization.

3.4.2.6 Rate Constants of Radiative Transitions. The natural radiative rate constant k_r of an electronic transition from a state S_i to a state S_f is related to the transition moment M and thereby to the oscillator strength f. It is convenient to factorize f to highlight the various factors which determine to what extent a transition is 'allowed' (f near 1) or 'forbidden' (f near 0). The transition moment includes the displacement of all particles of the molecule during the transition, nuclei as well as electrons. The heavy nuclei move much more slowly than the light electrons so that their motions can be considered to be independent. Within this approximation the transition moment is given as

$$\vec{M} = \langle \theta_i \theta_f \rangle \langle \psi_i | \vec{\mu} | \psi_f \rangle \tag{3.7}$$

For a one-electron transition the spin and orbital overlaps can be separated to yield

$$\vec{M} = \langle \theta_i \theta_f \rangle \langle S_i S_f \; \varphi_i | \vec{\mu} | \varphi_f \rangle \tag{3.8}$$

The vibrational overlap factor $\langle \theta_i \theta_f \rangle$ is also known as the 'Franck–Condon' factor and represents the probability of finding a common nuclear geometry in the initial and final states. If the nuclei do not move during the electronic transition, such a transition can take place only in such a common nuclear geometry. Figure 3.12, p. 39, provides an illustration of this Franck–Condon principle; electronic radiative transitions are vertical, non-radiative transitions are horizontal but they are followed by vertical vibrational deactivation.

Within the Born–Oppenheimer (B–O) approximation the oscillator strength can be expressed as a product of four independent factors

$$f = f_s \, f_p \, f_y \, f_v \tag{3.9}$$

f_s is the spin overlap factor which is theoretically 1 for a spin-allowed transition (*e.g.* singlet–singlet), or 0 for a spin forbidden transition (singlet–triplet); actual values of f_s depend on the magnitude of spin-orbit coupling, so it is never 0. f_p represents the spatial overlap of the orbitals; even if all other factors are favourable, an electronic transition between two orbitals widely separated in space is very unlikely. f_y is the symmetry factor related to the overlap integral of the wavefunctions; the transition is symmetry-forbidden if it implies a change in the angular momentum of the electron, as in a jump between an n and a π (or π^*) orbital in carbonyl compounds. This overlap integral includes only the electronic parts of the total wavefunctions. f_v is the vibrational overlap integral or Franck–Condon factor, as seen above.

Since the four factors which are combined to yield the oscillator strength f appear as factors, the transition will be 'forbidden' if any one of these factors is very small.

3.4.2.7 Breakdown of the Born–Oppenheimer Approximation. The B–O approximation is based on the independence of the motions of nuclei and electrons. This is generally a reasonable assumption, except at the crossing point of two electronic states where a minor nuclear displacement is linked to the transition between two electronic states (Figure 3.31).

3.4.2.8 Luminescence Kinetics, Luminescence Lifetimes. The Einstein coefficient A for spontaneous emission gives the probability of radiative transition. Since this probability is the same for all molecules of the same excited species, it follows that the decrease in the number of excited molecules within a differential time increment is simply proportional to the number of excited

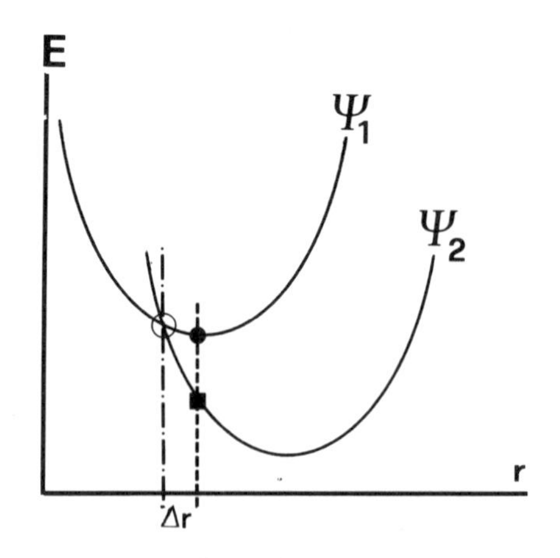

Figure 3.31 *Near the crossing point of two electronic energy states of a molecule, a very small change in nuclear geometry* r *can result in a very large change in the electronic wavefunction*

molecules

$$dN/dt = -k_r[N] \quad (3.10)$$

which integrates to

$$N(t) = N(0)\exp(-k_r\, t) \quad (3.11)$$

This is similar to a first-order reaction in chemical kinetics and follows the same law as radioactive decay. The rate constant k_r defined in this manner is the 'natural radiative' rate constant which also defines the natural radiative lifetime

$$\tau_r = 1/k_r \quad (3.12)$$

According to this definition the 'lifetime' is the time required for the luminescence intensity to drop from $I(0)$ to $I_t = I(0)/e$; it should not be confused with the 'half-life' often used for radioactive decays, this being the time $t_{1/2}$ required to decrease the intensity of emission from $I(0)$ to $I(0)/2$.

The natural radiative lifetime is the longest (average) lifetime of an excited molecule. This lifetime is seldom observable in practice because there are other deactivation processes which compete with the luminescence emission. These can be intramolecular, non-radiative transitions (internal conversion or intersystem crossing) or intermolecular quenching processes; these are considered in the next sections.

The observable kinetics of a luminescence can be derived from the energy diagram of Figure 3.32. This shows a simple example of two competing decay paths, fluorescence from S_1 to S_0, and a non-radiative transition from S_1 to T_1. These are two first-order processes of rate constants k_f and k_{isc} respectively so that the quantum yield of fluorescence is given by the branching ratio as

$$\Phi_f = k_f/(k_f + k_{isc}) \quad (3.13)$$

and the observed fluorescence lifetime is $\tau = 1/(k_f + k_{isc})$ (3.14) This is, of course, always shorter than the natural radiative lifetime.

3.4.2.9 Delayed Fluorescence. By definition the fluorescence emissions are spin-allowed radiative transitions of atoms or molecules; they have short lifetimes, of the order of ns to a few hundred ns. There are however some cases where molecules emit the very same fluorescence spectra but with much longer decays and often with complex non-exponential kinetics. These

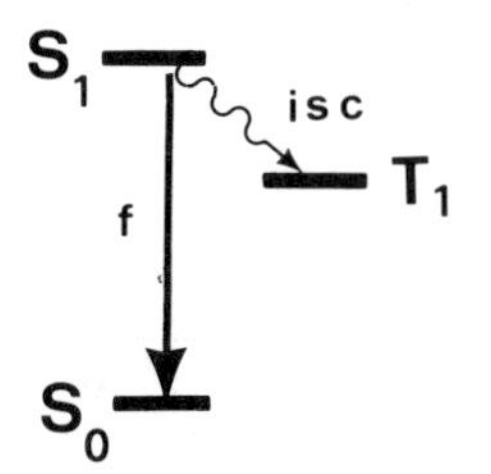

Figure 3.32 *Competition between fluorescence (f) and intersystem crossing (isc) in the kinetics of luminescence*

emissions of 'delayed fluorescence' result from two major photophysical processes, apart of electroluminescence and chemiluminescence which are discussed in section 4.8.

(1) 'E-type' delayed fluorescence: in some organic molecules the triplet state T_1 is close enough in energy to the singlet state S_1 for the thermal population of S_1 from T_1 to be significant according to the Boltzmann distribution law. The triplet state is long-lived and the fluorescence S_1–S_0 then follows the triplet state population.

(2) 'P-type' delayed fluorescence: this is also known as 'triplet–triplet annihilation' and depends on the energy transfer between two triplet excited states $^3M^*$ to form an excited singlet state $^1M^*$ and a ground state M

$$^3M^* + {}^3M^* \rightarrow {}^1M^* + M \tag{3.15}$$

This process depends greatly on the triplet state concentration and becomes important in some flash photolysis experiments. There is an obvious energetic requirement for this process, that the triplet state energy should be at least half that of the singlet state.

3.4.3 Non-radiative Transitions

The wavy arrows in the Jablonski diagram of Figure 3.23, p. 50, correspond to the non-radiative transitions of internal conversion (ic) and the short arrows to intersystem crossing (isc); the former are spin allowed, as they take place between energy states of the same multiplicity; the latter are spin forbidden and are therefore much slower. The rate constants of ic and isc span extremely large ranges because they depend not only on the spin reversal (for isc) but also on the energy gap between the initial and final states.

Figure 3.33 shows the two distinct steps involved in a non-radiative transition. The rate limiting step is the electronic transition without loss of energy (1) from the v = 0 vibrational level of S_i (initial state) to a vibrationally excited level v = n of S_f (the final state). From this v = n level the molecule can reach the v = 0 level of this state S_f through vibrational deactivation; the energy ends up as heat of the surroundings. This 'vibrational cascade' is a fast process, of the order of 10^{-12} s. The controlling step in the non-radiative deactivation is therefore the isoenergetic crossing from the initial state of wavefunction ψ_i to the final state of wavefunction ψ_f and can be expressed as follows:

$$k_{nr} = \frac{2\pi}{h}\langle\psi_i|H'|\psi_f\rangle^2\rho \propto \langle\phi_i|H'|\phi_f\rangle^2\sum_j\sum_k\rho\langle\theta_{ij}|\theta_{fk}\rangle \tag{3.16}$$

Here ρ is the density of vibrational levels of states S_i and S_f at the energy of the electronic transition E. The overlap of the electronic wavefunctions ϕ_i, ϕ_f and of the vibrational wavefunctions $\langle\theta_i|\theta_f\rangle$ are factorized according to the Born–Oppenheimer approximation just as in the case of radiative transitions. The density of vibrational levels is greater for the lower (final) state S_f

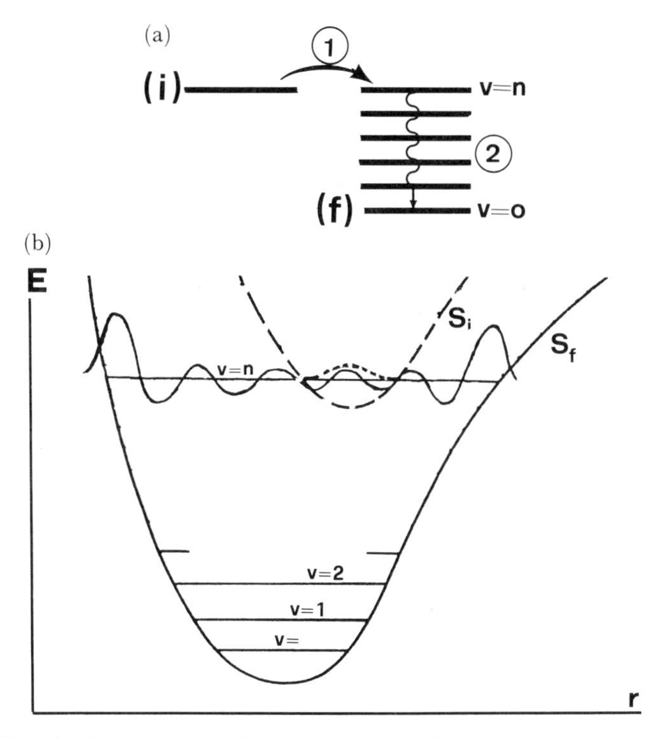

Figure 3.33 *(a) The two sequential steps of a non-radiative transition. 1 is the change in electronic wavefunction from state (i) to state (f); 2 is the vibrational cascade within state (f). (b) The vibrational overlap decreases rapidly with the energy gap of states S_i and S_f*

at the transition energy because the number of these levels increases with energy; in many cases the density ρ_i of the initial state is simply unity. The electronic overlap $\langle \phi_i | H' | \phi_f \rangle$ includes both symmetry and spin factors (as in radiative transitions) but the most important factor is the vibrational overlap $\langle \theta_i | \theta_f \rangle$ which is known as the Franck–Condon factor. Figure 3.33(b) shows that as the vibrational quantum number increases this overlap decreases rapidly because the higher vibrational level wavefunctions are increasingly spread out in space and the local amplitudes become smaller. This is the origin of the 'energy gap law' which states that the rate constants of non-radiative transitions decrease exponentially with the energy difference $E_i - E_f$ of the states S_i and S_f.

3.4.3.1 The 'Golden Rule' Equation. In a simple model of the radiationless transition from a state i to a state f the rate controlling step is the crossing from a single level of energy $E(\mathrm{i})$ to a number of levels of energies $E(\mathrm{f}) = E(\mathrm{i}) \pm n\varepsilon$ where n is an integer ($n = 0, 1, 2, \ldots$) and ε is the spacing of energy levels. The interaction energy which drives the crossing is given in terms of the wavefunctions of the initial and final states

$$v = \langle \psi_i | H | \psi_f \rangle \tag{3.17}$$

and the probability of finding the molecule in the initial state after a time t is

$$P_i(t) = e^{-(k_{nr}t)}$$

with the rate constant (3.18)

$$k_{nr} = 2\pi v^2/\hbar\varepsilon$$

This form of the rate constant of the radiationless transition is known as the 'golden rule'. It depends on two limiting conditions, which are:
(1) that the energy spacing ε should be small compared with the interaction energy v;
(2) that the observation time t should not approach the 'recurrence' time $\tau_r = \hbar/\varepsilon$.
This second point is quite an interesting one, for there is a theorem known as the Poincaré recurrence theorem which states that an isolated system (like our molecule left to itself) will in the course of time return to any of its previous states (*e.g.* the initial state), no matter how improbable that state may be. This recurrence can be observed with very small molecules but not with polyatomic molecules, because in the latter there are far too many levels of the final state; the recurrence time is then far longer than any practicable observation time.

3.4.3.2 Angular Momentum Conservation in Non-radiative Transitions. The very general law of conservation of the angular momentum of any isolated physical system (*e.g.* atom or molecule) applies to non-radiative as well as to radiative transitions. This is often described as the rule of spin conservation, but this is not strictly accurate since only the total angular momentum must remain constant. Electrons have two such angular motions which are defined by the orbital quantum number L and the spin quantum number S, the total

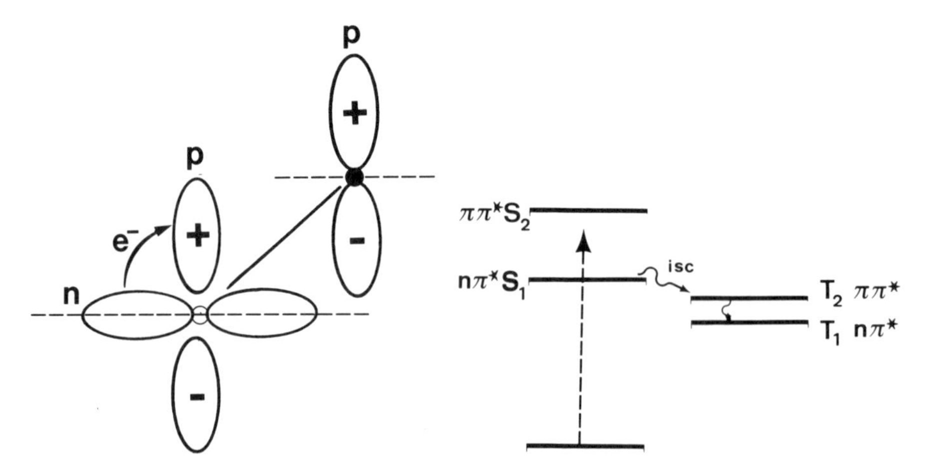

Figure 3.34 *The compensation between the changes of spin and orbital angular momenta can lead to allowed intersystem crossing (isc). In benzophenone crossing from $S_1(n-\pi^*)$ to $T_2(\pi-\pi^*)$ takes only approximately 20 ps*

angular momentum quantum number then being $J = L + S$. In any radiative or non-radiative transition the condition is $\Delta J = 0$. This allows S and L to vary, so long as there is a compensation between them, *e.g.* $\Delta S = 1$ and $\Delta L = -1$. This compensation of the changes in spin and orbital angular moments explains the very fast intersystem crossings in some molecules like aromatic carbonyl derivatives. The crossing from a singlet n–π^* state to a triplet π–π^* state is spin forbidden, but it implies also a change in orbital momentum since the n and π^* orbitals are orthogonal (Figure 3.34).

3.4.3.3 The Deuterium Effect on Non-radiative Transitions. In many organic molecules the C–H stretching frequency plays the main role as acceptor of vibrational energy, because of its wide energy spacing (of the order of 2800–3200 cm^{-1}). Figure 3.35 shows that the non-radiative deactivation of the triplet state T_1 to the ground state S_0 in a molecule like naphthalene is fairly efficient since the quantum number of the *n*th vibration of S_0 is still not very high at the level of T_1.

In molecules which contain only C–D bonds, the situation is different because the heavier D atom leads to much smaller spacings of vibrational levels, something like 2000 cm^{-1}. The crossing from T_1 (v = 0) to S_0(v = n) must reach a vibrational level of much higher quantum number so that the overlap of the nuclear wavefunctions is much smaller. The phosphorescence quantum yields and lifetimes are therefore greater in the deuterated compounds. To take one example, the observed phosphorescence lifetime of naphthalene-h_8 is 2.3 s, but that of naphthalene-d_8 is 18.4 s, both measured in a rigid glass at 77 K.

3.5 QUENCHING OF EXCITED STATES

Quenching is the non-radiative deactivation of an excited molecule M* by a molecule Q (the quencher), the excited state energy eventually becoming heat energy of the surroundings (*e.g.* liquid solvent or solid matrix)

$$M^* + Q \rightarrow M + Q \ (+ \text{ energy}) \tag{3.19}$$

This scheme shows only the overall process, and we shall see that there may be intermediate steps between these initial and final states. In the way it is

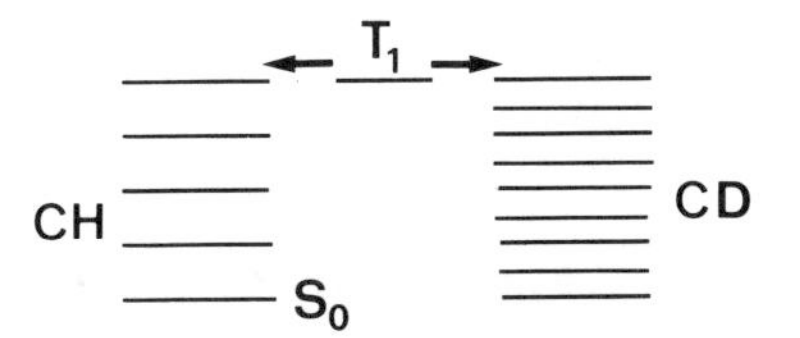

Figure 3.35 *The deuterium effect in non-radiative transitions. The spacing of the vibrational levels of CD (carbon–deuterium) is much smaller than that of CH, so that for the same energy gap a much more distorted geometry of RD must be reached*

written here, quenching could be defined as the catalytic deactivation of excited molecules without chemical reaction. It is obvious that a chemical reaction also results in the disappearance of the excited molecule M* but this is a photochemical process. We shall use here the definition of 'quenching' in the restricted sense of a photophysical process in which the molecule M* is restored unchanged to its ground state M.

3.5.1 Energy Transfer

If the quencher molecule Q has an excited state Q* lower than M* the excitation energy can be transferred according to

$$M^* + Q \rightarrow M + Q^* \qquad (3.20)$$

Radiative or non-radiative deactivation of Q* to Q then completes the quenching process. There are two major energy transfer processes.

(1) *Electron exchange: the Dexter mechanism.* Figure 3.36 shows a simple orbital diagram of an electron exchange between an excited molecule M* and a ground state molecule Q. The first requirement is of course that the excitation energies should be in the order $E(M{-}M^*) \geqslant E(Q{-}Q^*)$; the second requirement is that the orbitals should allow exergonic electron transfer from φ_M^* to φ_Q^* and from φ_Q to φ_M. The two electron transfers are simultaneous; both M and N remain neutral species throughout the electron exchange and no recognizable ions are formed as intermediates. This double

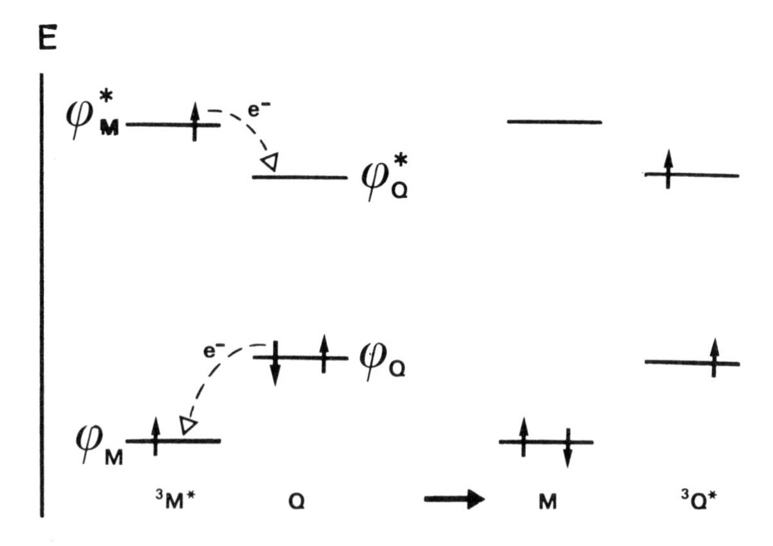

Figure 3.36 *The Dexter mechanism of energy transfer through simultaneous electron exchange. The HOMO and LUMO of the quencher Q must fall between the HOMO and LUMO of the energy donor*

electron transfer requires the spatial overlap of the orbitals so that the molecules M* and Q must be in close contact (van der Waals or hard sphere contact).

Although the overall spin quantum number of the system M*/Q must be kept in electron exchange, quenching can take place between singlet and triplet states in any combination:

$$^1M^* + Q \rightarrow M + {}^1Q^*;\ {}^3M^* + Q \rightarrow M + {}^3Q^*\ \textit{etc.} \quad (3.21)$$

(2) *Dipole–dipole interaction: the Förster mechanism* (Figure 3.37). This is in fact the interaction of the transition moments of the excitation Q → Q* and the deactivation M* → M. As the excited electron of M* falls to the lower orbital of M there is a change in dipole moment which produces an electric field; this field is proportional to the transition moment $\vec{M}$ and to the inverse cube of the distance. An electron in the molecule Q therefore experiences a force proportional to $\vec{M}/r^3$, and as it moves towards a higher orbital it produces its own electric field which results in a force being applied on the electron in molecule M*. In this way the 'downward' motion of the electron in M* and the 'upward' motion of the electron in Q are coupled by their electric fields, the rate constant for energy transfer being

$$k_q = \frac{c\Phi_e k^2}{n^4 \tau r^6} \int_0^\infty f_D(\nu)\varepsilon_A(\nu)\frac{d\nu}{\nu^4} \quad (3.22)$$

The important features of this equation are:

(1) the distance dependence, r^{-6}. This results from the coupling of the motions of electrons in the transitions M* → M and Q → Q*. This is a rather shallow distance dependence compared with the exponential decrease of the orbital overlap which governs the electron exchange (Dexter), energy transfer mechanism. Dipole–dipole interaction can be operative at distances of some 100 Å is some cases.

(2) the emission quantum yield of the energy donor, Φ_e. This is linked to the magnitude of the transition moment M* → M.

(3) the lifetime τ of the excited state of the donor M*; the probability of energy transfer is proportional to this lifetime since deactivation of M* competes with energy transfer.

(4) the spectral overlap of the emission spectrum of M and the absorption spectrum of Q. The law of energy conservation requires that the process of

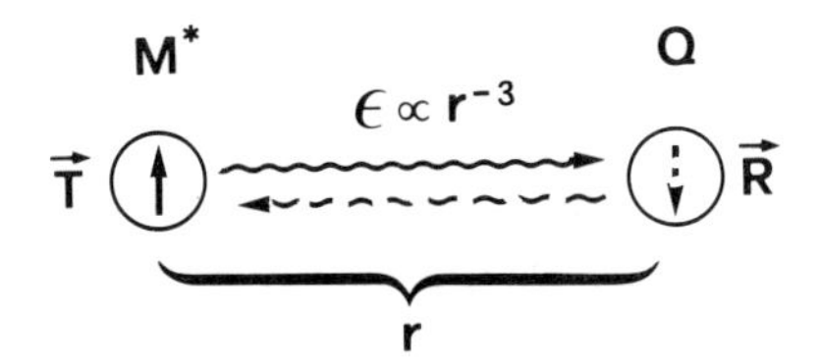

Figure 3.37 *The Förster mechanism of energy transfer through transition dipole interaction. Deactivation of the excited molecule M* creates an electric field T/r³ which promotes the excitation of Q through its transition dipole R*

energy transfer should be most efficient when the transitions $M^* \rightarrow M$ and $Q \rightarrow Q^*$ are isoenergetic; the integral in eqn. (3.22) represents this spectral overlap over all the frequency range, but in practice this range is quite small since it is restricted to the actual emission and absorption spectra of M and Q. k is an orientation factor to take into account the angle between the transition moment vectors of the molecules. For a random distribution it is equal to 2/3.

3.5.2 Quenching by Electron Transfer

In an electron transfer reaction two neutral molecules, for example, will form an ion pair; if one of the neutrals is electronically excited it will be deactivated, and the overall process may appear as a quenching if the ion pair recombines to return to the neutral ground state partners:

$$M^* + Q \rightarrow M^{\bullet +} + Q^{\bullet -} \rightarrow M + Q(+\text{ energy}) \qquad (3.23)$$

We arrive here at the limit of quenching considered as a purely photophysical process, involving no permanent chemical change. Electron transfer is in fact a chemical reaction which leads to new, distinct species $M^{\bullet +}$ and $N^{\bullet -}$; these may separate and react to form new molecules, in which case we enter the realm of photochemistry. Photoinduced electron transfer must be considered to be a photochemical reaction (see chapter 4), but in some cases it may appear to be a quenching reaction when the reactants are restored to their initial state (Figure 3.38).

3.5.3 The Heavy Atom Effect

We have seen that transitions between electronic states of different spin quantum numbers are in principle forbidden by the law of conservation of angular momentum. In practice these transitions take place only through the compensation of two simultaneous changes in angular momentum represented by the orbital quantum number L and the spin quantum number S; their sum $J = L + S$ remains constant while L and S vary in opposite directions.

The link between L and S is provided by the spin-orbit coupling which increases with the atomic number of the atoms; electrons move faster around nuclei which carry large positive charges, so that the interaction between the electron currents and the related magnetic fields increases with atomic number. This is the basis of the important processes known as the 'heavy atom effects' which enhance the rates of formally spin-forbidden radiative and non-radiative transitions.

The heavy atom effect can rely on the presence of an atom of high atomic number either within the molecule itself (the internal heavy atom effect) or in the solvent (external heavy atom effect). In both cases the fluorescence

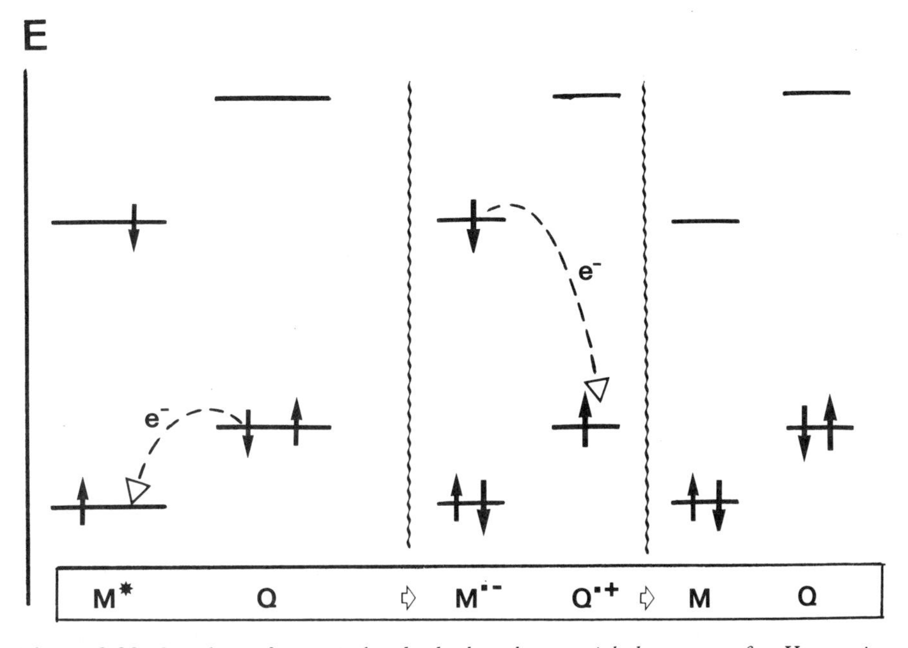

Figure 3.38 *Quenching of an excited molecule through sequential electron transfer. Here an ion pair is formed in the first place, within which back electron transfer leads to overall deactivation of the excited energy donor*

quantum yields decrease and the triplet yields increase, but the triplet lifetimes (*e.g.* phosphorescence) decrease as well.

Fluorescence lifetimes (ns)	Naphthalene	95
	2-Chloro-naphthalene	3.5
	2-Bromo-naphthalene	0.15
	9-Iodo-anthracene	0.035

The question then arises: what is a 'heavy' atom? Br and I can be considered as heavy atoms in organic molecules, but Cl is a borderline case. Most metals qualify as heavy atoms and the photophysical properties of metal complexes are related to this increased spin-orbit coupling. Some noble gas heavy atoms like Xe show important external heavy atom effects.

3.5.4 Paramagnetic Quenching

External magnetic fields can also enhance spin-orbit coupling, so that molecules which have a permanent magnetic moment have an action similar to the heavy atom effect. Molecular oxygen $^3O_2$ is a paramagnetic species with a triplet ground state and is the best known paramagnetic quencher. It must however be noted that its quenching action can be quite complex, for it can take part in energy and electron transfer reactions as well.

3.5.5 Concentration Quenching

It has been observed in many cases that the luminescence quantum yield in solution decreases as the solute concentration increases, and this applies also to the quantum yields of photochemical reactions. There must therefore be some process which can be written as

$$M^* + M \rightarrow M + M \ (+ \text{energy}) \tag{3.24}$$

whereby the ground state molecule M acts as a quencher for the excited molecule M*. This process of 'concentration quenching' seems to be quite general, but its mechanism has not been established; it is likely to involve the formation of an unstable excimer.

3.5.6 Static and Dynamic Quenching

Since any quenching action is a bimolecular process, it is essential that the molecules M* and Q should be in relatively close contact, but not necessarily in 'hard sphere' (van der Waals) contact. Theoretical models lead to the distance dependence of the quenching rate constants as exponentials or sixth powers of r, the centre-to-centre distance of M* and Q. Since these distance dependences are very steep, it is convenient to define a critical interaction distance r at which the quenching efficiency is $\frac{1}{2}$, this being the distance at which 50% of the molecules M* decay with emission of light (or undergo a chemical reaction) and 50% are quenched by some nearby molecule Q.

(1) In a rigid system such as a glass or a polymer, the molecules M and Q are distributed at random and do not move, at least within the lifetimes of excited states. The distance distribution follows the Perrin law which is based on a very simple model. Take any excited molecule M*, and ask if one quencher molecule Q happens to be within the 'volume of action' defined by the centre-to-centre distance r. Should any molecule Q be found within this action volume, the molecule M* is quenched instantaneously, but if there is no quencher Q within this space, then M* emits as if no quenchers at all were present. Figure 3.39 gives a picture of the Perrin model. The mathemat-

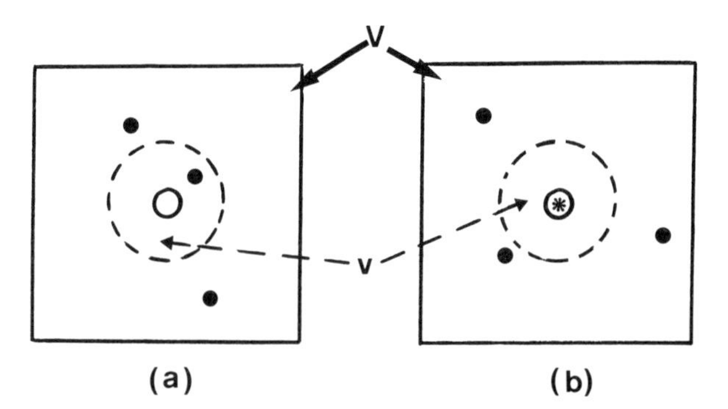

Figure 3.39 *The Perrin 'action volume' model of static quenching. Each excited molecule is surrounded by a sphere which can contain one (or several) quenchers (a), or no quencher (b). In (a) quenching is instantaneous, in (b) there is no quenching at all*

ical relationship between emission intensity (proportional to the concentration of unquenched excited molecules [M*]) and quencher concentration follows simply from the laws of probabilities. Take one molecule M* at the centre of an arbitrarily large volume V which contains n quencher molecules Q, and let v be the volume of action of any one quencher. The probability of finding the first quencher within the volume v is then v/V, neglecting the space occupied by M*. The probabilities of independent events being multiplicative, the probability of finding all n quenchers outside the volume v is $(1 - (v/V)^n)$.

The Perrin equation of static quenching is given in logarithmic form as

$$\ln I(\mathrm{Q}) = c[\mathrm{Q}] \qquad (I = 1 \text{ for } [\mathrm{Q}] = 0) \qquad (3.25)$$

and Figure 3.40 shows such a plot.

(2) The molecules M* and Q can come into contact (within the sphere of action) through their random diffusional motion. The rate constant k_D of diffusion controlled encounters is the upper limit for any bimolecular reaction. This must be multiplied by the probability of an encounter leading to reaction (quenching in the present case), and the luminescence quantum yield then follows the Stern–Volmer equation

$$\Phi_0/\Phi = 1 + k_q\tau[\mathrm{Q}] \qquad (3.26)$$

τ being the lifetime for $[\mathrm{Q}] = 0$.

Figure 3.41 provides an example of a Stern–Volmer plot. In this case the triplet excited state T_1 of an aromatic molecule is quenched by the external heavy atom effect. The Stern–Volmer plot must by definition have an

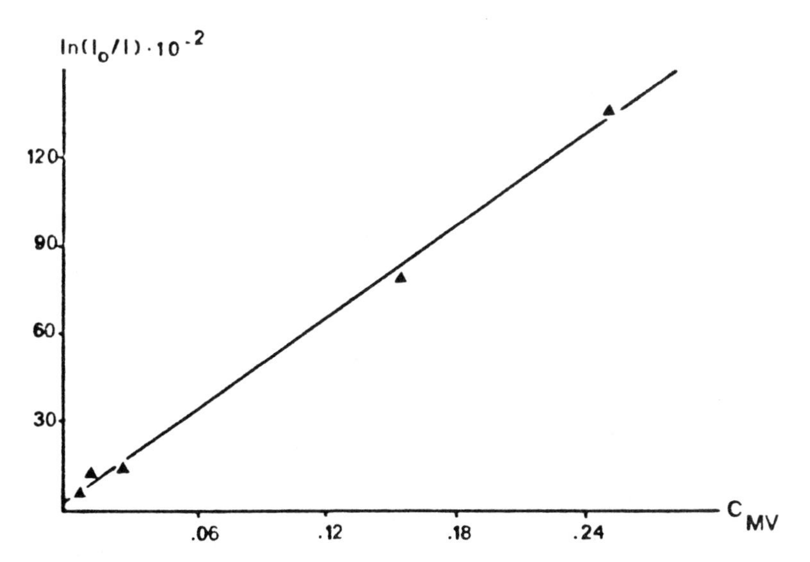

Figure 3.40 *Example of the Perrin plot of static quenching. Luminescence of a metal complex [Ru(bpy)$_3^{2+}$] in rigid glycerol in the presence of increasing concentrations of methylviologen (quenching by electron transfer)*

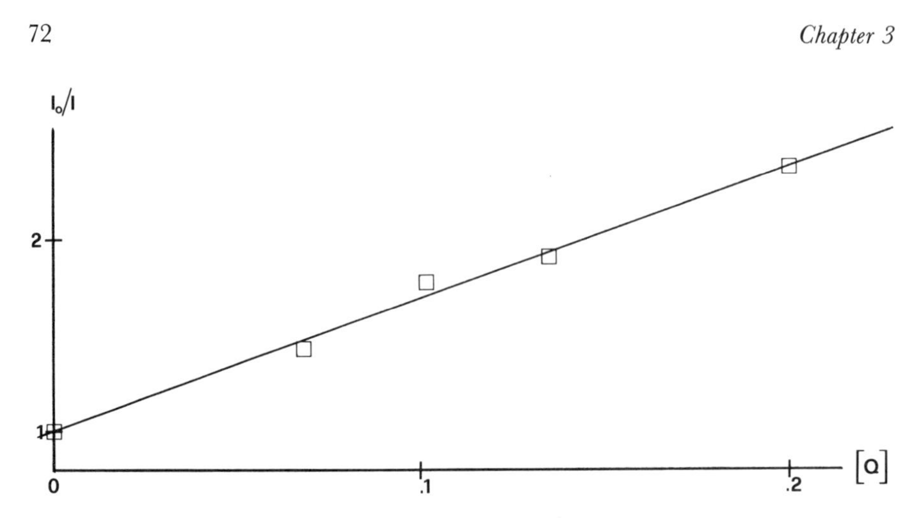

Figure 3.41 *Example of the Stern–Volmer plot of dynamic quenching. The fluorescence of anthracene is quenched by bromobenzene (heavy atom effect)*

intercept of unity at zero quencher concentration. When this plot is non-linear, the mechanism of quenching is more complex, and static quenching may become important at high quencher concentrations.

3.5.6.1 Kinetics of Luminescence Quenching. The dependence of the emission decay kinetics on quencher concentration is the best criterion to distinguish between static and dynamic quenching. In static quenching the emission lifetime remains constant, but the initial luminescence intensity decreases. However, this initial intensity remains constant in dynamic quenching and the lifetime decreases with increasing quencher concentration (Figure 3.42).

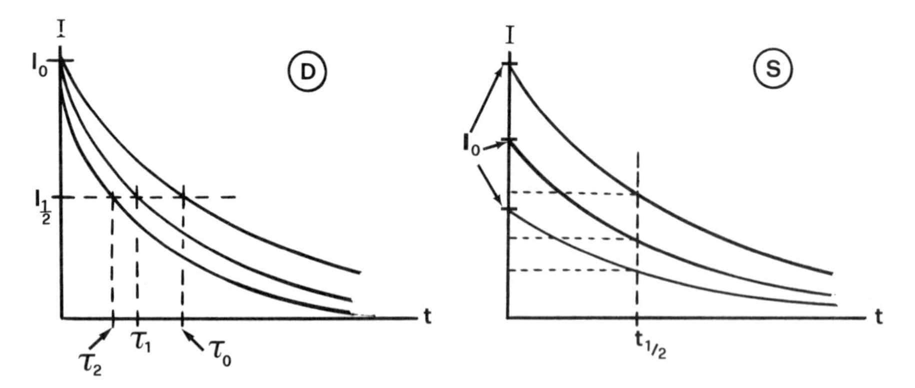

Figure 3.42 *Kinetics of dynamic (diffusional), D, and static, S, quenching. In dynamic quenching the excited state lifetime gets shorter with increasing quencher concentration, from τ_0 with no quencher to τ_1, τ_2 with added quencher. In static quenching excited state lifetime remains unchanged but the initial concentration of excited states is reduced*

3.6 ENERGY LEVELS IN SOLIDS

In a solid the atomic or molecular orbitals interact strongly to form broad energy bands (Figure 3.43) which are known as the valence band (VB) and the conduction band (CB). The valence band results from the interaction of filled bonding or non-bonding orbitals, while the conduction band results from the interaction of normally vacant antibonding orbitals. As their names imply, electrons are strongly bound to nuclei in the VB, but they can move freely in the CB over the whole solid lattice.

Three general cases can be distinguished according to the relative positions of the VB and the CB. In an insulator they are very far apart, so far that the CB cannot be populated significantly from the VB through thermal energy distribution. In a semiconductor the energy bands come close enough together to allow population of the CB from the VB either thermally or through the absorption of light. Finally, in a metal the VB and CB overlap and electrons move freely through the solid lattice at all temperatures (Figure 3.44).

The photophysical properties of insulators are of little interest and will not be considered here. Metals (electrical conductors) are characterized by a single property known as the 'work function' which is the energy required to

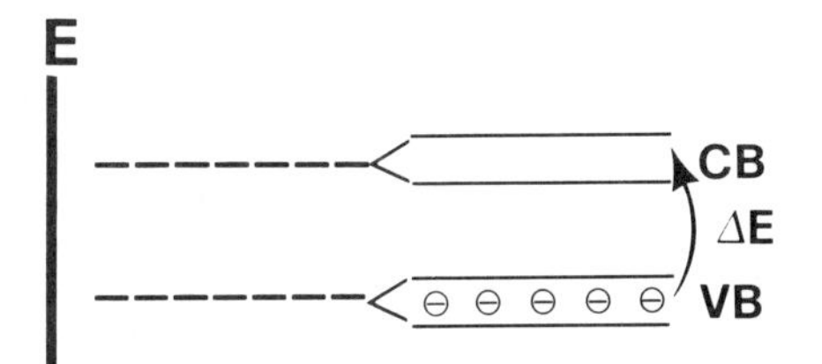

Figure 3.43 *In a solid the atomic or molecular orbitals interact to form broad energy bands known as the valence band, VB, and the conduction band, CB. The smallest energy between these bands is the band gap energy*

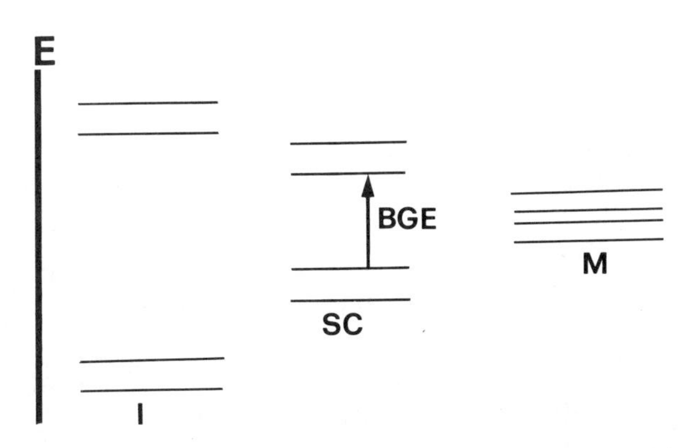

Figure 3.44 *In an insulator the band gap energy (BGE) is very large, while it is small in semiconductors (SC). In metals, M, the valence and conduction bands overlap*

remove the weakest bound electron from the solid to vacuum at 'infinite' distance; it is akin to the ionization potential of a molecule.

The most important case is that of the semiconductors. These show absorption spectra in the UV, visible and sometimes in the near IR; the onset of absorption corresponds to the 'band gap' energy $h\nu_{BG}$ between the highest level of the VB and the lowest level of the CB. A small selection of band gap energies of some semiconductors of practical importance is given in Appendix 3A.

3.6.1 Fermi Levels: Doped Semiconductors (Figure 3.45)

The 'Fermi level' is a theoretical energy of electrons in a semiconductor, such that the probability of occupation of the VB and CB is 50%. In an 'intrinsic' semiconductor this Fermi level is about half-way between the VB and the CB, but it can be displaced substantially in 'doped' semiconductors. An intrinsic semiconductor would be for example a crystal of pure Si or Ge, all tetravalent atoms being linked together in a three-dimensional array. In a doped semiconductor of 'n'-type some of the Si atoms are replaced by pentavalent atoms such as As, and these will release electrons into the CB. A 'p'-type semiconductor, however, contains some trivalent atoms like Al which are electron deficient. The Fermi level moves closer to the CB in the n-doped semiconductor, while it comes closer to the VB in the p-type semiconductor (Figure 3.46).

The Fermi level represents in a way the 'pressure' of electrons and is rather similar to the redox potential of an electrode.

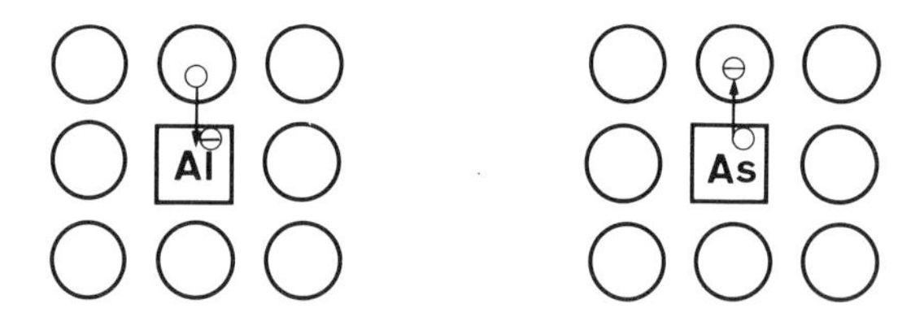

Figure 3.45 *In a doped semiconductor some tetravalent atoms are replaced by trivalent atoms (p-type) or pentavalent atoms (n-type)*

3.7 THE PHYSICAL PROPERTIES OF EXCITED MOLECULES

There are several important differences between ground state and excited state molecules with respect to the physical properties of shape, bond lengths and charge distribution. These properties are related also to the chemical properties of these molecules, so they have a direct bearing on photochemistry.

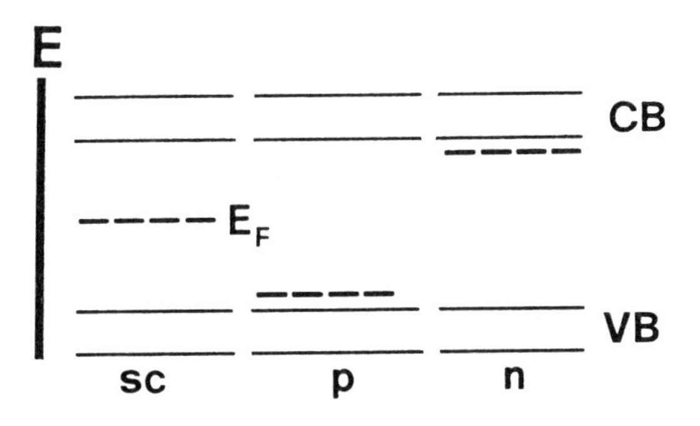

Figure 3.46 *In a p-type semiconductor the electrons are held by the trivalent atoms and their overall energy (the Fermi level) is lowered. In an n-type semiconductor electrons are released by the pentavalent atoms and the Fermi level comes closer to the conduction band. SC = semiconductor*

3.7.1 Geometrical Changes in Excited States

We will consider here two representative examples of major changes in molecular geometry following electronic excitation; it should not be forgotten that a molecule has several, sometimes many different, excited states, and each one of these will have its particular geometry. In practice only the lowest excited state of each multiplicity lives long enough to take part in chemical reactions, and upper excited states are in any case too short lived to reach their stable geometry. According to the Franck–Condon principle the molecular geometry remains unchanged during an electronic transition which lasts some 10^{-15} s, so that the rearrangement to the new shape must follow on a much longer time-scale, of the order of tens or hundreds of picoseconds.

3.7.1.1 A Planar Molecule Becomes Bent. In the ground state formaldehyde is planar, but in the n–π^* state an electron is promoted from the n to the π^* orbital. The excited molecule can be represented approximately as a biradical (Figure 3.47). The carbon atom now has four nearly equivalent 'bonds',

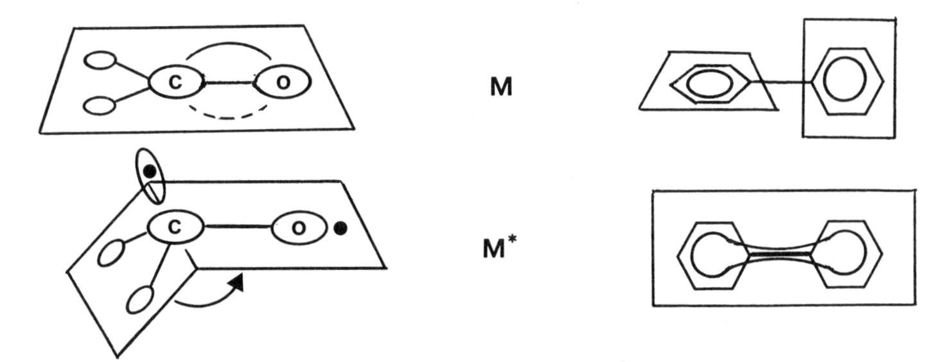

Figure 3.47 *In the ground state the molecule of formaldehyde (left) is planar, but it becomes bent in the S_1 excited state. Biphenyl (right) is twisted in the ground state but becomes planar in the S_1 excited state*

though one of these contains only a non-bonding electron. The C atom will then take the tetrahedral shape characteristic of the tetravalent species.

3.7.1.2 A Twisted Molecule Becomes Planar. Biphenyl can be seen as two benzene rings linked by a single C–C bond which forces the rings to remain very close together. In the ground state these two benzene rings cannot be coplanar for two reasons. In the first place, there is a small steric hindrance between the proximal hydrogen atoms in the *ortho* position with respect to the bond, but the most important factor is the repulsion between the closed electron shells of the rings. The interaction of these filled orbitals leads to an unstable overlap, so that the aromatic rings must find a non-planar geometry. This geometry is however very loose, the angle of the rings taking many different fluctuating values. The absorption spectrum of biphenyl is broad and structureless, as expected from an assembly of molecules of many different geometries. However, the fluorescence spectrum shows a vibrational structure which is characteristic of a nearly planar molecule; in the S_1 excited state biphenyl resembles an intramolecular excimer of benzene and its most stable geometry corresponds to maximum orbital overlap which can be reached only in a planar configuration since the actual folding is prevented by the C–C single bond.

3.7.2 Electron Distributions and Polarizabilities

The asymmetry of distributions of positive and negative charges in a molecule can have important effects on its physical properties as well as on its chemical reactivity. This asymmetry can have two quite different origins, illustrated by the 'permanent' and 'induced' dipole moments. A molecule such as HF has a permanent dipole moment which results from the electronegativities of the H and F atoms; but even a symmetrical molecule such as $H_2C{=}CH_2$ can acquire an induced dipole moment in an external electric field.

We have seen a few examples of charge transfer in excited states (section 3.3). There are also examples of charge recombination where a dipolar ground state leads to a non-polar excited state, as in the case of some betaine dyes (Figure 3.57, p. 86). Charge transfers also take place in formally symmetrical molecules, with or without symmetry breaking. When the symmetry is retained there can be no change in dipole moment, but there can be changes in the higher multipole moments (in particular in the quadrupole moment).

The *polarizability* of a molecule defines its response to an external electric field $\vec{E}$. The molecule's electrical charges (electrons and nuclei) are displaced and an induced dipole moment is created such that

$$\vec{\mu} = \alpha\vec{E} \tag{3.27}$$

α being the polarizability of the molecule in the direction of the external field $\vec{E}$. Experimental evidence shows that the polarizabilities of excited molecules remain rather similar to the ground state values, at least for the low energy excited states. There are however some exceptions in the case of long, conjugated chains which are the models for some important molecules in photobiology (carotenes) in which the polarizability can increase mostly in the direction of the long axis.

The polarizability is strongly related to the molecular volume. In simple electrostatic theory the polarizability of a perfectly conducting sphere of radius a is $\alpha = a^3$, but in real molecules of more complex shapes the average polarizability is still proportional to the size of the molecule; the larger molecules have the higher polarizabilities.

This brings us to consider the size of excited molecules. In the low-energy excited states this remains very close to that of the ground state species, but it increases gradually with the energy of the states; the higher excited states are more polarizable, until the limiting case of ionization is reached when the molecule becomes theoretically of infinite size.

3.8 THE EFFECT OF THE ENVIRONMENT ON THE ENERGY STATES OF MOLECULES

So far we have considered the various states of molecules as intrinsic molecular properties, as they would exist in isolated molecules in the gas phase at very low pressures. In practice most of chemistry (and all of biochemistry) concerns molecules in the condensed phase, as liquids, solids, or more or less in an organized state. The interaction of these condensed phase environments with a molecule is therefore of the greatest importance.

3.8.1 Non-specific Electrostatic Interactions

In a simple model a neutral molecule can be described through two properties related to its electron distribution, the permanent dipole moment μ and the average polarizability α. There are therefore four electrostatic interactions between a solute molecule and the surrounding solvent molecules, as shown in Table 3.1.

For practical purposes the solvent is described as a continuum, so that the dimensions of the solvent molecules do not appear explicitly in the interaction energy equations; the permanent dipole moments and the polarizabilities of the solvent are expressed as functions of macroscopic properties which are the dielectric constant D and the refractive index n; the interaction

Table 3.1 *Non-specific solute–solvent interactions*

Solvent	*Solute* Dipole moment	Polarizability
Dipole moment	Dipole–dipole interaction	Solvent Stark effect
Polarizability	Dipole–induced dipole interaction	Dispersion interaction

energy is then the product of a quantity which can be described as the solute polarity and another quantity which represents the polarity of the solvent.

3.8.1.1 Separation of Electronic and Nuclear Motions. The polarizabilities of the ground state and the excited state can follow an electronic transition, and the same is true of the induced dipole moments in the solvent since these involve the motions of electrons only. However, the solvent dipoles cannot reorganize during such a transition and the electric field which acts on the solute remains unchanged. It is therefore necessary to separate the solvent polarity functions into an 'orientation' polarization and an 'induction' polarization. The total polarization depends on the static dielectric constant D, the induction polarization depends on the square of the refractive index n^2, and the orientation polarization depends on the difference between the relevant functions of D and of n^2; this separation between electronic and nuclear motions will appear in the equations of solvation energies and solvatochromic shifts.

3.8.1.2 Dipole–Dipole Interaction. The first of the four terms in the total electrostatic energy depends on the permanent dipole moment $\vec{\mu}_M$ of the solute molecule of radius a (assuming a spherical shape) immersed in a liquid solvent of static dielectric constant D. The function $f(D) = 2(D-1)/(2D+1)$ is known as the Onsager polarity function. The function used here is $[f(D) - f(n^2)]$ so that it is restricted to the orientational polarity of the solvent molecules to the exclusion of the induction polarity which depends on the polarizability α_s of the solvent molecules, related to the slightly different Debye polarity function $\varphi(n^2)$ according to

$$\varphi(n^2) = (n^2 - 1)/(n^2 + 2) = (4N\delta_s\pi/3M_s)\alpha_s \quad (3.28)$$

$$\varphi(D) = (D - 1)/(D + 2) = (4N\delta_s\pi/3M_s)(\alpha_s + \mu_s^2/kT) \quad (3.29)$$

for a solvent of density δ and molecular weight M. The solvation energy is then

$$E(\text{solv.}) = -(\mu_M^2/2a^3)[f(D) - f(n^2)] \quad (3.30)$$

3.8.1.3 Solute Dipole–Solvent Polarization. In a solvent of refractive index n, a solute molecule of permanent dipole moment μ_M is stabilized by interaction with the dipoles induced in the solvent; this interaction energy is

$$E(\text{solv.}) = -(\mu_M^2/2a^3)f(n^2) \quad (3.31)$$

This is of course the major solvation energy of polar molecules in non-polar solvents, but it can be an important term in polar solvents as well, especially in highly polarizable solvents such as aromatic derivatives.

3.8.1.4 Non-polar Solutes in Polar Solvents: the 'Solvent Stark Effect'. At first sight a non-polar solute molecule cannot polarize the surrounding solvent since it develops no electric field. However, the solvent fluctuates around the non-polar solute, so that there is a small instantaneous electric field which acts on the solute to produce a fluctuating induced dipole which leads to some stabilization known as the solvent Stark effect. Experimental data show that this is at best a very minor effect, and it will not be considered further.

3.8.1.5 Dispersion Interactions. Last but not least in the range of solute–solvent electrostatic interactions come the dispersion forces which depend on the polarizabilities of the molecules. Any atom or molecule—non-polar or polar—has a small fluctuating dipole moment as the electrons move around the nuclei. These instantaneous dipoles induce dipole moments in all other polarizable molecules, so that the interaction energy is proportional to the product of the average polarizabilities α_M and α_s of the solute and solvent molecules

$$E(\text{disp.}) = -C\alpha_s\alpha_M/r^6 \tag{3.32}$$

The proportionality factor C is difficult to evaluate, but dispersion forces play a most important role in the stability of liquids.

3.8.2 Specific Solute–Solvent Associations

The four electrostatic interactions which involve the dipole moments and the polarizabilities of the solute and solvent molecules are termed 'non-specific' because they do not imply a fixed geometry or stoichiometry of the molecules. In a 'specific' association the molecules form a loose complex, of which hydrogen bonding is the most important example. Hydrogen bonding takes place between a protic (proton donor or acid) molecule such as water and a proton acceptor (base) such as an amine (Figure 3.48).

The hydrogen bond must be described as a mixture of a covalent bond and a purely electrostatic interaction. It links two molecules in a 1:1 stoichiometry with a fairly well defined bond angle and bond length, subject of course to minor fluctuations. Such specific associations can have important effects on the photophysics and photochemistry of solute molecules, and it must be realized that these interactions depend on the nature of the electronic states; major changes in hydrogen bonding can take place between the ground state and the excited states of many molecules of importance in biology as well as in industrial applications.

3.8.3 Solvatochromic and Thermochromic Shifts

The solvatochromic shift is the displacement of an absorption or emission band in different solvents. Figure 3.49 shows examples of such shifts, the transition energy being linear with the Onsager polarity function $f(D)$. The

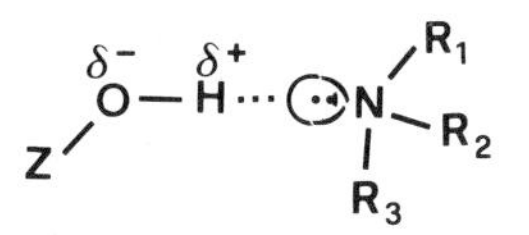

Figure 3.48 *Hydrogen bond formation between an amine and a protic molecule*

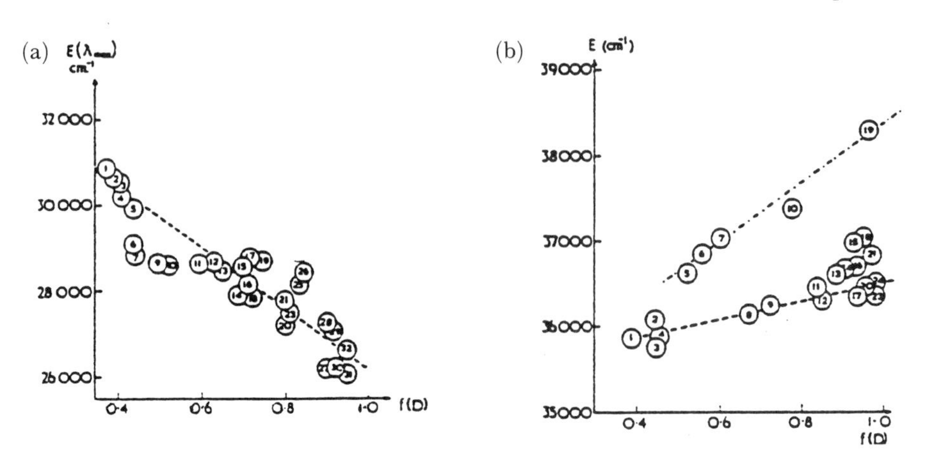

Figure 3.49 *Examples of solvatochromic plots of absorption spectra. (a) The first absorption band of 4-nitroaniline (charge transfer band). (b) The first absorption band of acetone (n–π* band). The ordinates are in units of 10^3 cm, measured at the band maximum; the abcissa are in units of Onsager solvent polarity function f(D)*

solvatochromic plot $h\nu$ vs. $f(D)$ is negative when the dipole moment of the molecule increases in the excited state (in this example we have chosen a 'charge transfer' state), but it can be positive when the dipole moment of the excited state is smaller or opposite to that of the ground state. An important example of this situation concerns the n–π^* states of carbonyl compounds, shown in Figure 3.49b. In the ground state the dipolar group C=O has a negative charge on the oxygen atom, but in the n–π^* excited state an electron is promoted from a non-bonding orbital of O to an antibonding orbital delocalized over C and O.

The equations of solvatochromic shifts represent the difference between the solvation energies of the initial and final states as a function of solvent polarity. For an absorption spectrum the initial state is of course the ground state and the final state is an excited state; this could be S_1 or S_2 or S_3, *etc.* For an emission spectrum the initial state is an excited state (almost always S_1 or T_1) and the final state is the ground state. The two main terms are shown below; these correspond to the dipole–dipole and dipole–induced dipole interactions.

$$(\Delta E_{ge})_{1-2} = -\vec{\mu}_g(\vec{\mu}_e - \vec{\mu}_g)a^{-3}[f(D) - f(n^2)]_{1-2} \tag{3.33}$$

$$(\Delta E_{eg})_{1-2} = -\vec{\mu}_e(\vec{\mu}_g - \vec{\mu}_e)a^{-3}[f(D) - f(n^2)]_{1-2} \tag{3.34}$$

These equations represent only the dipole–dipole interactions; the first one applies to an absorption spectrum, the second one to an emission spectrum (g for ground state, e for excited state).

The dipole–induced dipole solvatochromic shift is given by

$$\Delta E_{1-2} = -(\mu_e^2 - \mu_g^2)(2a^3)^{-1}\Delta f(n^2)_{1-2} \tag{3.35}$$

For a complete description of the solvatochromic shift the dispersion term should be added, but in many cases all terms in $f(n^2)$ can be neglected if the solvents have very similar refractive indices; in that case the shifts are described simply by eqns. (3.33) or (3.34). The solvent polarity function can be $f(D)$ instead of $f(D)-f(n^2)$ (Figure 3.50).

An important effect of solvatochromic shifts is the crossing of states of different electron distributions according to the polarity of the solvent. In polar solvents electronic states of large dipole moments (*e.g.* CT states) are stabilized with respect to states of small dipole moments; when these states are close in energy an inversion of their order can take place when the polarity of the solvent is changed. Since the photophysical and photochemical properties of most molecules are determined by the nature of the lowest excited state of a given spin quantum number, these properties can show very large variations with solvent polarity (Figure 3.51)

A thermochromic shift is the displacement of an absorption or emission band with the temperature of the solvent. These displacements result from the change in solvent polarity with temperature, the general rule being that the polarity decreases as the temperature increases. These shifts are small compared with solvatochromic effects and are unlikely to lead to state inversion (Figure 3.52).

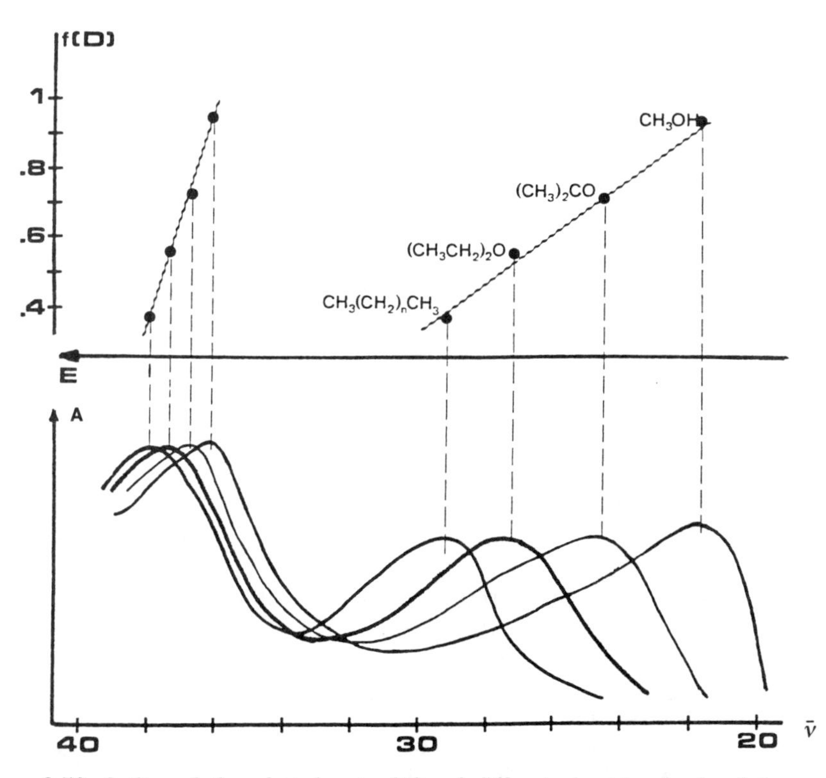

Figure 3.50 *Outline of the solvatochromic shifts of different absorption bands of the same molecule (e.g. 4-aminobenzophenone). In this case the first transition corresponds to a much larger change in dipole moment.* $\bar{\nu}$ *wavenumber in* 10^3 cm^{-1}; *$f(D)$ Onsager polarity function; A absorbance*

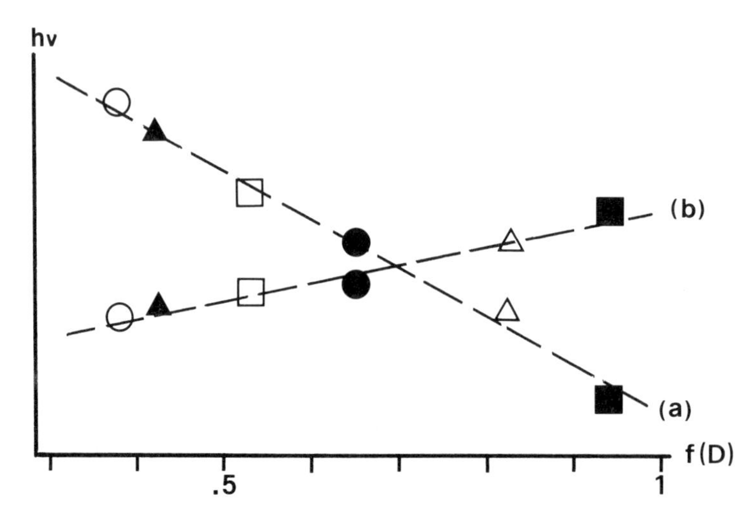

Figure 3.51 *Diagram of the crossing of two excited states of different dipole moments as a result of solvent stabilization. State (a),* e.g. *a CT state, is stabilized below state (b) in highly polar solvents. f(D) is the Onsager polarity function of the solvents shown by arbitrary symbols*

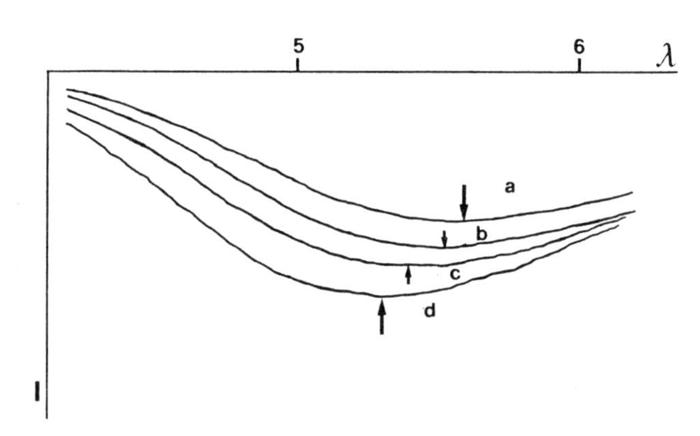

Figure 3.52 *Example of a thermochromic shift of the fluorescence spectrum of an exciplex in a solvent of medium polarity. [The temperature increases from (a) to (d) by 50 °C.]*

Finally we should mention the electrochromic shifts which are the changes in the energy of electronic transitions when an external electric field is applied to the sample. These effects are quite small but have proved useful for the measurements of the dipole moments and polarizabilities of excited molecules.

3.8.3.1 Effects of Hydrogen Bonding. In a solvatochromic plot of transition energy $h\nu$ *versus* solvent polarity [*e.g.* the $f(D)$ function], the effect of hydrogen bonding between the solute and the solvent is an anomalous blue shift or red shift, of which an example is shown in Figure 3.53. This is the

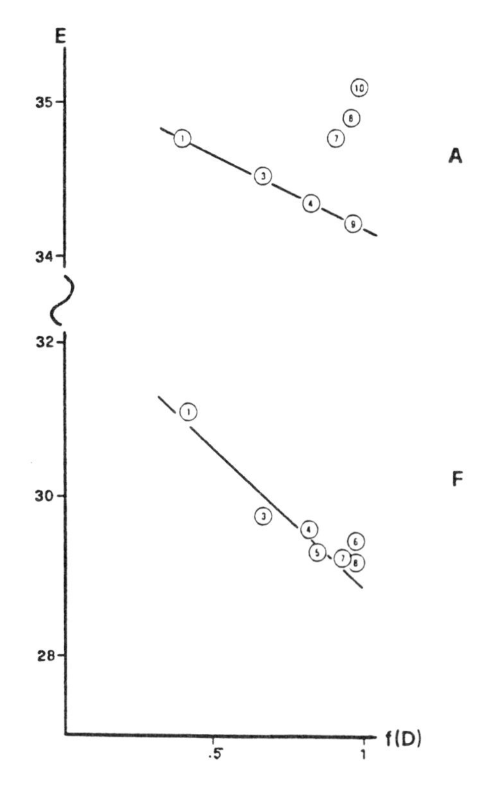

Figure 3.53 *Solvatochromic plots of the absorption (A) and fluorescence (F) spectra of an aniline, illustrating the anomalous blue shift in protic solvents (ethanol, methanol, water)*

absorption spectrum of aniline (S_0, S_1 transition) in which the ground state is stabilized when a protic solvent binds to the electron lone pair on the N atom.

The hydrogen bonding anomaly is related to the strength of the hydrogen bond although it is not equal to it in terms of energy, because the excited state reached in the vertical transition is destabilized before the solute–solvent structure is broken. In many fluorescence spectra there is a red-shift anomaly which shows that a new solute–solvent association is formed in the excited state. It is probable that these new hydrogen bonds link the protic solvent to electron-rich (or negatively charged) sites of the excited molecule and these produce a stabilization of the excited state as well as a corresponding destabilization of the ground state prior to solvent relaxation. Figure 3.54 shows a simple picture of the energy levels of a solute molecule M (excited molecule M*) in polar, non-protic and protic solvents.

3.8.3.2 Energy Shifts in Solvent Mixtures. In a mixture of two solvents of different polarities [*e.g.* the Onsager polarity function $f(D)$] the solute molecule often determines its own local environment; a process known as

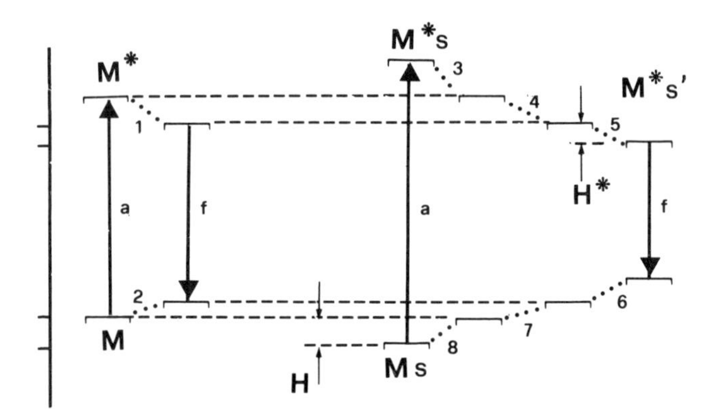

Figure 3.54 *Energy levels of a dipolar, hydrogen bonding molecule M (excited state M*) in a polar, non-protic solvent (left) and a polar, protic solvent (right). H is the H-bond energy in the ground state, H* in the excited state. (1), (2), (4) and (7) correspond to solvation relaxation, (3) and (6) to anti-H-bond dissociation, and (5) and (8) to H-bond formation*

dielectric enrichment. If the solute molecule is dipolar, then its solvent shell will be enriched in the more polar solvent. When the absorption or emission spectra of such molecules are measured in mixtures of increasing mole fraction of the polar solvent, the solvatochromic plots are not linear. Figure 3.55 provides two examples of this effect. The deviation from linearity at small concentrations of the polar solvent is much greater for the fluorescence spectrum than for the absorption spectrum, since the dipole moment of this molecule increases on excitation from 3.5 to 7.5 debye. When one of the solvents in the mixture can form hydrogen bonds with the solute (M or M*), the solvatochromic effects are much more complex (Figure 3.55b).

A simple model of a solute molecule of dipole moment μ and molecular radius a, surrounded by a mixture of two solvents N and P of polarity functions $f(D_N)$, $f(D_P)$ and bulk mole fractions x_N and x_P yields an equation for dielectric enrichment; as the local mole fractions change to y_N, y_P the equilibrium condition is

$$\frac{\partial E}{\partial y_P} - T\frac{\partial \Delta S}{\partial y_P} = 0 \tag{3.36}$$

and this gives the relationship

$$y_P/y_N = (x_P/x_N)e^Z \tag{3.37}$$

where Z is the index of preferential solvation. The more polar solvent (P) has a higher concentration around a dipolar solute molecule. Z is proportional to μ^2, a^{-6} and $\Delta f(D) = f(D_P) - f(D_N)$.

3.8.4 Ion Solvation: the Born Equation

Consider a spherical ion of radius a which carries a central electrical charge q. The work required to increase this charge by dq is the potential energy

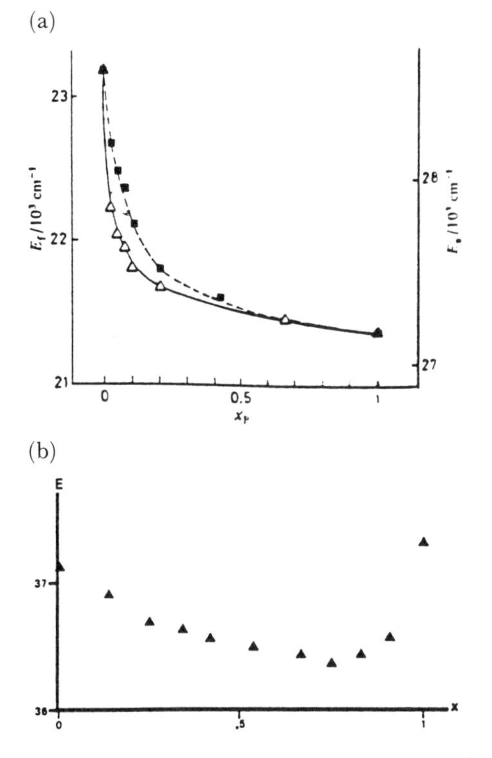

Figure 3.55 *Examples of solvatochromic plots in solvent mixtures. (a) The fluorescence of 4-aminophthalimide in ether/dimethylformamide shows a much steeper non-linearity than its absorption band because of the large dipole moment of the emitting state. (b) The absorption spectrum of an aminobenzene in dioxan/water displays a red shift followed by a steep blue shift at high water concentrations*

$(q\mathrm{d}q)/a$ in a vacuum, $(q\mathrm{d}q)/Da$ in a solvent of dielectric constant D, the difference being the solvation energy. Integration from $q = 0$ to $q = Ze$ leads to the Born equation of ion solvation

$$E = -[(Ze)^2/2a]\ F(D) \tag{3.38}$$

The solvent polarity function $F(D) = (1 - 1/D)$ having the same saturation property as the Debye and Onsager functions. As D goes to infinity all three functions approach the limit of 1.

3.8.4.1 Solvatochromic Shifts of Ions. There is at the present time no theory of the solvatochromic shifts of ions except for the effect of the change of volume in different electronic states. This follows from the Born equation, a_{i} and a_{f} being the ionic radii in states i and f (initial and final).

The Born equation is based on the simple model of a spherical ion with a single charge at its centre. Such an ion has no dipole moment and no higher multipole moments, but real molecular ions are of course much more complex. Since the electrical charge is distributed among all the atoms of the

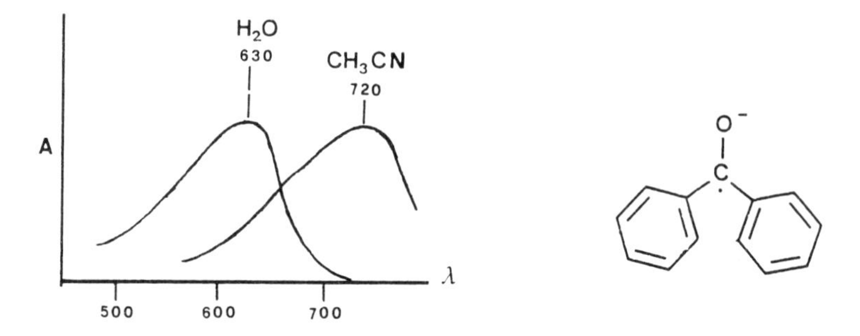

Figure 3.56 *Absorption spectra of the benzophenone radical anion in acetonitrile and water*

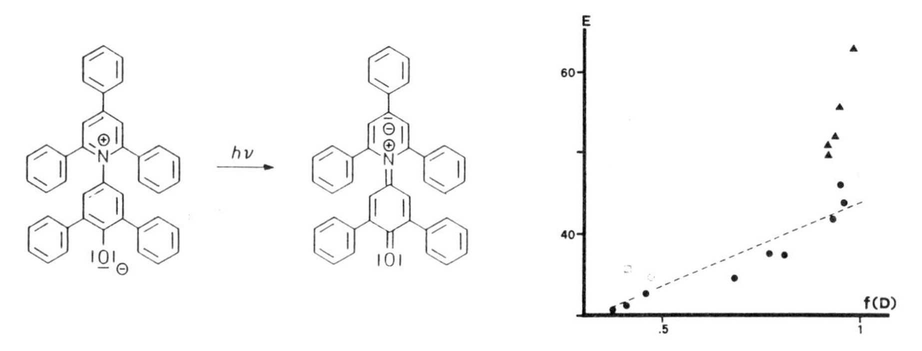

Figure 3.57 *Solvatochromic plot of the first absorption band of a betaine dye. The transition corresponds to a CT from O to the aromatic system. Triangles are protic solvents. E is in kcal mol^{-1} and f(D) is the Onsager function*

ion, there will in fact be multipole moments which will change in various electronic states, even though the formal charge itself does not change. Assuming that the various solvation interactions are independent, the solute dipole–solvent dipole interaction is then the same as for an identical neutral molecule.

Experimentally the absorption spectra of molecular ions often show important solvatochromic shifts. We take here the example of the benzophenone ketyl radical anion (Figure 3.56) which involves both dielectric and hydrogen-bonding effects, both being blue shifts with increasing solvent polarity. In the ground state of this ion the negative charge is localized essentially on the carbonyl group, the first excitation corresponding to a charge transfer towards the aromatic rings.

CHAPTER 4

The Chemistry of Excited Molecules

4.1 'DARK' CHEMISTRY AND LIGHT-INDUCED CHEMISTRY

In the preceding chapters it has been seen that the chemical reactions in thermal ('dark') chemistry originate from the ground state of a molecule, whereas photochemical reactions originate from its excited states. Photochemistry is therefore different from thermal chemistry because electronically excited molecules are different from ground state molecules.

(1) As a result of electronic excitation the molecule finds one of its electrons promoted to a higher energy orbital, leaving a vacancy (a 'positive hole', so to speak) in the lower orbital.

In Figure 4.1 the frontier orbitals of a molecule are shown, with respect to some arbitrary electrode potential E_a. The process of *oxidation* in electrochemistry is the removal of an electron from M to the electrode, and this

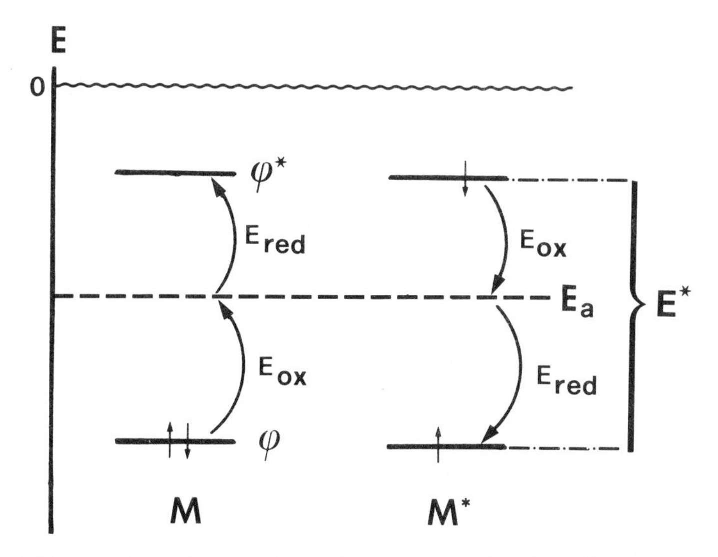

Figure 4.1 *Orbital energies, redox potentials and excited state (HOMO–LUMO) energy. The zero level is the energy of an electron in a vacuum, $E_{(a)}$ is the potential of a reference electrode*

requires an energy E_{ox}. Similarly, the electrochemical *reduction* of M is the addition of an electron from this electrode to form the negative ion $M^{\bullet -}$, and this again requires some input of energy shown as E_{red}. In this illustration both 'redox' processes go uphill—in proper chemical terms they are 'endergonic'—but this need not always be the case (if the electrode potential E_a was set below the occupied orbital of M, the oxidation would be exergonic, but then the reduction would also be that much more endergonic). If this same molecule is in an electronically excited state M*, both the oxidation and the reduction become more exergonic, the oxidation potential E_{ox} and the reaction potential E_{red} decreasing *both* by the excitation energy E^*.

An excited molecule is therefore at the same time a stronger reducing agent (an electron donor) and a stronger oxidizing agent (an electron acceptor) than the ground state molecule. Not surprisingly, we shall find that there are many redox photochemical processes which cannot occur in the dark, and the primary reaction of photosynthesis is one of these.

(2) The promotion of an electron to an antibonding orbital means that one (or several) bonds are weakened; such bonds may dissociate, or may become very reactive.

The rotation around a double bond, for instance, is practically impossible in a ground state molecule M, since this would require the breaking of the π bond. Figure 4.2 shows that such a rotation can however readily take place in the excited molecule M*, for in this case the π bond is indeed practically broken in the π–π^* excited state, although the σ bond is still there; but rotation around a single bond is virtually free, unless there is some steric hindrance when bulky substituents are attached to the carbon atoms.

(3) The distribution of electrical charges can be very different (in particular as a result of charge transfer between various parts of the molecule) and this can result in quite different electrostatic activation barriers, *e.g.* towards 'electrophilic' reactions which involve the attack of negative charge centres, or 'nucleophilic' reactions which involve the attack of positive charge centres.

4.1.1 Pathways of Dark and Light-induced Chemical Reactions

A photochemical reaction is not simply an 'activated' dark reaction, as shown in Figure 4.4. In the transition-state kinetic model, the initial and final states (reactants R and products P) of the chemical system are linked through a transition state (TS) which can be reached only through thermal activation (Δ). It would be tempting to see a photochemical reaction as being simply a light-activated ground state reaction, such that the excited state (R*) would decay to the transition state TS, thus overcoming the activation barrier. Such a point of view is however totally incorrect. If light had no other effect than to provide thermal activation for a ground state reaction, then the product distributions would be identical for dark and light-induced processes; experimentally this is not the case. (For some specific examples, see for instance section 4.6.3.)

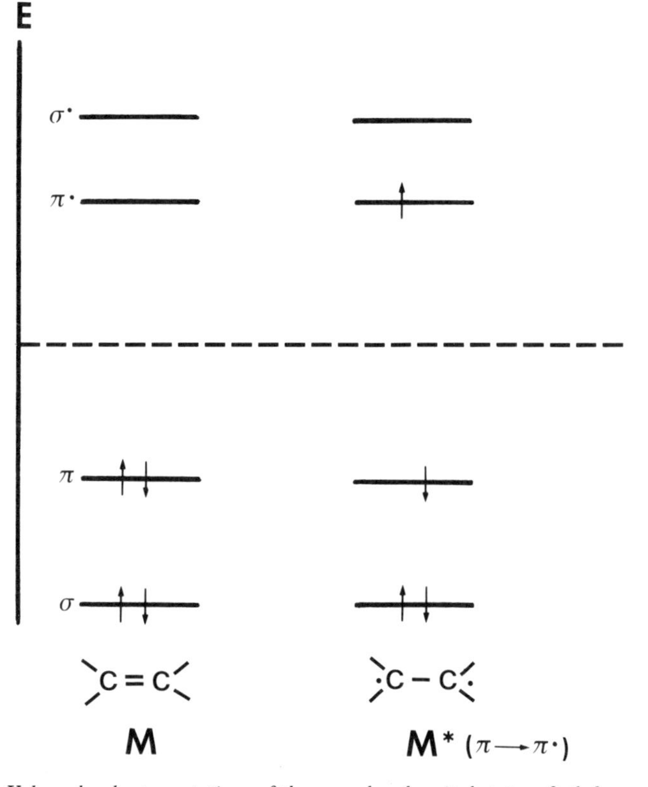

Figure 4.2 *Valence bond representations of the ground and excited states of ethylene*

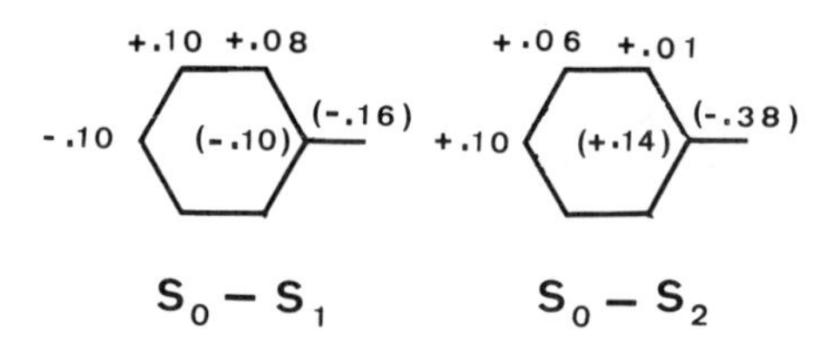

Figure 4.3 *Electron density changes in aniline from the ground state to the S_1 and S_2 excited states. Values in brackets are calculated only, as they are not accessible experimentally*

4.1.2 Adiabatic and Non-adiabatic Processes: the Role of the Energy of Excited States

An 'adiabatic' process is one in which the energy is conserved within the reactive system, whereas in a 'non-adiabatic' process the energy is lost in the form of heat to the surrounding medium. A general question arises in the case of photochemical reactions: are these processes adiabatic or non-adiabatic, in other words what happens to the energy of the photon, or of the excited state, in the course of the reaction?

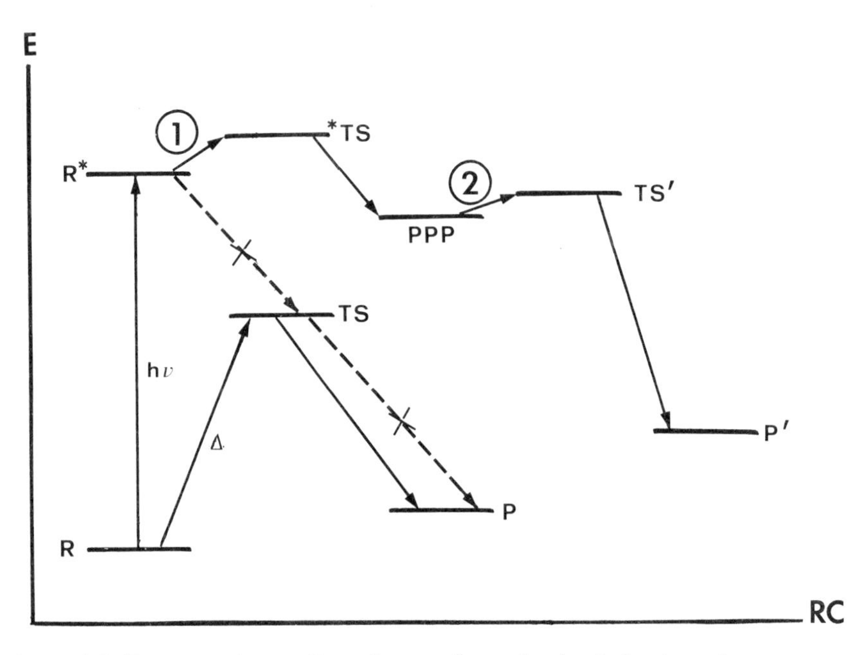

Figure 4.4 *Energy–reaction coordinate diagram of ground and excited state reactions*

In Figure 4.4 for example, the direct reaction from R* to P would be a non-adiabatic process. Although there is no simple and general answer to this question, most *primary* photochemical reactions can be considered to be adiabatic when the primary photoproduct (PPP) retains a large part of the excitation energy. In some cases this is fairly obvious, when the photoproduct is formed in an excited state; for instance in a reversible proton transfer reaction (see section 4.3).

$$M^* + H^+ \rightleftharpoons (MH^+)^* \tag{4.1}$$

In other cases the adiabacity of the primary photochemical process may not be immediately apparent, for instance when a pair of free radicals is formed by dissociation or by a bimolecular reaction such as hydrogen atom transfer (see section 4.4.4)

$$M^* + HZ \rightarrow HM^{\bullet} + Z^{\bullet} \tag{4.2}$$

However, the pair of radicals $HM^{\bullet}$, $Z^{\bullet}$ is a 'high energy' primary photochemical product, since it still retains the energy of the broken bond. In the energy–reaction coordinate diagram (Figure 4.4) it is readily seen that the excitation energy is finally lost only when the high-energy PPP react to form the final product P′ so that the primary process 1 is in fact adiabatic; it is the *thermal reaction* of the excited molecule, going through a transition state *TS.

4.1.3 Is There a Temperature Effect in Photochemical Reactions?

In a thermal reaction R → TS → P, as shown in Figure 4.4, the transition state TS is reached through thermal activation, so that the general observation is that the rates of thermal reactions increase with temperature. The same is in fact true of many photochemical reactions when they are essentially adiabatic, for the primary photochemical process is then a thermally activated reaction of the excited reactant R*. A non-adiabatic reaction such as R* → (TS) → P is in principle temperature independent and can be considered as a type of non-radiative transition from a 'state' R* to a 'state' P of lower energy, for example in some reactions of isomerization (see section 4.4.2).

'Catalysis' by light is an expression that is still used from time to time, but it is fundamentally incorrect because light does not act like a chemical catalyst in a thermal reaction. By definition, a catalyst must be recovered unchanged after the reaction, whereas light must be *absorbed* in order to have any chemical effect (light which is simply transmitted through transparent matter does not lead to photochemistry.)

Although the concept of 'catalysis by light' is quite wrong, *photoactivated catalysis* does exist; but here the catalyst is not light, it is an electronically excited molecule which is indeed recovered unchanged (but in its ground state) after the reaction. Such a catalyst is active only in its excited state.

The following scheme illustrates the process of photoactivated catalysis, C being the catalyst for a reaction R → P

$$C + h\nu \rightarrow C^*; \qquad R \xrightarrow{C^*} P \qquad (4.3)$$

Sensitized photo-oxidation reactions are an example of such photoactivated catalysis (Section 4.5) in which C would be a dye which absorbs light and forms excited oxygen molecules (P) through energy transfer; these excited oxygen molecules then lead to various oxidation reactions. There are other important examples of photoactivated catalysis in photoelectrochemistry (section 4.5.1).

4.1.4 Monophotonic and Multiphotonic Processes

When a photochemical reaction is written as

$$M^* \rightarrow P \text{ (unimolecular)}; \quad \text{or} \quad M^* + N \rightarrow P \text{ (bimolecular)} \qquad (4.4)$$

it is implied that the excited state M* has been formed by the absorption of *one* photon; the entire process is then said to be monophotonic. There are however processes which depend on the absorption of two photons (biphotonic processes), and in principle there could be reactions resulting from the absorption of any number of photons. Such multiphotonic processes become however increasingly improbable as the number of photons required increases, and it is sufficient to consider the biphotonic case of which there are indeed many important examples.

There are two distinct ways in which a biphotonic process can take place.

(1) By the sequential absorption of two photons by the same molecule, according to

$$M \xrightarrow{h\nu} M^* \xrightarrow{h\nu} M^{**} \quad (4.5)$$

whereby the molecule M is promoted to an excited state of very high energy; this process is discussed further on (section 4.2).
(2) By the encounter of two excited molecules

$$M^* + M^* \rightarrow P \quad (4.6)$$

The emission of delayed fluorescence through triplet–triplet annihilation (see section 3.4.2) can be taken as an example of this type of biphotonic process.

Biphotonic processes depend greatly on light intensity, and on the lifetimes of the excited states M*. Their importance increases markedly in conditions of very intense, pulsed light excitation, *e.g.* in laser beams or in flash light.

4.1.5 Primary and Secondary Photochemical Processes

These take place in chemical reactions in which the excited molecule M* takes part in the 'primary' photochemical process. This may lead directly to the final products (*e.g.* in isomerizations), or more often to unstable or reactive chemical species (*e.g.* free radicals or radical ions) which then react further in 'secondary' processes through dark reactions which lead ultimately to the final photoproducts.

The sequence of a photochemical reaction can therefore be given as a succession of steps.

(1) Absorption of light $M + h\nu \rightarrow M^*$ (4.7a)
(2) Fluorescence $M^* \rightarrow M + h\nu_F$ (4.7b)
(3) Non-radiative deactivation $M^* \rightsquigarrow M$ (+ heat) (4.7c)
(4) Intersystem crossing $M^* \rightsquigarrow {}^3M^*$ (4.7d)
(5) Primary reaction, unimolecular $M^* \xrightarrow{\text{light reaction}} P$ (4.7e)
bimolecular $M^* + N \xrightarrow{\text{light reaction}} P$ (4.7f)

P is the 'primary photochemical product(s)'; these can then react further in a variety of dark reactions.

(6) Secondary processes $P \xrightarrow{\text{dark reaction}} F$ (final products) (4.7g)

In the case of closed-shell organic molecules M* can be an excited singlet or triplet state. M* can react on its own in unimolecular reactions (dissociations, isomerizations) or it can react with another (ground state) molecule; N in bimolecular processes (*e.g.* additions, substitutions, *etc.*).

For a reaction originating from the (lowest) triplet state the radiative and non-radiative deactivations of the molecule ($^3M^*$) must also be considered

(2a) Phosphorescence ${}^3M^* \rightarrow M + h\nu_P$ (4.7h)
(3a) Non-radiative deactivation ${}^3M^* \rightsquigarrow M$ (+ heat) (4.7i)

4.1.5.1 Yield, Quantum Yield and Rate Constant of Photochemical Reactions. The overall efficiency of a thermal reaction is given by its 'yield', that is the fraction of reactants convertible into products; this is usually expressed in %.

In the case of a photochemical process the overall efficiency is given by the 'quantum yield' which is defined as the number of molecules of photoproducts formed for each photon absorbed

$$\varphi = \frac{\text{number of product molecules obtained}}{\text{number of quanta of light } \textit{absorbed}} \tag{4.8}$$

In the same manner a quantum yield of reactant disappearance can be defined as the ratio of the number of reactant molecules which have disappeared to the number of photons absorbed.

The quantum yield of a primary photochemical process must be a number between 0 and 1, since it represents the probability that the excited state will undergo the reaction. The quantum yield of overall (that is, primary and secondary) processes can on the other hand be greater than 1, depending on the secondary reactions; it can be very large in the case of chain reactions.

The quantum yield of a primary photochemical process is related to the actual *rate constant* for the reaction and to the rate constant(s) for all other deactivation processes of the excited state, as well as to the population of the reactive state (from the state reached directly by the absorption of the photon). If the reactive state is not populated directly by the absorption process it must be populated through the non-radiative decay of higher state(s). If these higher state(s) can also decay without going to the reactive state, the population yield of this reactive state will be less than 1, and the quantum yield of the primary photochemical process cannot be greater than this yield.

It is important to bear in mind the fact that the intrinsic reactivity of an excited state is given by the rate constant, not by the quantum yield. An excited state can be very 'reactive' in terms of the rate constant, yet the reaction quantum yield can be very low if it is of very short lifetime as a result of other deactivation or quenching processes.

The relationships between rate constants and quantum yields can be derived from a Jablonski diagram such as the one shown in Figure 4.5. Let

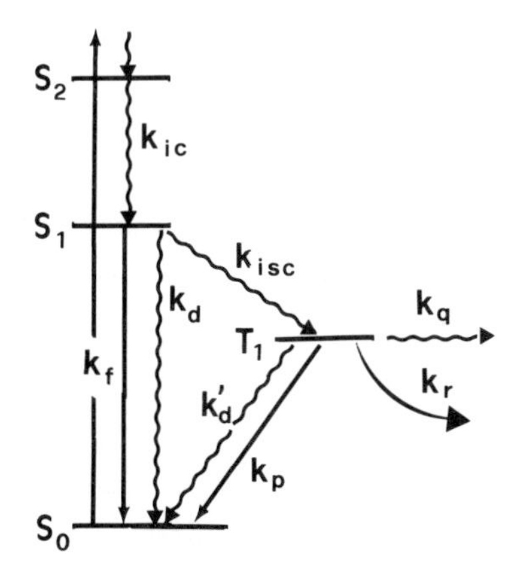

Figure 4.5 *Jablonski diagram of electronic transitions in a polyatomic molecule which reacts chemically from the triplet state T*

us assume that absorption of a photon promotes the molecule to some high excited state M** (this could be a higher vibrational level of S_2, for example), but the chemical reaction itself takes place only from the lowest triplet state T_1. After excitation the molecule will undergo very fast internal conversion to S_1; here we assume that intersystem crossing between the higher states (*e.g.* S_2 to T_2) can be neglected compared with internal conversion. The population of S_1 states is then unity, but now the triplet state T_1 is formed in competition with other deactivation paths of S_1 such as fluorescence (rate constant k_F) and internal conversion $S_1 \rightarrow S_0$ (rate constant k_d) to which bimolecular quenching processes could be added as well. In a general form the yield of T_1 will be given by the branching ratio of these rate constants

$$\Phi_T = k_{isc}/(k_{isc} + k_F + k_d) \quad (4.9)$$

The molecules which have reached T_1 will now react with a rate constant k_r (unimolecular reaction) or $k'_r[N]$ (bimolecular reaction with a ground state partner N) in competition with radiative (phosphorescence of rate constant k_P), non-radiative (k'_d) deactivations as well as quenching processes ($k'_q[Q]$) so that the final reaction quantum yield of the primary process is

$$\Phi_r = \Phi_T\{k'_r[N]/(k'_r[N] + k_P + k'_d + k'_q[Q])\} \quad (4.10)$$

Note that all the rate constants are unimolecular or pseudo-unimolecular (in the case of bimolecular processes like quenching or chemical reaction $M^* + N \rightarrow P$).

4.1.6 Kinetics of Photochemical Reactions

In the gas phase, encounters between atoms and molecules are controlled by their velocities related to the temperature T,

$$\langle\sigma\rangle = \sqrt{8kT/\pi m} \quad (4.11)$$

for a particle of mass m.

A useful concept is the 'mean free path', the average distance travelled by a particle between two collisions

$$\lambda = (1/\sqrt{2}\sigma)(kT/p) \quad (4.12)$$

where σ is the molecule's cross-section and p is the gas pressure. The actual collision time—the time during which the molecules are in contact—is usually very short in the gas phase, unless there is some strong interaction energy; even so, the presence of a 'third body' M is necessary to take up the kinetic energy of the reactants, otherwise the collision complex cannot be stabilized and will dissociate to reform the reactants.

In the liquid phase reactions are controlled by the diffusion of the reactants, except when they are formed from a pre-existing complex. Intermolecular reactions of independent species cannot be faster than the rate of diffusion given in a simple form as

$$k_D = 8RT/3\eta \quad (4.13)$$

This 'diffusion controlled' rate constant depends only on the viscosity η of the solvent and on its temperature T; the size of the molecules does not appear in it, and this can be something of a surprise at first sight. The reason for this is that although larger particles move slower, their encounter cross-section is that much larger, and in a simple model these two effects cancel out.

When a reaction is diffusion controlled its actual rate constant cannot be determined by simple kinetic experiments. All that can be said is that it must be greater than the diffusional rate constant, since diffusional encounters become the rate limiting step.

When two molecules come together in a liquid they are held in a solvent cage for some appreciable time; they may undergo some 100 collisions during the encounter time, a situation quite different from the gas phase. If the molecules are dipolar, or if they form a dipolar complex on encounter (*e.g* an exciplex), then the stabilization in a polar solvent can result in a long encounter time and enhanced opportunity for chemical reactions.

The diffusion equation for the rate constant, eqn. (4.13), assumes that the molecules must come within van der Waals (hard sphere) contact for reaction to take place. This is of course a reasonable assumption when chemical bonds are made or broken in the reactants, but does not apply to long range processes such as energy transfer and possibly electron transfer.

The situation for reactions in solids is much more complex and is treated in a separate section (4.7.4, p. 153). Physical diffusion of molecules can be neglected within the lifetimes of excited states, but exciton interactions can become important. These have no counterpart in dark reactions and can lead to unusual photochemical properties in crystals and polymers.

Excitons in solids can move with velocities far beyond the diffusional rates in liquids at comparable temperatures. It should therefore never be assumed that photochemical processes can be neglected in the solid phase.

4.2 PHOTOIONIZATION, LIGHT-INDUCED ELECTRON CAPTURE AND ELECTRON TRANSFER REACTIONS

Electrons are the lightest particles of matter and they play the most fundamental role in chemical reactions. The loss or the gain of an electron by a molecule is at the borderline of 'physical' and 'chemical' processes, if it is held that a *chemical* process requires the making and/or the breaking of formal chemical bonds.

Photoinduced ionization (or simply 'photoionization') is the complete separation of an electron from a molecule. In the gas phase (isolated molecules) this requires considerable energy. Since there are many electrons in a molecule the 'ionization potential' (IP) refers to the energy needed to separate 'to infinity' the least tightly bound electron. The energies are such that only light in the far UV or in the 'vacuum UV' (this refers to wavelengths below 180 nm) can directly ionize molecules in the gas phase.

In condensed media ionization can be greatly helped by the solvation of the molecular radical cation and the ejected electron. The polarity of the solvent or matrix plays a most important role, and water in particular is an excellent medium for photoionization processes. In water the ejected electron is solvated in the form of a 'hydrated electron' (denoted e^-_{aq}) which is a powerful reducing agent.

In other polar solvents such as alcohols and acetonitrile (CH_3CN) the ejected electron can be trapped as a 'solvated electron', shared between several solvent molecules, or as a negative ion by attachment to a solvent molecule. Many aromatic molecules such as naphthalene, anthracene, *etc.*, undergo such photoionizations with low quantum yields. The ions eventually recombine on a time-scale of microseconds, and there is no overall chemical effect (Figure 4.6).

If an electron scavenger such as bromobenzene is added to the solution, then the electron can be transferred from the solvent to the scavenger which dissociates with elimination of an ion

$$(MeCN)^{\bullet -} + C_6H_5Br \rightarrow (MeCN)_n + C_6H_5^{\bullet} + Br^- \qquad (4.14)$$

and this results in an irreversible photochemical degradation.

The photoionization of amines, which are strong electron donors, is a process of significance in the photochemical degradation of living matter (see section 5.5).

In conditions of very intense light (flash light and laser light in particular) photoionization can occur by a sequential biphotonic process even when one quantum does not provide enough energy to eject an electron. The first photon promotes the molecule to some short-lived intermediate state (in general the lowest triplet excited state), the second photon is absorbed by this intermediate state, and the combined photon energies bring the molecule to an energy level beyond the condensed phase ionization potential (Figure 4.7).

Since an electronically excited molecule is at the same time a stronger oxidizing agent and a stronger reducing agent than the ground state molecule, photoionization can be seen as a process of oxidation. The opposite reaction is the reduction of the molecule M to the radical anion $M^{\bullet -}$ by *electron capture* from the surroundings (liquid solvent or solid matrix). Such processes are relatively rare, but some examples are known of electron capture by photoexcited quinones in polar solvents; such quinones are among the strongest electron acceptors and play a considerable role in many biological processes (Figure 4.8).

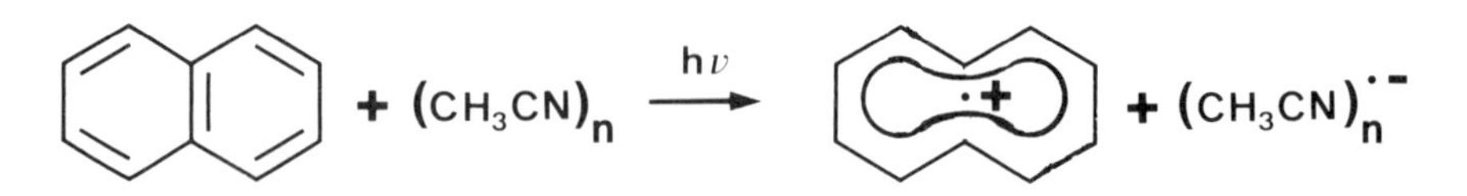

Figure 4.6 *Photoinduced electron transfer from naphthalene in acetonitrile*

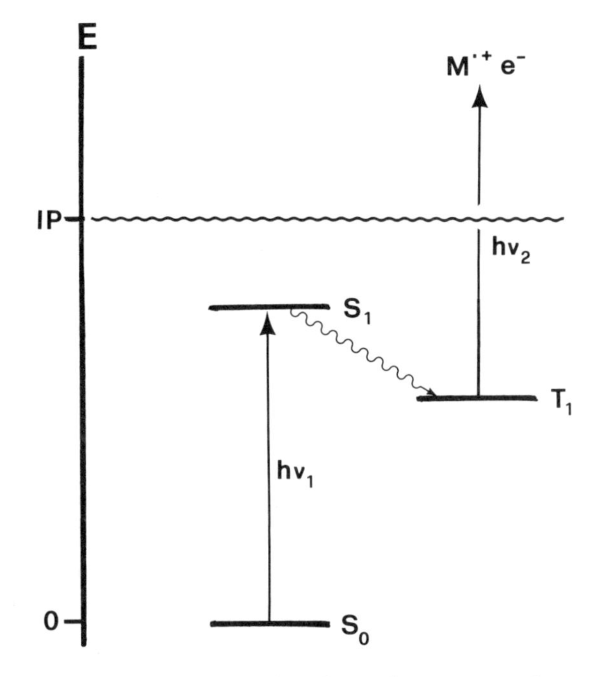

Figure 4.7 *Sequential biphotonic ionization through a triplet state intermediate*

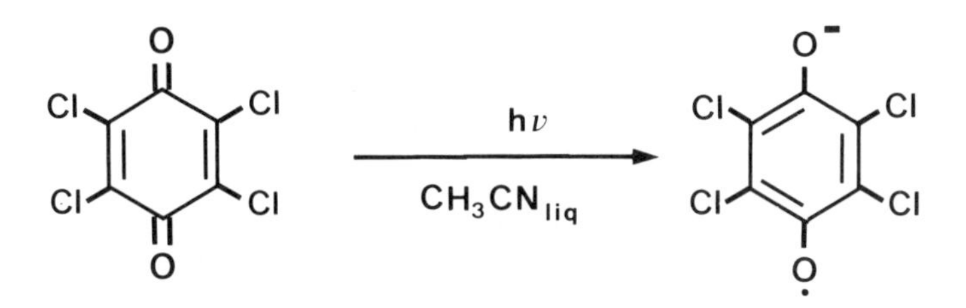

Figure 4.8 *Photoinduced electron transfer between chloranil and acetonitrile*

4.2.1 Intramolecular and Intermolecular Electron Transfer

Electron transfer between two molecules, *e.g.* M* and N, is one of the most fundamental and widespread of all photoinduced chemical reactions. It is at the basis of photosynthesis (section 5.1) in nature, and of photography (section 6.1) in industrial applications.

$$M^* + N \rightarrow M^{\cdot+} + N^{\cdot-} \qquad \text{or} \qquad M^* + N \rightarrow M^{\cdot-} + N^{\cdot+} \tag{4.15}$$

In the first case, M* is the electron *donor* and N is the electron *acceptor*, these roles being reversed in the second case. These properties of 'donor' and 'acceptor' are relative, the same molecule M* being a donor towards some species N and an acceptor towards other partners N. It will be seen presently that the direction of electron transfer depends simply on the energy balance of the reactions.

The loss of an electron by M, $M \rightarrow M^{\bullet+} + e^-$, is the process of oxidation in electrochemistry. The electron is then accepted by an electrode of well defined potential, so that the *oxidation potential* E_{ox} is the free energy of the reaction, as was seen in Figure 4.1. Similarly the *reduction potential* E_{red} is the energy of the reduction reaction, *e.g.* $N + e^- \rightarrow N^{\bullet-}$. By definition the molecule, which is oxidized, is the donor (M in this case), and the molecule, which is reduced, is the acceptor. The electron transfer from M to N is therefore equivalent to the combined oxidation of the donor and reduction of the acceptor, so that the energy balance is

$$\Delta G = E_{ox}(D) - E_{red}(A) + C - E^* \quad (4.16)$$

Here E^* is the energy of the excited state involved in the light-induced process. It has been noted in section 4.1 that both the oxidation and the reduction potentials of an excited molecule are lowered by the excited state energy, compared with the ground state molecule. In the present example the relevant excited state energy is that of the reactive state of M*, which can be in general either the lowest singlet state S_1 or the lowest triplet state T_1, in the case of an organic molecule. The additional term C in this 'Rehm–Weller' equation is called the 'Coulomb term' and represents the electrostatic energy gained when the two product ions, $M^{\bullet+}$, $N^{\bullet-}$, are brought from 'infinite separation' to the actual encounter distance in electron transfer (usually van der Waals, or 'hard sphere' contact). This correction is necessary because the oxidation and reduction potentials are measured independently, so that the difference $E_{ox}(D) - E_{red}(A)$ would apply to electron transfer between independent donor and acceptor molecules, as if they were separated to infinity. If the product ions carry opposite charges, as in this example, then an electrostatic stabilization will make the electron transfer process easier at short distances.

In a point charge model the Coulomb term is simply $C = qq'/rD$ where q and q' are the charges of the ions which are separated by a distance r in a solvent of dielectric constant D. This usual form of the Coulomb term may not be valid when the molecules are very close together, because there is then no solvent between the point charges, and a modified expression is then $C = qq'/4r$.

The positive and negative ions formed by electron transfer are held together strongly by the force of electrostatic attraction. They may however separate to form free, solvated ions in polar solvents and then various secondary reactions can take place. Radical cations often undergo cycloadditions with the neutral (Figure 4.9).

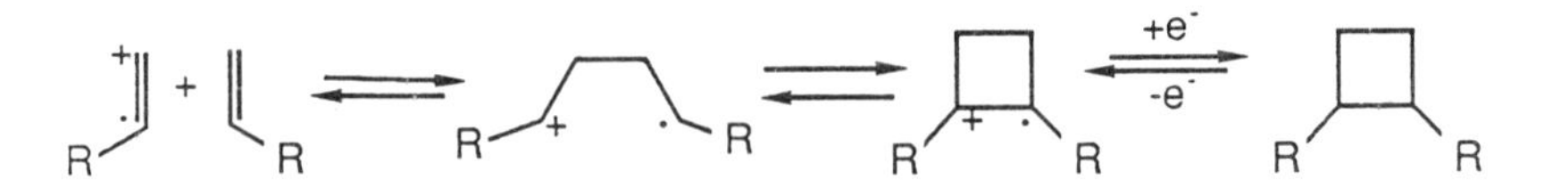

Figure 4.9 *Cycloaddition of a radical cation with the neutral molecule*

In the presence of oxygen the radicals can react to form oxidation products such as carbonyl derivatives, carboxylic acids, *etc.*

The importance of electron transfer as a primary chemical reaction has become increasingly recognized in recent years, and its detailed mechanism is a subject of active research. In most cases the direct contact of the donor and acceptor molecules seems to be necessary for efficient electron transfer, particularly for photoinduced electron transfer which is inevitably limited by the lifetime of the excited state molecule M*. In principle, electron transfer could also take place between distant molecules by the mechanisms of 'electron hopping' or of 'electron tunnelling'.

Electron hopping is in fact a primary charge transfer to solvent (a 'photoionization'), followed by charge separation by solvation of the radical cation $M^{\bullet +}$ and the solvent radical anion $S^{\bullet -}$. In general the electron may not be attached to any one solvent molecule but may exist as a 'solvated electron' delocalized over several solvent molecules. Such solvated electrons have a very high chemical reactivity (they are powerful reducing agents). The motion of a solvated electron can be seen as a sort of hopping of the electron from one group of solvent molecules to another, so that it does not require the physical diffusion of molecules and can be much faster than a diffusion-limited process.

Superexchange is another mechanism of electron transfer over relatively large distances in which the solvent or matrix acts as a bridge between the donor molecule D and the acceptor molecule A. It differs from electron hopping in that the electron is at no time actually localized on a molecule of the medium; there is an interaction between the orbitals of the molecules A, B and D which form a sort of very loose supermolecule over which the electron is delocalized (Figure 4.10). This mechanism seems plausible when the relevant orbitals of A, B and D are rather close in energy. This is similar to the requirement for the interaction of atomic orbitals to form a molecule.

Electron tunnelling is an electron transfer between distant molecules through

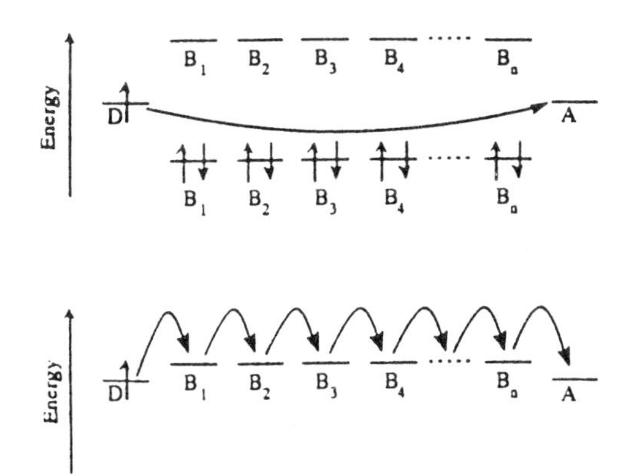

Figure 4.10 *Energy diagram of the superexchange electron transfer between a donor D and an acceptor A in a medium B*

the overlap of the donor and acceptor orbitals. In classical physics the dimensions of a microscopic object such as an atom or molecule can be defined precisely and such particles can be treated as 'hard spheres' with a well defined boundary. This ceases to be true in quantum mechanics since the orbital wavefunctions show an exponential decrease of amplitude with distance. Although it is a very rapid decrease it implies that the probability of presence of the electron is never quite zero, even at large distances from the centre of the atom or molecule. The electron in orbital Ψ_a of A will therefore be found from time to time near B, and at such times it may go into the orbital Ψ_b (of B) if it is of lower energy than Ψ_a. Such far-away visits of the electron in Ψ_a occur very infrequently and the rate of electron tunnelling goes down exponentially with distance (Figure 4.11).

When the donor and the acceptor are linked together by covalent bonds so that they are formally part of a single molecule, the process of *intramolecular* electron transfer can be observed. Some rigid spacers made up of saturated bonds seem to act somewhat like electrical conductors and electron transfer can then take place over quite large distances (20 Å) as a result of the 'through-bond' interaction (Figure 4.12a).

Similar molecules with flexible spacers show two different types of interactions.

(1) When a quinone is linked to a porphyrin, as in Figure 4.12(b), the rate of electron transfer depends on the nature of the linkage, even if it is too short to allow a folding back to give direct contact between the chromophores. This is strong evidence for the through-bond character of the electron transfer, since through-space interaction between identical pairs of chromophores should be independent of the linkage.

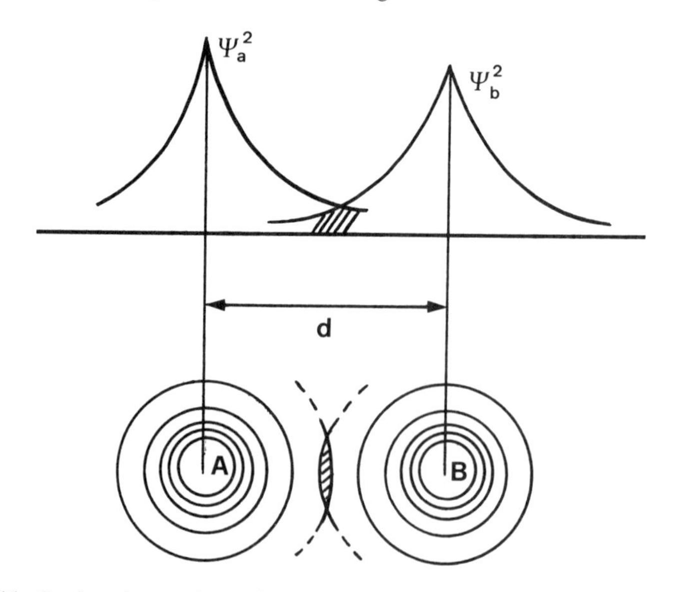

Figure 4.11 *Overlap of squared wavefunctions (probabilities of the presence of electrons) in the s orbitals of two atoms, A and B, separated by an internuclear distance d*

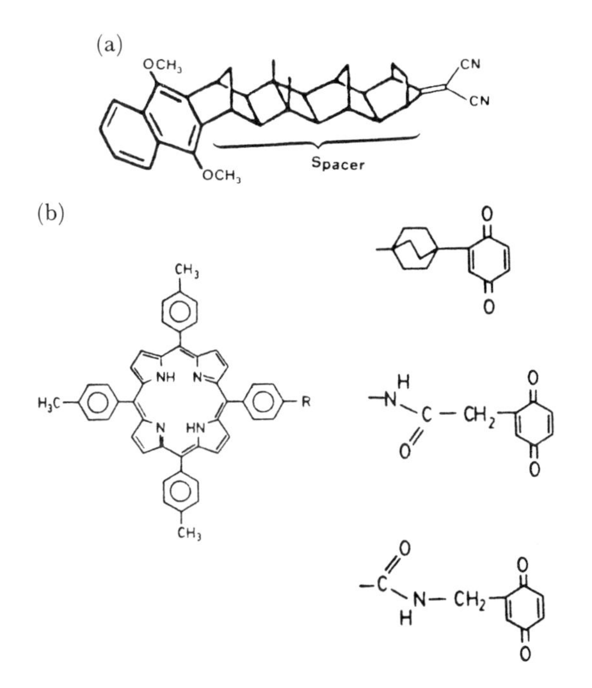

Figure 4.12 *Examples of bichromophoric molecules used for the study of intramolecular electron transfer. (a) Dimethoxynaphthalene electron donor, dicyanoethylene electron acceptor with rigid spacers; (b) porphyrin donor and quinone acceptors separated by flexible spacers*

(2) Longer linkages between the same chromophores may increase the rate of electron transfer, but this is not expected for a through-bond interaction of any kind. The flexibility of such chains however allows a folding of the molecule which brings the chromophores into van der Waals contact. Thus the conditions for 'through-space' interaction are restored, but there is really no 'space' at all between the donor and the acceptor.

The general conclusion seems to be that in most cases the direct contact of the partners is necessary for efficient electron transfer, unless they are linked by some suitable 'conducting' chain which allows through-bond interaction.

4.2.2 The 'Marcus–Hush' Model of Electron Transfer

In this model the reactants R and products P are represented by two potential energy wells (Figure 4.13). It must, of course, be realized that such a two-dimensional representation is only one section in a multi-dimensional space, so that the 'reaction coordinate' which is the abscissa can have many different meanings; it can be an intermolecular distance, but it can also be the state of polarization of the solvent, *etc.*

In a classical picture the reaction from R to P takes place when the system reaches the saddle point X at the intersection of the wells; the activation energy is then $\Delta G^{\#}$, the reaction free energy being ΔG^0. If the energy wells

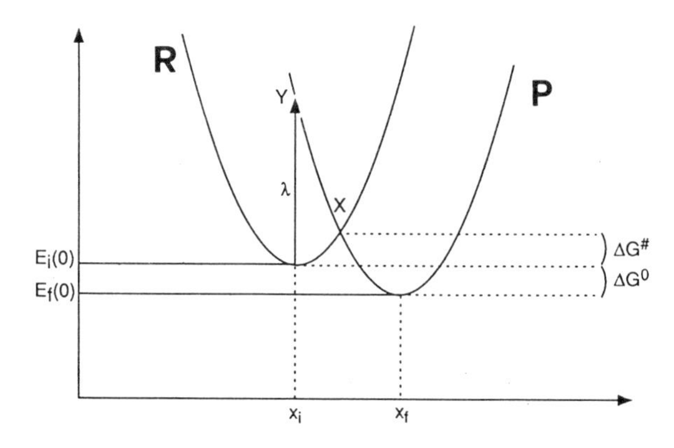

Figure 4.13 *Potential energy diagram of the intersection of two parabolic energy wells as a simple model of electron transfer. The reactants R can cross to the products P at the point X of common energy and common coordinate x*

are identical parabolic functions within the model of harmonic oscillators the activation energy can be expressed according to ΔG^0 and the 'reorganization energy', λ, which is defined as the vertical energy required to move from the reactant R to the product P at constant 'geometry' (*e.g.* internuclear distances, positions of solvent molecules).

$$E_a = \Delta G^\# = \frac{(\Delta G^0 + \lambda)^2}{4\lambda} \tag{4.17}$$

This quadratic dependence of the activation energy on the reaction free energy leads to the prediction of an 'inverted region' in which the reaction rate constant (which depends on $\Delta G^\#$) falls when the overall reaction free energy becomes more favourable. This is readily seen from the simple picture shown in Figure 4.14. When the intersection point of the wells leads to $\Delta G^\# = 0$ the reaction becomes free of an activation barrier, but as the products well sinks deeper the point of intersection rises again.

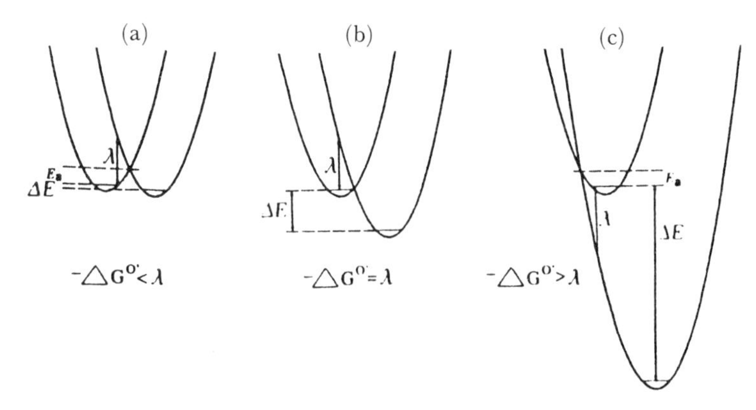

Figure 4.14 *Schematic representation of the classical activation energy $\Delta G^\#$ for different reaction energies ΔE. (a) is the normal thermodynamic region, (b) is the activationless reaction and (c) is the 'inverted' region*

The reorganization energy λ is the sum of an 'internal' reorganization λ_i which is related to the change in molecular geometry on going from the neutral species to the ions and an 'external' reorganization of the solvent shell which becomes polarized around the charged product molecules, λ_0.

This external or 'outer-sphere' parameter λ_0 can be expressed according to a modified form of the Born equation to take into account only the solvent's orientation polarization. (As in the theory of solvatochromic shifts, it is assumed that the induction polarization can follow electronic motion.)

$$\lambda_0 = e^2\left[\frac{1}{2r_A} + \frac{1}{2r_B} - \frac{1}{r_{AB}}\right]\left[\frac{1}{\varepsilon_{op}} - \frac{1}{\varepsilon_s}\right] \quad (4.18)$$

ε_s and ε_{op} are the static and optical dielectric constants of the solvent, r_A and r_B are the radii of the molecules, r_{AB} is their centre-to-centre separation, and e is the electronic charge.

There is no such simple expression for the internal reorganization energy λ_i, which must be given as a sum of terms related to the vibrational states of the reactants and products

$$\lambda_i = \sum_j (f_j^R f_j^P/(f_j^R + f_j^P))(\Delta x_j)^2 \quad (4.19)$$

f is the force constant of the jth normal mode of the reactants R and products P, and Δx_j is the displacement from the position of equilibrium.

4.2.2.1 'Marcus Behaviour' and 'Rehm–Weller Behaviour' in Electron Transfer Reactions. The prediction of the Marcus model of intersection of energy wells is that the rate constant k_{et} of electron transfer should show a bell-shaped function against the reaction free energy ΔG^0. In practice this behaviour is observed only with dark reactions such as ion recombinations, but not in photoinduced electron transfer reactions which do not display the inverted region. The rate constant rises as the reaction becomes more exergonic (the normal region), and reaches the diffusional limit k_D in the case of bimolecular reactions. As the exergonicity of the reaction increases further, there is no sign of the drop in rate constant as forseen by the Marcus model. This observation, which is common to all known photoinduced electron transfers, is known as 'Rehm–Weller' behaviour.

Figure 4.15 shows a graph of the Marcus and Rehm–Weller functions of reaction kinetics *versus* reaction energetics, and Figures 4.16, 4.17 and 4.18 give illustrations pertaining to these two cases. It is not yet known why the photoinduced reactions do not show the Marcus inverted region, one possibility being that in such reactions one of the product ions may be formed in an excited state

$$M^* + N \rightarrow M^{\bullet +} + N^{\bullet -*} \quad (4.20)$$

The actual ΔG^0 of the primary process would then be greater by the ion's excited state energy, so that the reaction would not reach the inverted region. This explanation is quite reasonable since it is well known that open-shell species such as radical ions have generally excited states of very low energy.

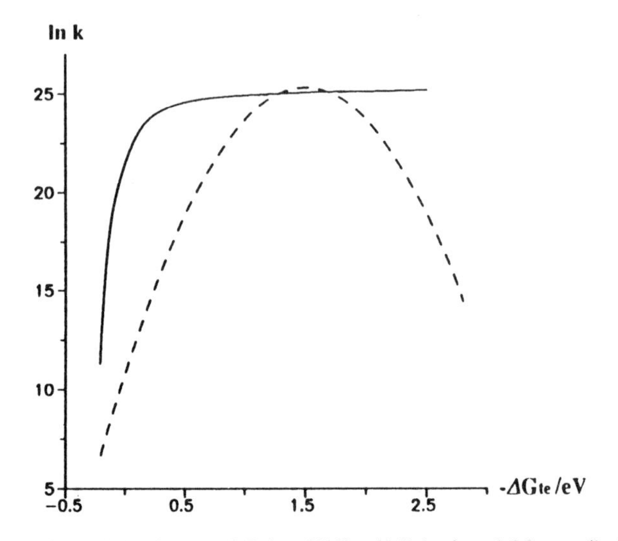

Figure 4.15 *Outlines of the theoretical Rehm–Weller (full line) and Marcus (broken line) plots of the reaction rate constant as a function of reaction free energy change*

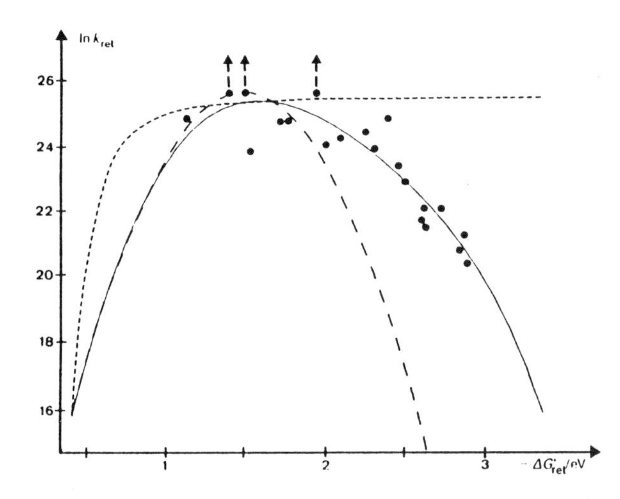

Figure 4.16 *Example of the inverted region observed in the geminate recombination of an ion pair. Theoretical plots are shown from the top as Rehm–Weller, Marcus with quantum effect correction, and Marcus classical model*

4.3 ELECTRONICALLY EXCITED SUPERMOLECULES: EXCIMERS AND EXCIPLEXES

In chemistry the word 'complex' is used to describe an association of molecules; a simple case is the formation of a hydrogen-bonded complex, *e.g.* between the carboxylic acids shown in Figure 4.19.

Other complexes are made up of many molecules, like the organometallic complexes that are considered in section 4.6.3. The important thing is that

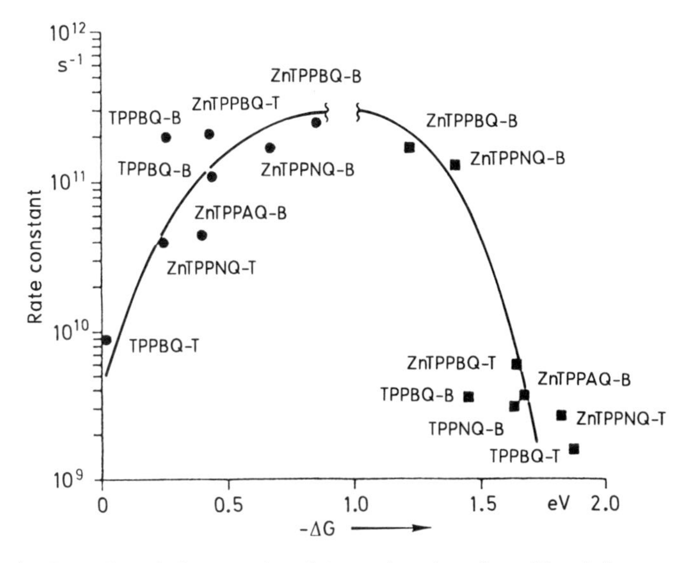

Figure 4.17 *Examples of the normal and inverted region plots. The circles correspond to a photoinduced electron transfer in which an ion pair is formed, the squares to the geminate recombination of these ion pairs*

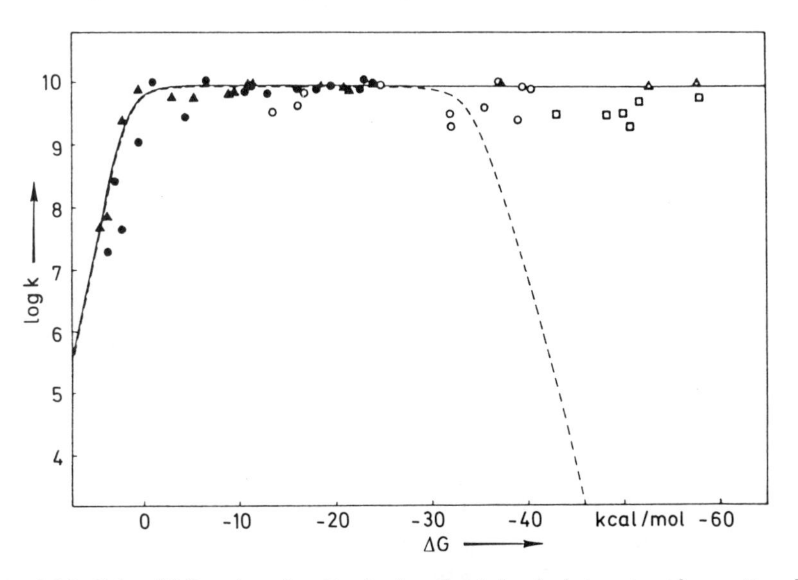

Figure 4.18 *Rehm–Weller plot of a bimolecular photoinduced electron transfer reaction. The horizontal limit is the rate constant of diffusional encounters. The broken line shows the classical Marcus plot*

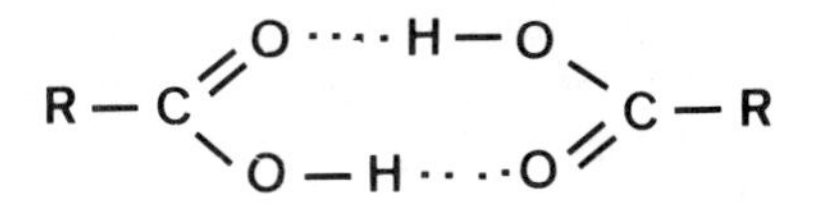

Figure 4.19 *Intermolecular hydrogen bonding in carboxylic acids*

these molecules are associated according to a precise stoichiometry and a fairly rigid geometry, *i.e.* the distances and angles between the various molecules of the complex are well defined (the loose association of a solute molecule with its surrounding solvent shell is *not* a 'complex' according to this definition.) Just as a molecule is an association of atoms, so a complex is an association of molecules and the word 'supermolecule' may be appropriate when the association energies are relatively large.

The association of an electronically excited molecule M* with a ground state molecule N leads to an *exciplex* (*exci*ted com*plex*) or to an *excimer* (*exci*ted di*mer*) if N = M,

$$\underset{\text{Exciplex}}{M^* + N \rightleftharpoons (MN)^*}\,; \qquad \underset{\text{Excimer}}{M^* + M \rightleftharpoons (MM)^*}\ (\text{or } M_2^*) \tag{4.21}$$

We shall use here the word 'exciplex' as the general term for all such electronically excited complexes, the 'excimer' being merely a special case when the excited and ground state species happen to be identical.

Exciplexes (including excimers) play a major role in a number of photophysical and photochemical processes, so they deserve a fairly detailed study. Here we must consider the questions:

(a) what is an exciplex; why is it formed?

(b) what happens when an exciplex is deactivated?

(c) how does an exciplex behave chemically?

(a) Perhaps the simplest way to understand the nature and the formation of an exciplex is to consider the helium molecule. When two He atoms come within encounter distance, they do not form a helium molecule He_2 because their orbitals are either filled (with two electrons of opposite spins), or are empty. As shown in Figure 4.20, the molecular orbitals formed through the interaction of the atomic orbitals are then also either filled or just left empty. The stabilization energy $-\Delta E$ of Ψ_b is offset by the destabilization energy $+\Delta E$ of Ψ_a, so that the ground state molecule He_2 is not stable. This simply illustrates the fact known to all chemists, that noble gas atoms do not form diatomic molecules because they have 'closed shells', fully occupied or totally vacant orbitals. It is not so widely known that the excited species $He_2{}^*$ and the ionic species He_2^+ are however stable, because one of the atoms is of *open-shell* type.

In Figure 4.20(b), the encounter of a closed-shell species M with the open-shell *excited* species M* shows that the occupation of the very same molecular orbitals leads to a net stabilizing situation: two electrons sink to the lower Ψ_b orbital, with an energy gain $-2\Delta E$, while one electron is raised by $+\Delta E$ to Ψ_a; but the 'excited' electron now *sinks* into Ψ_a^*, so that the energy balance is approximately $E = -2\Delta E + \Delta E - \Delta E = -2\Delta E$. This simple picture of excimer formation applies to closed-shell organic molecules as well as to He atoms. The excimer is indeed an electronically excited dimer which is stable only so long as its electrons follow the distribution of M_2^*.

It is clear that the excimer is really an excited state of the M_2(He_2) dimer, and that like any excited state it has a lifetime limited by its radiative deactivation.

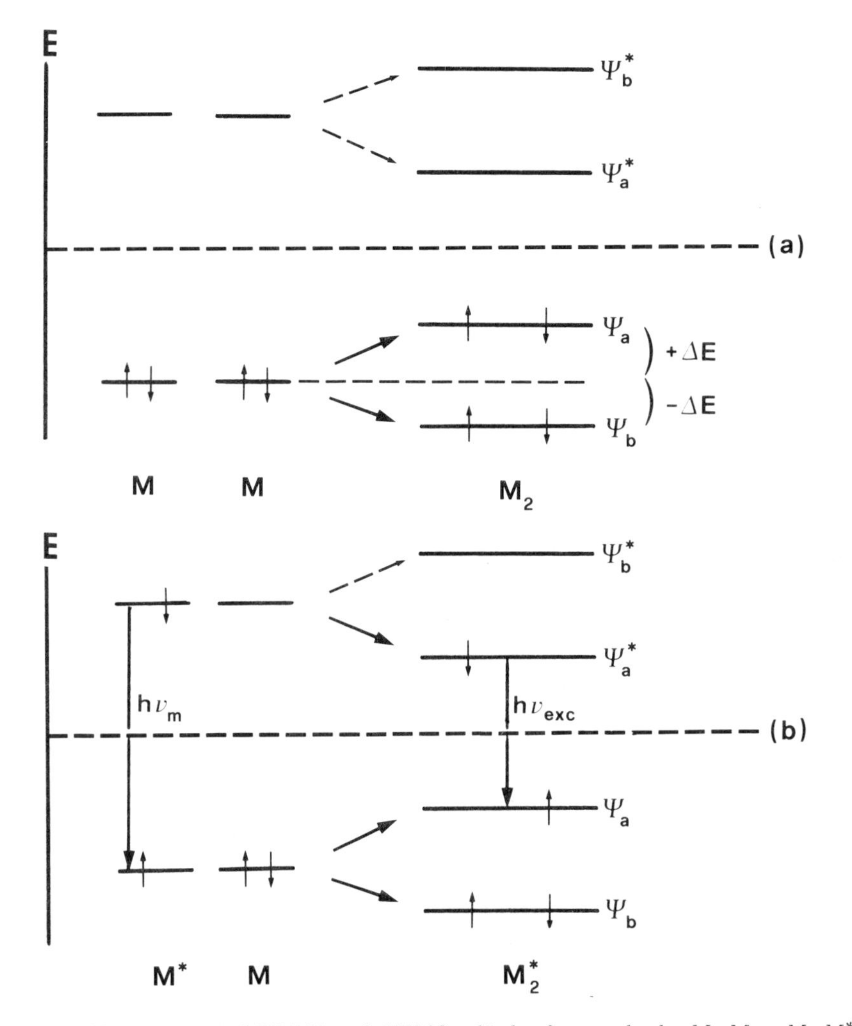

Figure 4.20 *Interaction of HOMO and LUMO orbitals of two molecules M, M or M, M*. $h\nu_m$ is the molecular fluorescence of M* and $h\nu_{exc}$ is the excimer fluorescence*

(b) This radiative deactivation corresponds to the return of the excited electron from Ψ_a^* to Ψ_a; the unstable situation of the ground state complex (Figure 4.20a) is then restored, and this leads to immediate dissociation

$$M_2^* \xrightarrow{h\nu} M_2 \rightarrow M + M \qquad (4.23)$$

The radiative deactivation is known as *excimer fluorescence*. Figure 4.21 shows that it must come at lower energy than the molecular fluorescence $h\nu_m$, since the supermolecular orbitals Ψ_a and Ψ_a^* are closer together than the molecular orbitals of M, M*. Excimer fluorescence is observed as a structureless band on the long-wavelength side of the molecular fluorescence in concentrated solutions, as illustrated in Figure 4.22.

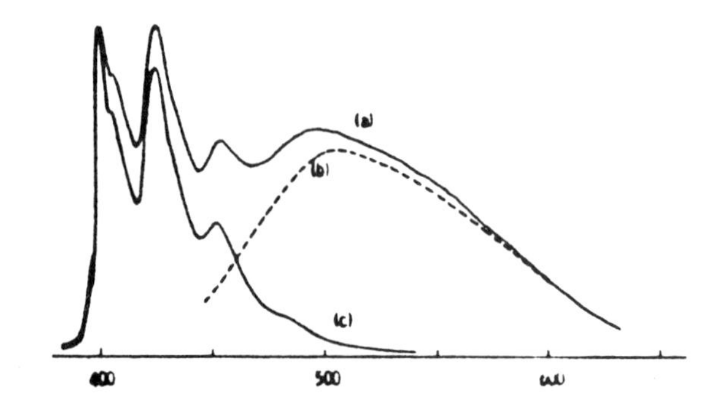

Figure 4.21 *Example of molecular and excimer fluorescence spectra. (a) is the composite spectrum observed at high concentrations, (b) the excimer emission when the molecular fluorescence (c) is substracted*

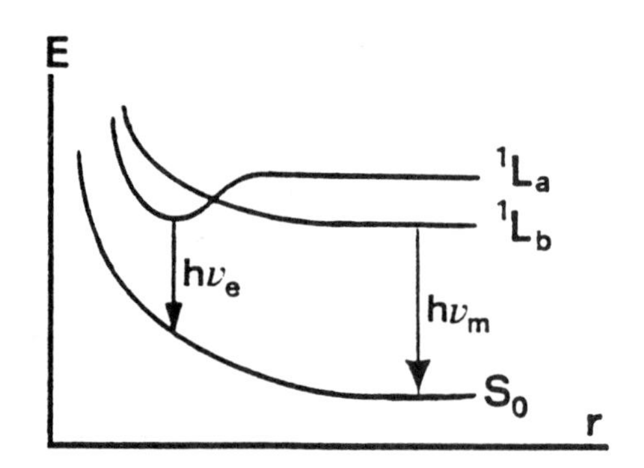

Figure 4.22 *The potential energy of excimer formation and emission of pyrene. The excimer ground state S_0 is dissociative and association takes place only from the L_a excited state when the transition to L_b is forbidden. $h\nu_m$ is the molecular emission, and $h\nu_e$ the excimer emission*

The lack of vibrational structure is proof that the ground state excimer is dissociative, *i.e.* it does not 'exist' at all; remember that the molecular fluorescence of naphthalene, for example (Figure 3.28, p. 55) shows a structure which is related to the transitions from the $v = 0$ level of the S_1 excited state to the various discrete vibrational levels of the S_0 ground state.

We have been concerned so far with excimers in which the partner molecules are identical. An exciplex in which $M \neq N$ is different, since the orbital energies of M* and N do not coincide. One of the molecules will now act as an electron donor and the other as an electron acceptor, as illustrated in Figure 4.23.

Figure 4.23 shows one example of the interaction of LUMOs (Ψ^*_m and Ψ^*_n) to form new supermolecular orbitals that are delocalized over the complex. If the HOMOs are too far apart in energy to interact, as in this example, then they will of course remain localized on the partner molecules; excitation of an electron from a *localized* HOMO to a *delocalized* LUMO results in partial

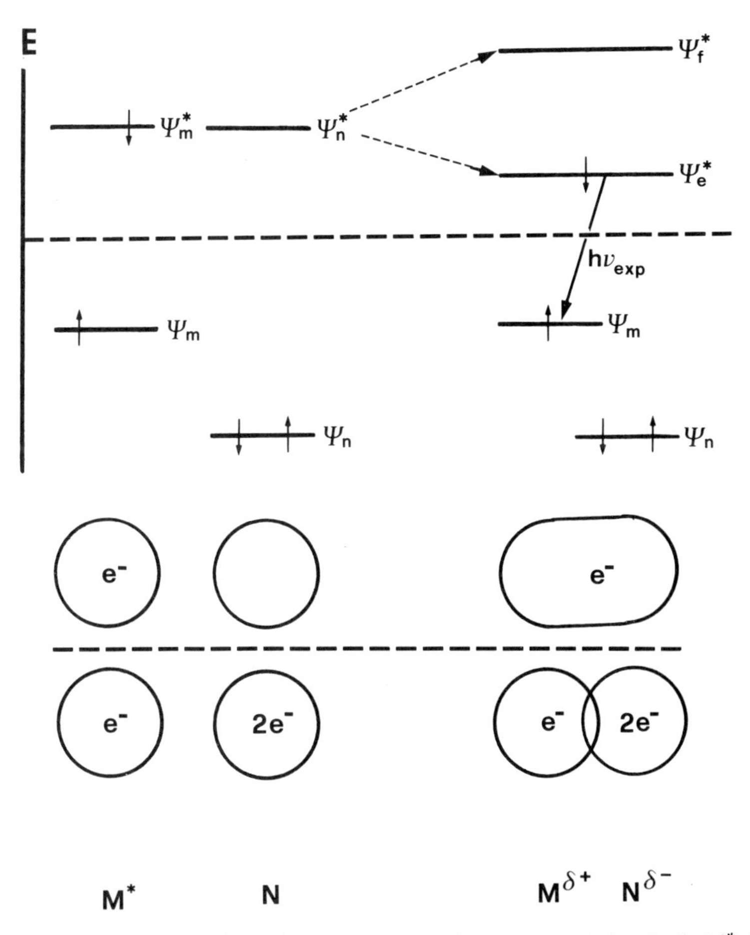

Figure 4.23 *Orbital interactions in exciplex formation between an excited molecule M^* (electron donor) and a ground state molecule N. $h\nu_{exp}$ is the exciplex emission*

charge transfer. Exciplexes made up of different partners, M ≠ N, are dipolar species, resembling sometimes ion pairs. In highly polar solvents such as water, alcohols, acetonitrile, *etc.*, they can dissociate to form free, solvated ions:

$$(M^{\delta+}N^{\delta-})^* \rightarrow M^{\bullet+}_{s} + N^{\bullet-}_{s} \tag{4.23}$$

This represents a second mode of deactivation of exciplexes, equivalent to an electron transfer reaction. Dipolar exciplexes are in fact often intermediates in such reactions.

(c) The chemical results of dipolar exciplex formation and dissociation are therefore similar to those of an electron transfer reaction. Things are however different when a non-polar excimer M_2^* is formed of two identical molecules, for separation into ions does not then take place (as a rule). Non-polar excimers are often intermediates in cycloaddition reactions, and these are considered in section 4.4.4.

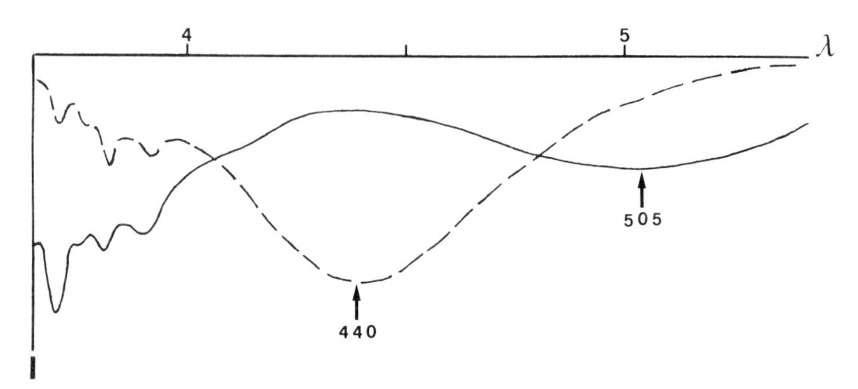

Figure 4.24 *Example of the solvent dependence of an exciplex fluorescence: the pyrene/N,N-dimethylaniline exciplex in cyclohexane (broken line) and tetrahydrofuran (full line). Wavelength λ(×100) nm*

4.4 ORGANIC PHOTOCHEMISTRY

Organic molecules are made up mostly of the lighter elements of the periodic table: carbon, of course, and hydrogen, nitrogen, oxygen, and sulfur. This restriction has two important consequences for their photochemistry.

(1) The lowest of all excited states is a triplet state of long lifetime. In the ground state, organic molecules have 'closed shells'; *i.e.* all bonding and non-bonding orbitals are filled (by two electrons of opposite spins), and all antibonding orbitals are empty. The total spin is then zero, the ground state is a singlet state (S_1). An excited state which results from the promotion of one electron to an antibonding orbital can therefore have a total spin of either 0 or 1; excited states come in pairs of singlet ($S = 0$) and triplet ($S = 1$) states. Since the triplet state is always lower in energy than the corresponding singlet state, the lowest of all excited states is necessarily a triplet state (labelled T_1 in a Jablonski diagram, *e.g.* in Figure 3.23, p. 50). The crossing from that triplet state to the ground (singlet, S_0) state is very slow, so that the T_1 state has a very long intrinsic lifetime, of ms to s in the absence of quenching. The lowest triplet excited state of organic molecules therefore acts as an energy reservoir, and it is in many cases the only excited state of sufficiently long lifetime to lead to efficient bimolecular reactions.

(2) Upper excited states are extremely short-lived. When the molecule is promoted to an excited singlet state beyond S_1, the non-radiative deactivation by internal conversion is much faster than the spin-forbidden intersystem crossing to any triplet state. Therefore, the first excited singlet state is formed with near unit quantum yield. If an upper triplet state could be reached, it would also deactivate very rapidly to T_1, and no singlet excited state would be formed. The extremely short lifetime of all upper excited states $S_n(n > 1)$ and $T_n(n > 1)$ means that luminescence emission and chemical reaction are, as a rule, not observed from such states. There are some exceptions to this rule, but there are many more mistaken reports of chemical reactions from short-lived upper excited states. Any such report

must be treated with caution and a simple wavelength dependence of the photoproducts or of the reaction quantum yield is not, in principle, sufficient evidence for an upper excited state reaction.

The heavy-atom effect in organic photochemistry occurs when a molecule contains one or more 'heavy' atoms such as Br, I, *etc.* All spin-forbidden transitions become faster and upper excited triplet states can be populated significantly. The quantum yield of formation of the longer-lived S_1 and T_1 states can then vary with the excitation wavelength, as illustrated in Figure 4.25. Assuming that crossing from S_2 to T_2 is now fast enough to compete with internal conversion from S_2 to S_1, with an excitation wavelength shorter than $h\nu_2$, the S_1 excited state is populated less efficiently and the quantum yield of a photochemical reaction which originates from that state would show a similar dependence on irradiation wavelength. Conversely, a reaction of the T_1 triplet state would become more efficient with short-wavelength irradiation. Note that the photochemical reactions still come from the lowest excited states S_1 and T_1, and not from any upper excited state. This example illustrates the fact that the wavelength dependence of a photochemical reaction is not, on its own, evidence for an upper excited state reaction.

4.4.1 The Mechanisms of Photochemical Reactions: Quenching, Sensitization and Wavelength Effects

The nature of the reactive state (or states) is the first important point in the overall mechanism of a photochemical reaction. It is not merely of academic

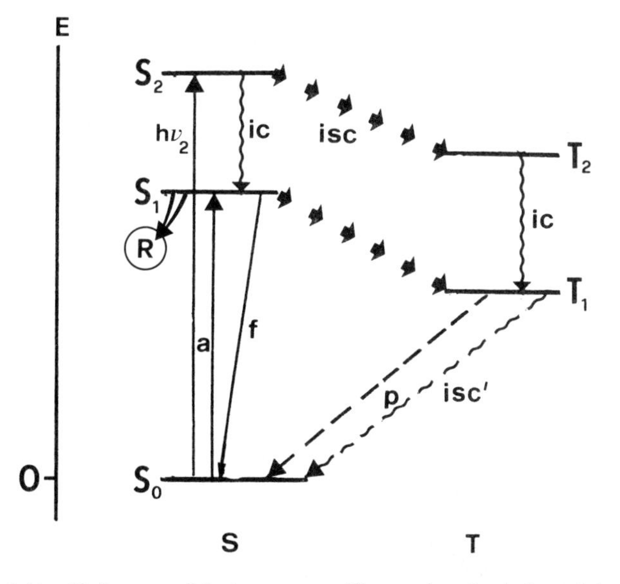

Figure 4.25 *Jablonski diagram of the heavy atom effect on photochemical reactivity. If excitation to S_2 ($h\nu_2$) is followed by intersystem crossing (isc) to T_2, the quantum yield of reaction R decreases at higher excitation energies. ic = internal conversion, a = absorption, f = fluorescence, p = phosphorescence*

interest: information about the nature of the reactive state can be used to improve the conditions of reaction, or to modify the conditions to prevent an unwanted reaction. Here the 'conditions' mean the physical parameters such as temperature, pressure or viscosity, as well as irradiation wavelengths and irradiation times, and the choice of chemical additives, solvents, *etc*.

Quenching is observed as the lowering of a quantum yield of reaction with increasing concentrations of an additive Q. In the first place, this is simply an experimental observation which implies no particular quenching mechanism. Before the actual quenching of the reactive state by energy transfer can be established, it is necessary to eliminate certain apparent 'quenching' effects:

(a) The *inner filter effect* of Q. Consider a reaction $R \rightarrow P$, in which a chemical species of the reactants R absorbs the irradiating light. It may happen that the additive Q absorbs also at the irradiation wavelength(s), so that the buildup of the product P is slowed down simply because of the reduction in light intensity. This inner filter effect can be a serious problem when polychromatic light is used for irradiation, for it is then difficult to ensure that *none* of the light is absorbed by Q, throughout the entire wavelength range absorbed by R. When two species (*e.g.* R and Q) absorb light at a wavelength λ, the absorbed light is shared between them according to their absorbances; the absorbance A_m of the mixture being the sum of the individual absorbances

$$A(\lambda)_m = A(\lambda)_R + A(\lambda)_Q \qquad (4.25)$$

Remembering that the absorbance is proportional to the concentration of each species, it is readily apparent that the neglect of the inner filter effect may be dangerous when the 'quencher' is present in concentrations much higher than the reactant. It should not be forgotton that the additive may also contain absorbing impurities, even if it does not absorb itself.

(b) The *chemical actions* of Q, in particular as a scavenger of free radicals formed in the photochemical reaction. Consider for example the process of hydrogen abstraction by a carbonyl compound RR′–C=O from some molecule HZ (where Z is an aliphatic carbon residue, *e.g.* C_6H_{11} if HZ is cyclohexane)

$$RR'\text{–}CO \xrightarrow{h\nu} (RR'\text{–}CO)^* \xrightarrow{HZ} RR'\text{–}C^{\bullet}OH + Z^{\bullet} \qquad (4.26)$$

The free radicals $RR'\text{–}C^{\bullet}OH$ then react further to form the final addition products (see section 4.9 for details). If an additive Q such as acridine is present in the solution, it may react with the radical by removing its H atom to reform the ketone (Figure 4.26).

If the photochemical reaction is followed only by measuring the concentration of the light-absorbing ketone, it will indeed appear that the additive (acridine) acts as a 'quencher'; in reality it acts as a *radical scavenger* through a thermochemical reaction which is quite unrelated to the nature of the reactive excited state.

Once these two major potential quenching artefacts have been eliminated,

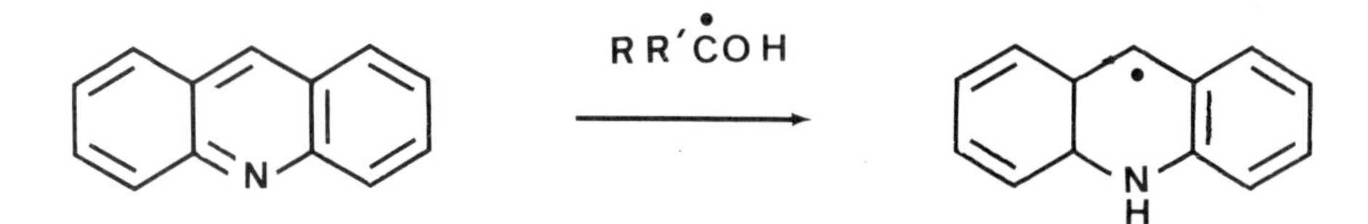

Figure 4.26 *Chemical sensitization through hydrogen atom transfer between a ketyl radical and acridine*

it still remains to prove that the observed quenching action does result from energy transfer

$$R^* + Q \rightarrow R + Q^* \tag{4.27}$$

rather than from electron transfer or a paramagnetic interaction. The observation of Q^* as a transient species in flash photolysis provides such conclusive proof; this is discussed further on in section 7.5. When quenching takes place by electron transfer, *e.g.*

$$R^* + Q \rightarrow R^{\bullet +} + Q^{\bullet -} \tag{4.28}$$

one of the radical ions (or sometimes both) may be detected in the same way. A mechanism of paramagnetic quenching as postulated for molecular oxygen is difficult to establish directly, but can be inferred only when other mechanisms have been ruled out.

4.4.2 Unimolecular Reactions

An isolated molecule can only undergo two types of chemical change:
(1) the *rearrangement* of its atoms into a new molecule which keeps the general formula unchanged (the product is then an *isomer* of the reactant, so that the term *isomerization* is equivalent to that of rearrangement);
(2) the *dissociation* into two or more 'fragments', which can be individual atoms, free radicals, radical ions, closed-shell ions, biradicals, smaller molecules, protons, electrons, *etc.*
A distinction must be made between truly *isolated* molecules which react in the absence of any collision with other molecules, as in the gas phase at very low pressures or in molecular beams, and molecules in liquid or solid environments. A condensed phase medium, liquid or solid, imposes a 'cage effect' which can prevent large geometrical changes in rearrangement reactions, and the separation of fragments in dissociation reactions.

4.4.2.1 Dissociation Reactions. A glance at the list of fragments which can be produced from the dissociation of a molecule shows that these apparently simple reactions can in fact follow several mechanisms. Of these, two are treated separately: the loss of an electron, which is the process of photoionization; and the loss of a proton, which is one side of the acid–base equilibria considered in section 4.4.3.

Homolytic dissociations. The dissociation of a diatomic molecule is probably the most simple of all chemical reactions; thus the chlorine molecule Cl_2 leads to atomic chlorine $Cl^{\bullet}$

$$Cl_2 \xrightarrow{hv} 2\,Cl^{\bullet} \tag{4.29}$$

Although this type of dissociation is quite common in small molecules, it is by no means universal; so that the question does arise, why do some molecules dissociate after electronic excitation, and others not?

We have seen in section 3.2 that excited states of small molecules can be either associative or dissociative (Figure 3.11, p. 38). Excitation of the molecule to the latter leads to 'instant' dissociation, often within the time of a single vibration (some 10^{-13} s), but excitation to an associative state can also lead to eventual dissociation when its vibrational energy is raised to the crossing point of the dissociative state, or by the dissociation energy E^* of the excited state; in this case the fragments will have some excess energy which will be distributed as translational and (in polyatomic molecules) vibrational energies.

There are two limiting models of dissociation known as the 'impulsive' model and the 'statistical' model. In the former, bond dissociation occurs during a single vibration and this can be represented by the breaking of a spring (Figure 4.27), in this case for the example of a triatomic molecule. The kinetic energy will be distributed according to the reduced masses of the fragments. In Figure 4.27, when the photon is absorbed, a repulsive potential exists between A and BC. E_{avl} [eqn. (4.29)] is the translational energy available intially between A and B, E_t being the total translational energy of the fragments A and BC.

$$E_t = \frac{\mu_{A-B}}{\mu_{A-BC}} E_{avl} \tag{4.29}$$

The laws of conservation of energy and momentum imply that the lighter fragments move away from the zero position faster than the heavier fragments; in particular hydrogen atoms which are formed by the disso-

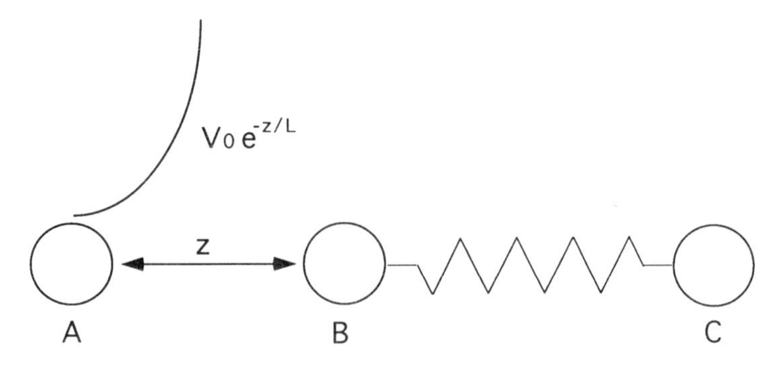

Figure 4.27 *The 'impulsive' model of the photodissociation of a triatomic molecule: ABC→ A + BC*

ciation of a Z–H bond carry a considerable momentum $p = mv$ (mass × velocity), so that they move with extremely high velocities. These 'hot' H atoms are therefore very reactive in gas phase collisions and can lead to troublesome secondary reactions in gas phase photochemistry. These problems can be minimized by the use of 'moderators' in the gas mixture, for instance some rare gas atoms such as He, Ar, *etc.* which will slow down the fast H atoms through elastic collisions.

As a rule, an excited diatomic molecule can be deactivated only through luminescence or dissociation, in the absence of collisions. Dissociative states show up in the absorption spectra as continuum absorptions, devoid of the sharp spectral lines characteristic of transitions between associative states.

In some molecules there is another, slower dissociation path known as *predissociation*. In this case the crossing to the dissociative state is the rate-limiting step, and this may take place after many vibrations; in the absorption spectrum the vibrational sub-levels remain sharp, but the rotational levels are blurred (Figure 4.28).

As the molecule gets larger, so the homolytic dissociation becomes less likely, unless there is a weak bond such as the peroxide linkage O–O, the sulfide bridge S–S, or some carbon–halogen, C–X, bonds (X = I, Br, sometimes Cl, but never F). In a polyatomic molecule, the general rule is that the weakest bond breaks preferentially, and this is to be expected on the grounds of the statistical distribution of the excitation energy among the vibrational levels of the molecule; but there are exceptions, particularly in the case of aromatic halides.

The energy requirement of photodissociation: the photochemistry of haloaromatics in solution. The law of conservation of energy requires that the dissociative excited state should be higher than the bond dissociation energy. This follows from the fact that when the fragments recombine to restore the original molecule the bond energy is released, and can be used (in principle, at least) to drive other energy-consuming processes. The photochemistry of a series of halonaphthalenes, X–Np (Np = naphthalene) provides a good illustration of this general energy requirement.

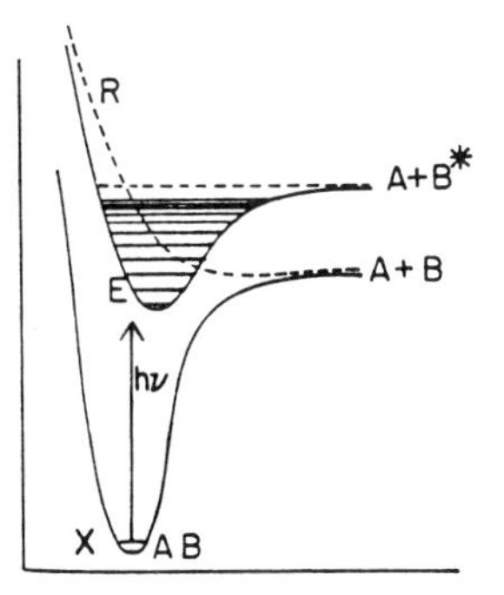

Figure 4.28 *Potential energy diagram of predissociation. The dissociative state crosses an associative state of lower energy*

The halogens F, Cl, Br and I are very weak π electron donor substituents which leave the energies of the aromatic π–π^* excited states of the parent naphthalene practically unchanged; in Figure 4.29 the S_2, S_1 and T_1 levels therefore apply to all four X–Np derivatives. However, the aromatic C–X bond energies vary greatly, as shown on the right-hand side of Figure 4.29. It is indeed observed that 1-iodo-Np dissociates from its triplet state T_1 when this is populated through sensitization by energy transfer, but that 1-bromo-Np dissociates only from its higher energy S_1 state, and then with a much lower quantum yield; 1-chloro-Np and 1-fluoro-Np do not dissociate at all, although they can take part in substitution and oxidation reactions.

The photodissociation of aromatic molecules does not always take place at the weakest bond. It has been reported that in a chlorobenzene, substituted with an aliphatic chain which holds a far-away Br atom, dissociation occurs at the aromatic C–Cl bond rather than at the *much weaker* aliphatic C–Br bond (Figure 4.30). This is not easily understood on the basis of a simple picture of the crossing to a dissociative state, and it is probable that the reaction takes place in the π–π^* S_1 excited state which is localized on the aromatic system. There are indeed cases in which the dissociation is so fast ($< 10^{-12}$ s) that it competes efficiently with internal conversion. 1-Chloromethyl-Np provides a clear example of this behaviour, its fluorescence quantum yield being much smaller when excitation populates S_2 than when it reaches S_1. Figure 4.31 shows a comparison of the fluorescence excitation spectrum and the absorption spectrum of this compound. This is one of the few well-documented examples of an upper excited state reaction of an organic molecule which has a normal pattern of energy levels (*e.g.* unlike azulene or thioketones). This unusual behaviour is related of course to the extremely fast dissociation, within a single vibration very probably. We must now

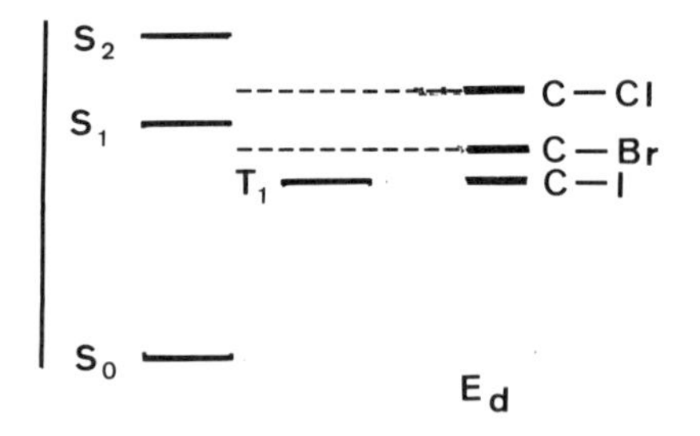

Figure 4.29 *Excited state energies of halonaphthalenes, and bond dissociation energies of C–X*

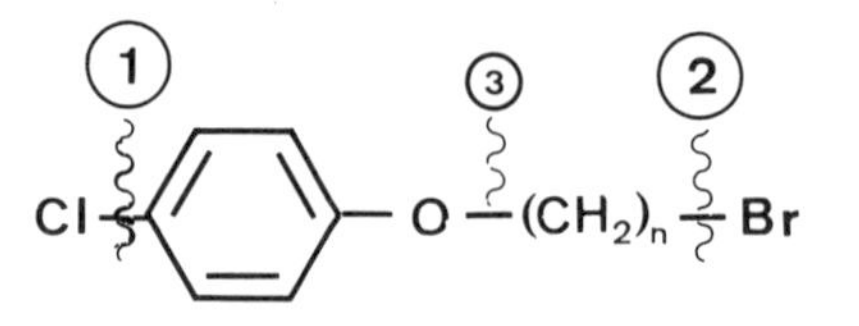

Figure 4.30 *Points of dissociation in an ω-bromochlorobenzoyl ester*

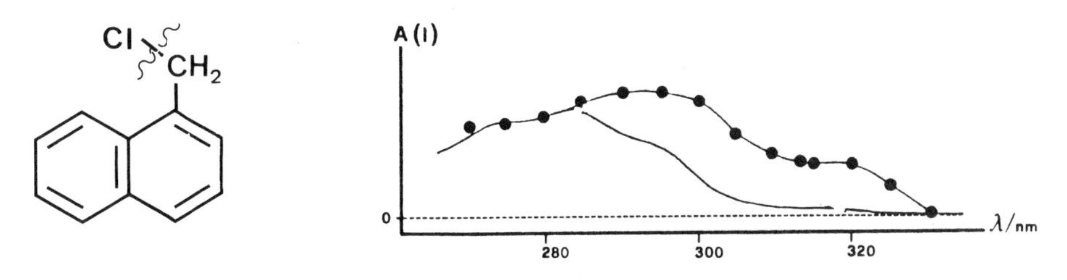

Figure 4.31 *Absorption spectrum (full line) and fluorescence excitation spectrum (filled circles) of 1-chloromethylnaphthalene*

consider another question: why should 1-chloromethyl-Np dissociate so efficiently when 1-chloro-Np does not dissociate at all from excited states of similar energies?

The energy balance of photodissociation: the importance of stabilization of the free radicals. When chlorobenzene or chloro-Np loses the halogen atom, a phenyl or a naphthyl radical is formed with the odd electron localized in an sp^2 orbital which is orthogonal to the aromatic π orbitals; such a radical is not stabilized through resonance, unlike the benzyl- or the methyl-Np radicals for which several resonance structures can be drawn (Figure 4.32).

These resonance stabilization energies of free radicals can be quite large, *e.g.* 50 kJ mol^{-1} for benzyl-, and 70 kJ mol^{-1} for methyl-Np. These must be included in the overall energy balance of the reaction, and can make all the difference between a fast, highly exergonic process, and an endergonic process which in practice does no take place at all.

The dissociation of carbonyl compounds. Many carbonyl derivatives undergo a homolytic dissociation between the CO group and a saturated hydrocarbon chain (*e.g.* Figure 4.33). The stability of the free radical $R^{\bullet}$ plays a very important role in the efficiency of this reaction. In a series of aromatic carbonyl compounds dissociation is not observed with benzophenone ($R = C_6H_5$), benzaldehyde ($R = H$) or acetophenone ($R = CH_3$), but is observed with the longer R chains.

Dissociation is more efficient when the R chain contains electron donors (*e.g.* an amino group), and this can be rationalized on the basis of the

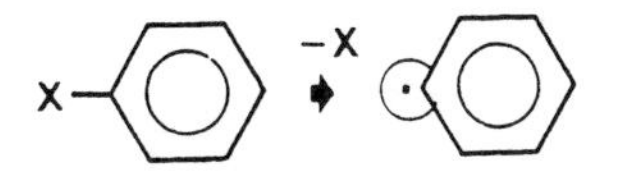

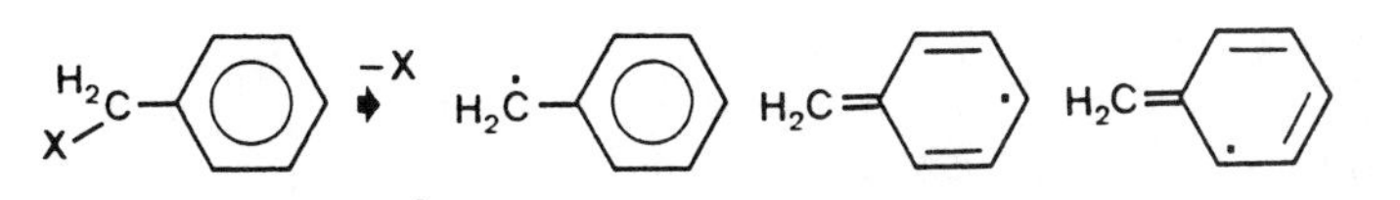

Figure 4.32 *Structures of a phenyl radical formed from the dissociation of a halobenzene, and of a benzyl radical from dissociation of a halomethylbenzene*

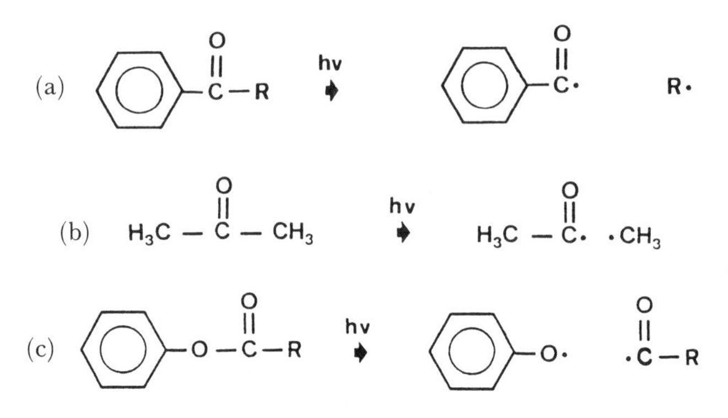

Figure 4.33 *α-Dissociation of carbonyl compounds. (a) An aromatic ketone or benzaldehyde, (b) acetone, (c) an ester*

electron distribution in the excited state which is in this case generally the $^3(n-\pi^*)$ state in which the electron density is high at the carbonyl C atom. These dissociations of carbonyl compounds have found important applications in photoinitiation processes of polymerization (section 6.2).

Many aliphatic carbonyl compounds show the same dissociation reaction, even acetone (Figure 4.33b). When the carbonyl group is separated from the benzene ring by a suitable substituent such as O, the dissociation takes a different course, as shown in Figure 4.33(c). The resonance stabilization of the phenoxyl radical is much greater than that of the aliphatic radical R$^\bullet$, so that splitting of the O–CO bond is favoured.

Photoelimination reactions. In these reactions one of the fragments is a small, stable, closed-shell molecule such as N_2, CO_2, CO, HX (X = halogen), *etc.* The name 'photo-extrusion' is also used for these reactions. Since the reactant molecule is of closed-shell type in the ground state, the second fragment must be either another closed-shell molecule or a *biradical*, that is a species with two unpaired electrons held in distinct orbitals.

If the eliminated fragment (*e.g.* N_2) is attached to *two* distinct carbon atoms, then the biradical can reorganize rapidly by making a bond between these atoms (provided the odd electron spins are antiparallel, the biradical being formed through a photochemical reaction originating from a *singlet* excited state, because then a *singlet* biradical is the primary product). If the photochemical reaction starts from a *triplet* state (*e.g.* formed by sensitization), then the primary product will be a *triplet* biradical. This cannot rearrange to a closed-shell molecule prior to spin inversion, so it has a longer lifetime which gives it the chance to take part in addition reactions with other molecules, in particular oxygen, O_2 which is itself a triplet ground state biradical (Figure 4.34).

This provides an illustration of the way in which the chemical outcome of a light-induced reaction can depend on the nature of the reactive excited state: triplet states lead to long-lived primary products which can take part in further bimolecular processes, whereas singlet states lead to fast uni-

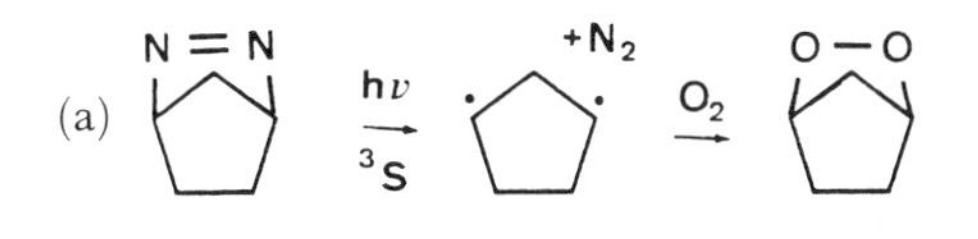

(b) $R_2C{=}N_2 \xrightarrow{h\nu} R_2C{:} + N_2$ $\quad RN_3 \xrightarrow{h\nu} RN{:} + N_2$

Figure 4.34 *Photoelimination reactions of nitrogen. (a) Formation of a carbene through triplet state sensitization, and addition of molecular oxygen. (b) Formation of nitrenes through photodissociation of azo compounds and azides*

molecular reorganization. This should however only be considered as a general rule, which suffers in fact from many exceptions.

These reactions are important in synthetic organic chemistry, because they lead to highly reactive chemical species called *carbenes* and *nitrenes*. A carbene has two unpaired electrons on a carbon atom, a nitrene has two unpaired electrons on a nitrogen atom. They are formed through the elimination of N_2 from azo-compounds, RN_2, and from azides, RN_3 (Figure 4.34b).

4.4.2.2 Reactions of Rearrangement (Isomerization). In reactions of *isomerization* (or 'rearrangement') the molecule keeps its overall formula but undergoes a structural change which either keeps the bonding pattern of the atoms unchanged but modifies the geometry of the molecule (stereo-isomerism), or produces a new bonding pattern (valence isomerization).

Stereo-isomerizations are quite common photochemical processes with unsaturated organic molecules (the primary photochemical reaction of vision is of this type).

A substituted ethylene molecule can have two stereo-isomers in the ground state, which can be labelled as the *cis* and the *trans* forms. Rotation around the double bond is practically impossible and the stereo-isomers do not interconvert thermally. In the excited states, S_1 or T_1, the promotion of an electron from the bonding π orbital to the antibonding π^* orbital reduces the π bond order to near zero. The molecule can be represented in the form of a biradical (this is a highly simplified view of an excited state, but reasonably accurate in this case), the central C atoms being linked by a *single bond* which allows free rotation (subject only to steric hindrance of the substituents). The photochemical process can be pictured as the excitation of the molecule without change of geometry, to the excited state S^* potential energy surface shown in Figure 4.35 (this could be the S_1 or the T_1 state). Now the molecular geometry is unstable in this state, and relaxation brings it down to the most stable excited state geometry, in which the two unpaired electrons have the least possible interaction; this corresponds to the twisted form, the orbitals of the two unpaired electrons being orthogonal. This molecular shape however is also that of the least stable ground state geometry, so here the ground and excited state potential energy surfaces come very close together; this is described as a 'funnel', through which deactivation can take place by non-radiative crossing from S^* to S_0.

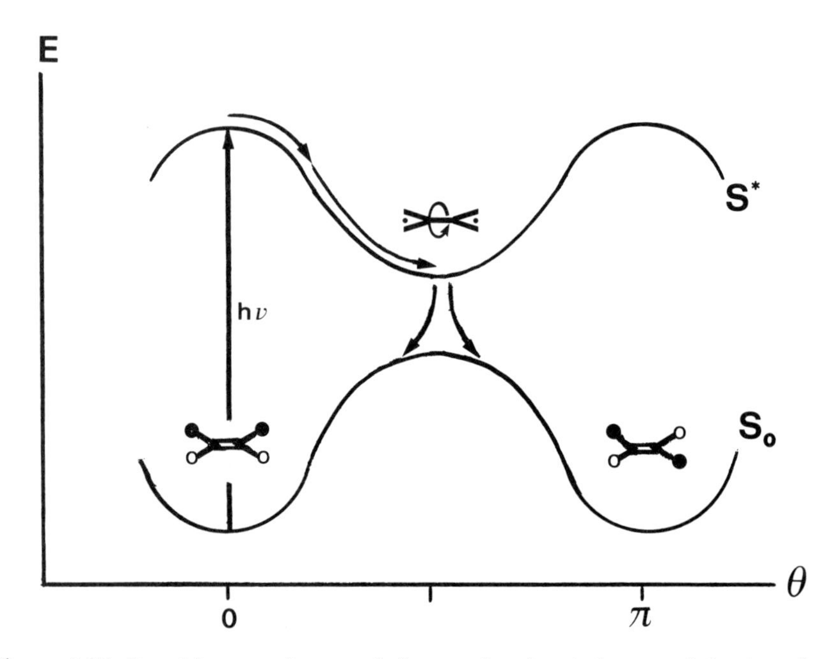

Figure 4.35 *Potential energy diagram of the ground and excited states of the cis and trans isomers of an olefin. The energy is shown as a function of the bond angle*

Valence isomerizations are rearrangement reactions in which the 'connectivity' of the atoms is changed: for example, two atoms that were linked together in the starting molecule become separated in the product molecule (Figure 4.36).

There are indeed many photochemical processes of valence isomerization, and these may follow quite different mechanisms which we shall consider in turn, with the help of actual examples.

A first general question must now be discussed, concerning the mechanism of any chemical reaction in which a new bond (or several new bonds) is made while other bonds are broken.

Concerted and unconcerted reaction pathways: intermediates in chemical reactions. When the making and breaking of bonds take place simultaneously, the process is 'concerted'; there is then no actual 'intermediate' in the reaction, which simply passes through a 'transition state' which is the point of highest energy in an energy–reaction coordinate diagram (Figure 4.37).

However, there is in principle another pathway for the same reaction, if first the AB bond is broken but the new bonds are not yet formed. In this

Figure 4.36 *Valence isomerization of cyclobutene and butadiene*

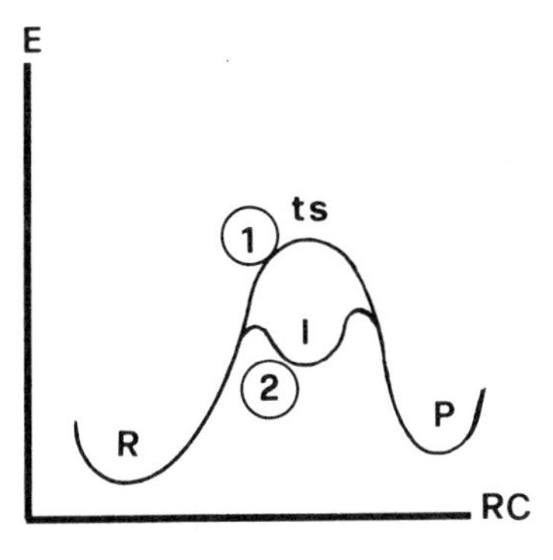

Figure 4.37 *Potential energy diagram of concerted and unconcerted reactions. The former pass through a transition state ts (1), the latter through an intermediate I (2)*

case the first reaction product is a biradical, a chemical species which has a well defined structure and (in principle, at least), a measurable lifetime. This is an 'intermediate', as distinct from a transition state. In the energy–reaction coordinate diagram it corresponds to a minimum between the points of the reactant and product.

In the general discussion of the differences between thermal and photochemical reactions (section 4.1), it was mentioned that the primary product of the latter is often a high-energy product which retains a large part of the electronic excitation energy of the reactant. This is precisely the case when a biradical is formed in an unconcerted isomerization reaction, and such a biradical is of course a genuine chemical intermediate.

An unconcerted photoisomerization: the 'di-π-methane' rearrangement. This reaction is of interest in synthetic organic chemistry, because it can form the small, strained three-carbon cyclopropane ring. Molecules of this type are rather unstable thermally, and therefore are useful reactants for further (thermal) processes.

The di-π-methane reaction is fairly general with molecules which contain two double bonds separated by two single bonds of the same carbon atom (Figure 4.38).

The π–π^* excited state can be pictured as a biradical which rearranges to

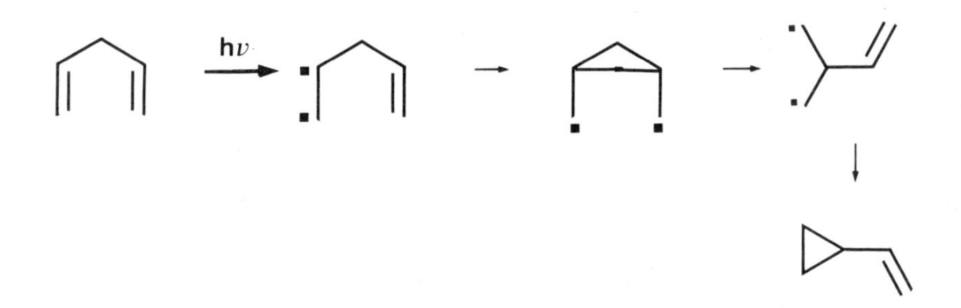

Figure 4.38 *Reaction sequence of the di-π-methane rearrangement*

an actual biradical through cyclization, followed by further rearrangement to the final product.

The experimental evidence for the existence of the biradical as a genuine chemical intermediate comes from the detection of other products which are characteristic of addition reactions of such biradicals with various unsaturated molecules.

Concerted rearrangements: sigmatropic and electrocyclic processes and the principle of orbital symmetry conservation. In a concerted reaction the changes in bonding take place simultaneously, so that the transition state corresponds to a molecule which has 'half-bonds' between the atoms involved in the overall change in bonding pattern. In an *electrocyclic* process this transition state can be drawn as a molecule with half-bonds, delocalized over a ring of carbon atoms (a 'cycle'; Figure 4.39). The ring-closure and ring-opening reactions are in principle reversible. Such electrocyclic isomerizations can be drawn for all conjugated molecules that can form a ring of carbon atoms through bonding of the terminal C atoms.

There is another class of isomerization reaction which proceed through a cyclic transition state, known as a *sigmatropic* reaction. This is formally the shifting of a single (σ) bond between the terminal carbon atoms of a linear conjugated molecule, according to the example drawn in Figure 4.40.

The common feature of these electrocyclic and sigmatropic rearrangements is that a change of bonding takes place at the terminal carbon atoms of a linear conjugated molecule. The word 'linear' should not be taken in a strict geometrical sense, for although such molecules are usually drawn as lines of alternating single and double bonds in a simple chemical formula, this does

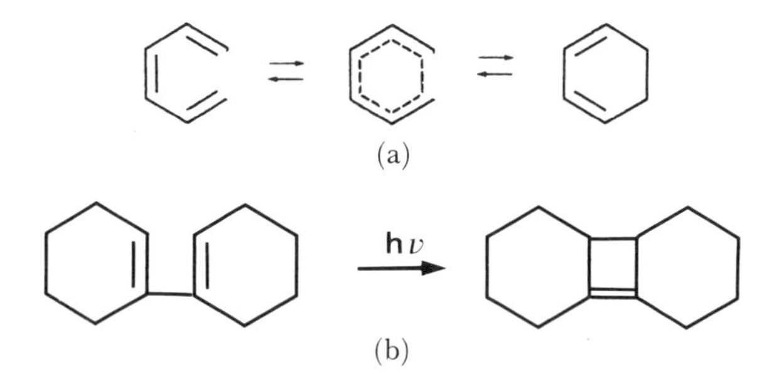

Figure 4.39 *(a) The principle of a concerted electrocyclic reaction and (b) an example of a photochemical process*

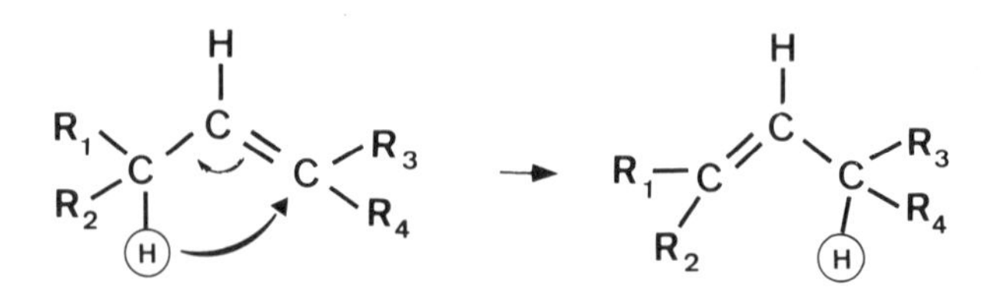

Figure 4.40 *The principle of a sigmatropic reaction*

not represent their actual shape. Indeed, these molecules are flexible and can therefore assume many different shapes, in particular the folded, quasi-cyclic patterns involved in sigmatropic and electrocyclic rearrangements. In Figure 4.41 only the p orbitals of the terminal atoms are therefore shown, the rest of the conjugated chain being represented simply as a string. In an electrocyclic ring-closure process, there are two distinct ways in which the terminal p orbitals can overlap to form the new σ bond: in a 'conrotatory' motion they twist in the same direction, whereas in a 'disrotatory' motion they turn in opposite directions.

Depending on the relative symmetry of the p orbitals, a bonding (that is, a positive overlap) is obtained either for the conrotatory or for the disrotatory motion, and these result in different stereochemistries if the terminal atoms carry distinct substituents, as shown in Figure 4.41(b).

In a dark reaction the relevant frontier orbital is the HOMO, while it will be the LUMO for a photochemical reaction. In the case of linear conjugated molecules there are then two simple rules:

(1) there is an inversion of stereochemistry between ground and excited state reactions;

(2) there is an inversion of stereochemistry between molecules which have 4n π electrons and those which have 4n + 2 π electrons (*e.g.* butadiene and hexadiene).

Similar arguments apply to sigmatropic rearrangements. Here the migration of a terminal s orbital can take place with or without the crossing of the surface defined by the quasi-cyclic conjugated system (Figure 4.42). The preferred mode is the one which results in positive (bonding) overlap of the s and p orbitals which will form the new bond.

The approach of the two orbitals on the same side of the surface is called 'supra', that on opposite sides is called 'antara'. The above rules, 1 and 2, also apply in the case of sigmatropic rearrangements, since the changes of orbital symmetries are the same on going from a HOMO to a LUMO or from a chain of 4n to a chain of 4n + 2 π electrons.

The rules of orbital symmetry conservation apply only to concerted reactions; in photochemical processes these are usually those of singlet excited states, since the triplet states often lead to long-lived biradical intermediates.

Secondary rearrangements: apparent isomerizations through radical recombination reactions. In the rearrangement reactions considered so far, the isomerization step is the primary photochemical process, except when a biradical is formed as an intermediate; for in that case the primary photochemical process is really a dissociation, even though the fragments cannot separate. There are however cases of overall isomerizations which result from the recombinations of separated free radicals formed through a process of photodissociation. The 'photo-Fries' reaction is an important example of this mechanism, and is illustrated in Figure 4.43.

The primary process is the photoinduced homolytic dissociation of an

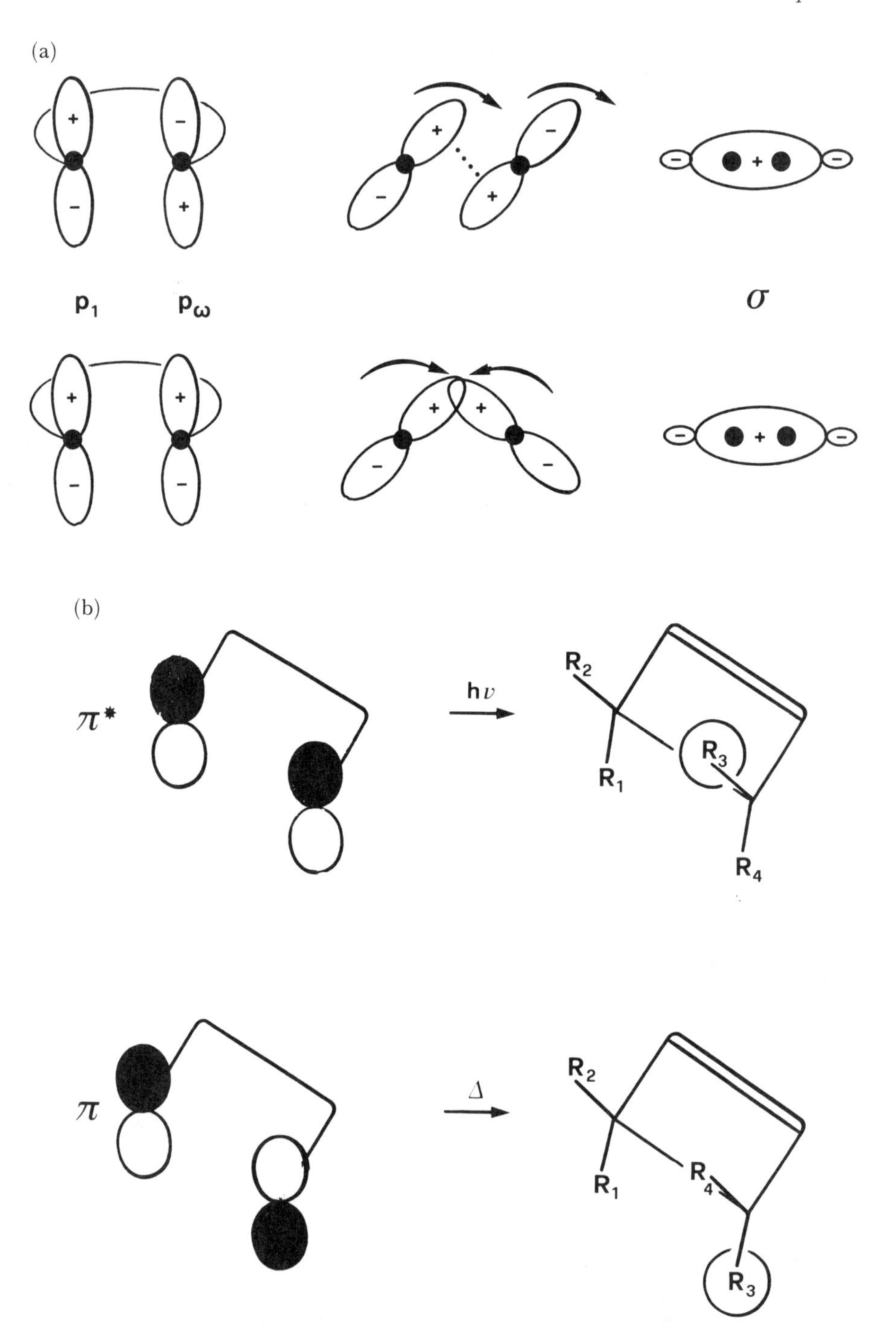

Figure 4.41 *(a) Orbital overlaps in conrotatory and disrotatory σ bond formation in a ring-closure reaction. (b) Stereochemistry of preferred products in photochemical and thermal (dark) processes*

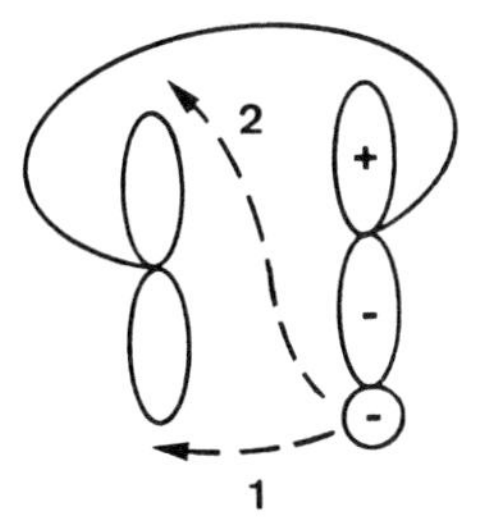

Figure 4.42 *Supra (1) and antara (2) pathways of hydrogen atom migration as an example of a sigmatropic isomerization*

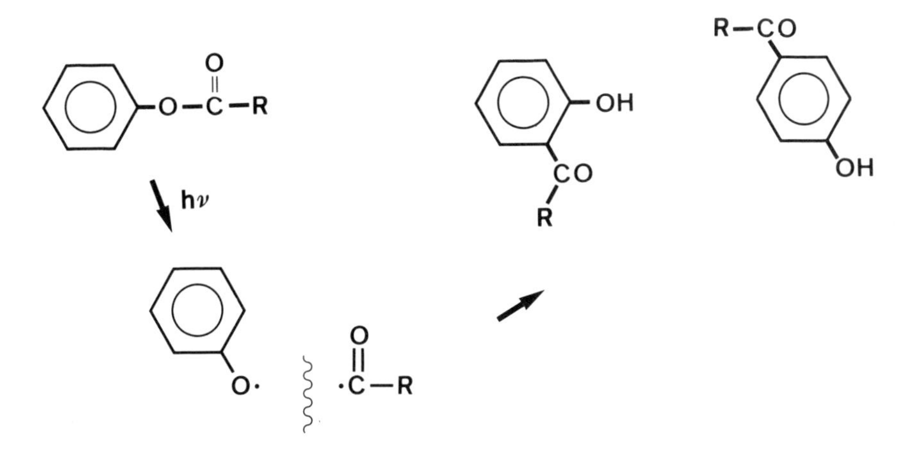

Figure 4.43 *Mechanism of the photo-Fries 'rearrangement'*

aromatic ether or amide. The free radicals then separate but may come together again in various thermal (secondary) reactions, some of which form addition products which are then formally isomers of the reactant molecule.

The potential complexity of photoinduced isomerizations: the many isomers of the simple benzene molecule. Some molecules can be drawn in several, indeed sometimes many, isomeric structures. These will in general have different stabilities, so that under given experimental conditions (temperature, pressure, environment, *etc.*), one of the isomers should be the species of lowest energy (greatest stability), and all others should interconvert to it (in the course of time). The molecule of benzene C_6H_6 provides a fascinating example of such multiple isomerizations, the various isomer photoproducts being shown in Figure 4.44. These are known as 'Dewar benzene', prismane, fulvene and benzvalene from left to right.

Although they are all formed from benzene under irradiation, they rearrange thermally back to the stable cyclic form.

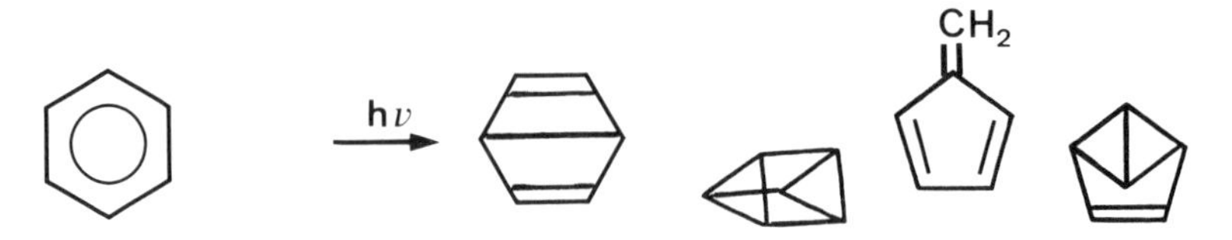

Figure 4.44 *Photochemical isomers of benzene. From left to right: 'Dewar benzene', prismane, fulvene, benzvalene*

4.4.3 Protolytic Equilibria (Acid–Base Reactions)

Acid–base equilibria are important chemical processes both in thermal and in photochemical reactions, since many molecules have an 'acid' form AH, and a 'base' form A, which are connected by the transfer of a proton (eqn. 4.30 and an example in Figure 4.45)

$$AH^+ \rightleftarrows A + H^+ \tag{4.30}$$

The equilibrium constant of this reversible proton transfer reaction is

$$K = [A][H^+]/[AH^+] \tag{4.31}$$

and this is usually given by the 'pK' of the reaction, which is the pH of the solution in which the concentrations of the acid and base forms are equal (remembering that the 'pH' of a solution is defined as, $pH = -\log[H^+]$). These acid-base equilibria are important since the acid and base forms of many molecules have quite different physical and chemical properties.

The 'acidity' (as defined by the pK; *NB* a strong acid has a low pK) of a molecule is strongly influenced by its electrostatic charge distribution, and this is not surprising since the partial negative charge on the O atom in phenols, naphthols, *etc.*, determines the electrostatic energy of the OH bond ($O^{\delta^-}-H^{\delta^+}$). Taking the example of such hydroxy-substituted molecules, it is an experimental fact that they are very weak acids in the ground state; *i.e.* the (partial) negative charge is localized on the oxygen atom. In the lowest singlet excited state these molecules become *strong* acids because of the charge transfer from the hydroxy oxygen to the aromatic ring(s). This O atom is now less negative, consequently its electrostatic attraction for the proton is smaller (Figure 4.46).

4.4.3.1 The Spectroscopy of Proton Transfer and the Förster Cycle. The absorption and emission spectra of the acid and base forms of a molecule are of course different, since these forms represent distinct chemical species. The excitation energies (*e.g.* $S_0 \rightarrow S_1$) are however related according to a very simple scheme of energy levels known as the 'Förster cycle'; this is shown in Figure 4.47(a).

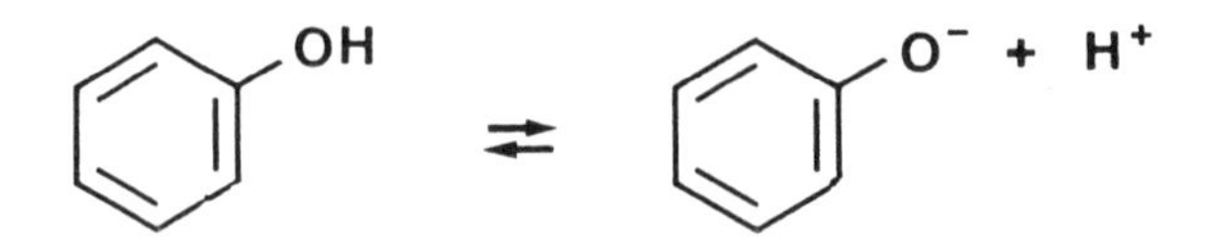

Figure 4.45 *Acid–base equlibrium of phenol phenolate*

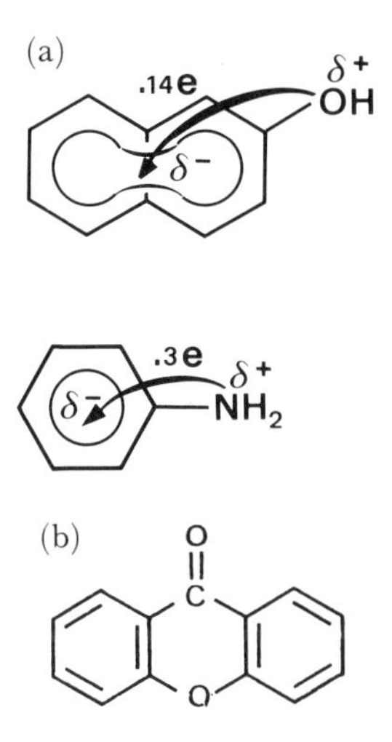

Figure 4.46 *(a) Charge transfer in excited states of hydroxy- and amino-aromatics. (b) Xanthone offers two potential protonation sites*

An equilibrium constant K is related to the change in free energy G between the products and reactants, $\Delta G = -RT \ln K$. ΔG is here the chemical *free* energy of the reaction, but it can be equated with the reaction *enthalpy*, ΔH, if the change of entropy ΔS can be neglected, or if, as in the present case, it can be assumed to be the same for the ground state (thermal) and excited state (photochemical) equilibria. The enthalpy change ΔH is then related to the ground state acid–base equilibrium, and ΔH^* is similarly related to the equilibrium in the excited state. The species A/A*, on the one hand, and AH^+/AH^{+*}, on the other hand, are connected through the excitation energies $h\nu_{OH}$ and $h\nu_{O^-}$ in this example, so that the energy cycle becomes:

$$\Delta H + h\nu_{OH} = \Delta H^* + h\nu_{O^-} \qquad (4.32)$$

The Förster cycle therefore permits the calculation of the excited state pK^* (from ΔH^*) simply from spectroscopic data (assuming of course that the ground state pK is known, from usual titration measurements).

The pK^* value derived from the Förster cycle is however a 'theoretical' value, two important assumptions being involved in its derivation:

(a) The free energy ΔG is replaced by the reaction enthalpy (potential energy) ΔH; this is valid if the reaction entropies are the same in the ground and excited state reactions. In practice, this restriction means that the geometry of the molecule should change little between the ground and excited states, and most importantly that the site of protonation should be the same. This condition must be considered carefully when the molecule contains several chromophoric sites which may be available in principle for proton transfer. Xanthone provides an example of such a molecule, its carbonyl group and oxygen bridge being able to accept protons (Figure 4.46b).

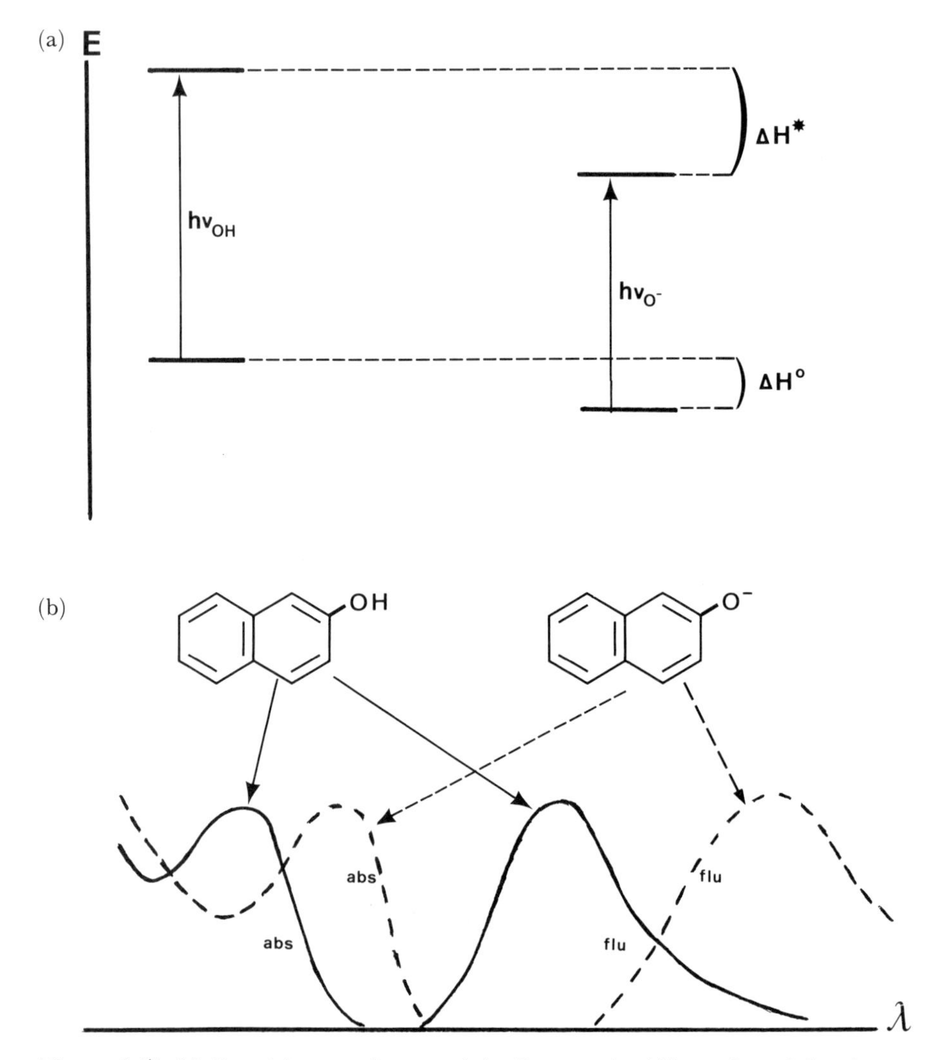

Figure 4.47 *(a) Potential energy diagram of the Förster cycle. ΔH are the enthalpy changes between the acid and base forms. (b) Absorption and fluorescence spectra of a naphthol and a naphtholate*

(b) The equilibrium defined by the theoretical pK may not be established during the lifetime of the excited state. In general the rate of proton attachment is simply diffusion controlled with a second-order rate constant of the order of 10^{10} $\text{M}^{-1}\,\text{s}^{-1}$. The first-order rate constant of proton dissociation is then often quite small, as readily calculated from the equilibrium constant

$$K = k_{\text{dis}}/k_{\text{D}} \tag{4.33}$$

The deprotonation time is therefore often so long that protolytic equilibrium cannot become established during the lifetime of singlet excited states (say

10^{-8} s). However, many triplet states live long enough ($> 10^{-6}$ s) to reach equilibrium.

This question of equilibration of the protonation and deprotonation processes leads to another fundamental problem in the case of excited state reactions: between which states can a protolytic equilibrium be at all established? A molecule has only one ground state, so there can be no ambiguity about the thermal protolytic equilibrium which connects of course the ground states of the acid and base forms. However, there are many excited states of both these forms, excited states which can differ greatly in electron distribution (*e.g.* nπ^* and $\pi\pi^*$ states) or even in multiplicity (*e.g.* singlet and triplet states).

One thing is clear: there can be no acid–base equilibrium between states of different multiplicities; thus it is correct to consider only the 'pK^* of the singlet state', or the 'pK^* of the triplet state'. However, the question of the protolytic equilibrium between an nπ^* singlet and a $\pi\pi^*$ or charge transfer (CT) singlet remains open. This problem is illustrated in Figure 4.48 for the case of 4-hydroxybenzophenone, in which there is a reversal in the order of nπ^* and CT states between the acid and base forms. Excitation of the protonated molecule in ethanol for example leads to the ground state deprotonated form, but the detailed mechanism of this process is not known.

4.4.3.2 Quenching of Excited Molecules Through Proton Transfer. An aromatic molecule such as benzene cannot be protonated in the ground state even in strong acids

$$M + H^+ \rightleftharpoons MH^+ \tag{4.34}$$

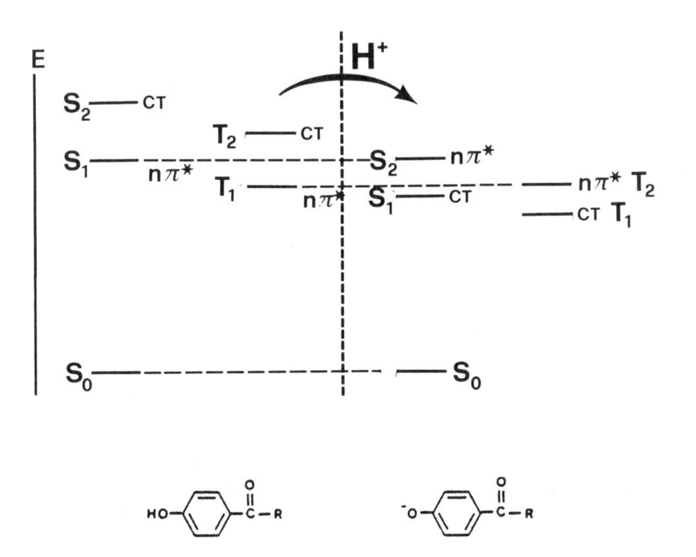

Figure 4.48 *Jablonski diagram of the acid and base forms of 4-hydroxybenzophenone, illustrating the reversal of the ordering of n–π^* and CT states*

Electronically excited benzene, however, can accept a proton to form a kind of complex which eventually dissociates back to the ground state molecule (Figure 4.49).

Deactivation of the S_1 excited state takes place because a new C–H bond must be formed in the σ complex. This bond requires two electrons, and since the bare H^+ proton does not provide any electrons, both must come from the aromatic system. The excited electron will therefore fall from the π^* orbital into the new σ orbital, hence the electronic excitation energy is lost through protonation. In the scheme shown in Figure 4.49 above there would be no detectable change in the benzene molecules. If however this same process is carried out in D_2O instead of H_2O, then an H/D isotope exchange would take place when the σ complex dissociates by losing either H^+ or D^+. The experimental evidence for this type of quenching of aromatic molecules by protonation of the aromatic ring is indeed found in the H/D isotope exchange related to the quenching of the fluorescence of the aromatic molecule in acidic solutions.

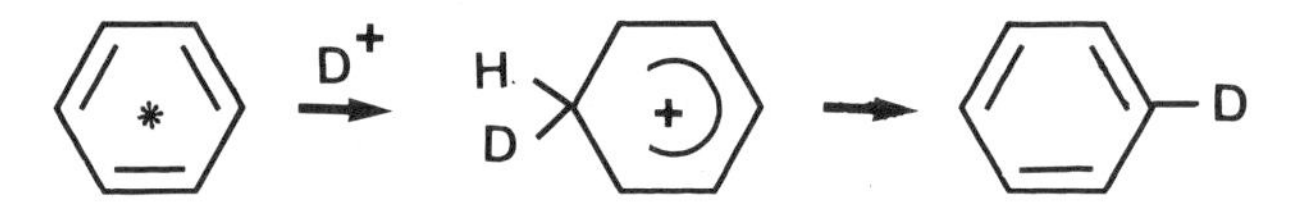

Figure 4.49 *Deuterium exchange through substitution in the excited state*

4.4.4 Bimolecular Reactions

All these reactions result from the interaction of an electronically excited molecule M* with a ground state molecule N. The most important processes of this type can be classified as:

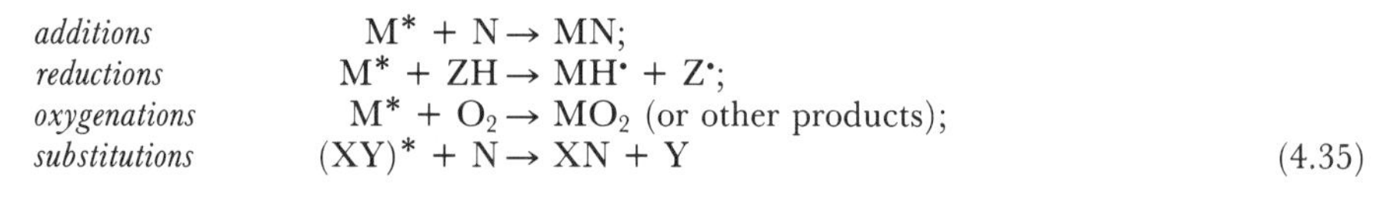

additions $M^* + N \rightarrow MN$;
reductions $M^* + ZH \rightarrow MH^{\bullet} + Z^{\bullet}$;
oxygenations $M^* + O_2 \rightarrow MO_2$ (or other products);
substitutions $(XY)^* + N \rightarrow XN + Y$ (4.35)

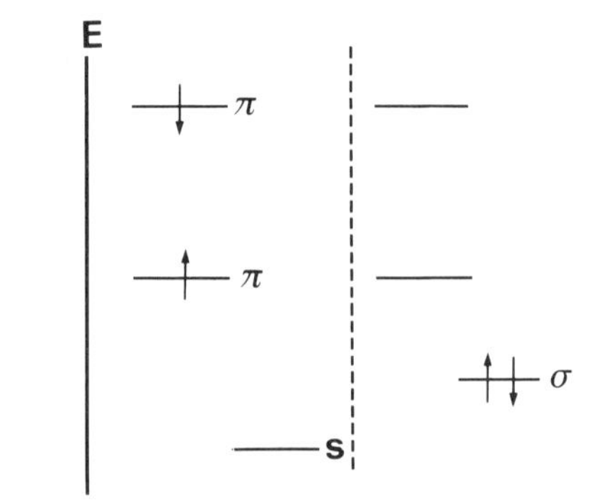

Figure 4.50 *Orbital energies in the protonation reaction*

In *addition* reactions an excited, unsaturated molecule uses its weakened π bond to form two new σ bonds; when these σ bonds form a new ring in the molecule MN the reaction is one of 'cycloaddition'. Thus alkenes can add photochemically to alcohols, as shown in Figure 4.51(a) to form a non-cyclic adduct; or to other alkenes in cycloaddition processes Figure 4.51(b).

Anthracene undergoes a photochemical 9,10,9′,10′-cycloaddition which goes through the excimer as intermediate. Many aromatic molecules follow similar cycloaddition paths. The close approach of the molecules in the excimer is essential for bond formation, and steric hindrance can prevent the reaction; unsubstituted anthracene dimerizes so fast that no excimer fluorescence can be detected, 9,10-dimethylanthracene shows both excimer fluorescence and photodimerization, but 9,10-diphenylanthracene shows neither excimer emission nor photodimerization (Figure 4.52).

Another important class of cycloaddition reactions is the formation of oxetane rings between a photoexcited carbonyl compound and an unsaturated molecule. These reactions also occur probably through an exciplex although these exciplexes are non-fluorescent as they are formed from the triplet state of the ketone or aldehyde. The formation of the four-membered oxetane ring is an interesting example of a typical photochemical reaction

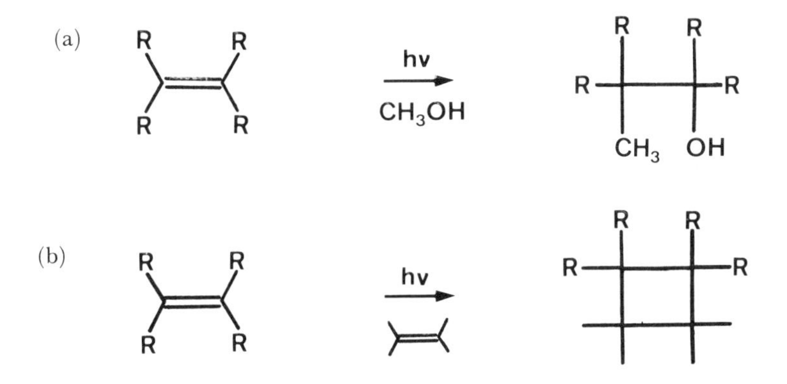

Figure 4.51 *Examples of photoinduced addition reactions of double bonds: (a) acyclic addition of methanol, (b) cycloaddition of an olefin*

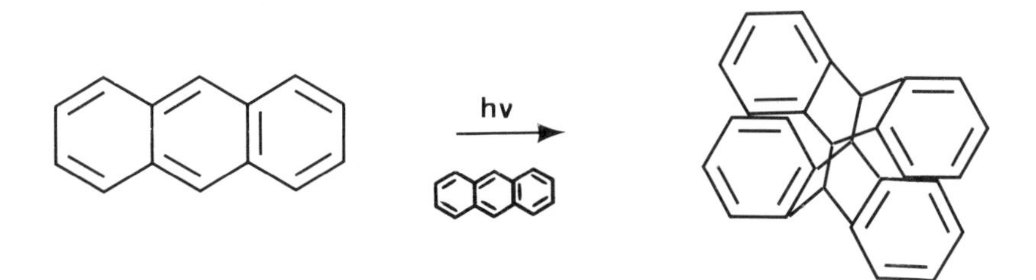

Figure 4.52 *Cycloaddition of anthracene to form a photodimer*

with no obvious equivalent in the ground state. There are indeed other examples of formation of small, strained rings through photochemical processes, when such rings are quite difficult to form by dark reactions (Figure 4.53).

4.4.4.1 Cycloadditions of Enones and Pyridinones. When the carbonyl group is conjugated to a C=C double bond, cycloaddition takes place at that double bond and a new ketone is formed. These reactions are fairly general with enones and their derivatives, as illustrated in Figure 4.54. The cyclobutane products are often useful synthetic intermediates.

These photochemical processes originate in most cases from the lowest triplet excited state of the carbonyl reactant, so that they are seldom stereo-selective. However, they show some regio-selectivity when the partner ethylene is substituted with different groups on its C atoms; the regio-selectivity can be explained by the relative stabilities of the biradicals which can be formed in the primary photochemical step.

4.4.4.2 Orbital Symmetry Conservation in Bimolecular Cycloadditions. The cycloaddition reactions of carbonyl compounds to form oxetanes with ethylenes, as well as those of enones and their derivatives to form cyclobutanes, are examples of reactions which originate from triplet excited states and lead in the first step to biradical intermediates. Such reactions are of course not concerted, and they show little or no stereo-specificity.

Concerted cycloadditions can exist in principle when the photochemical process originates from a singlet excited state. Such reactions are rather exceptional in bimolecular cycloadditions, simply because singlet excited states have short lifetimes (in the ns time-scale), so that encounter with a

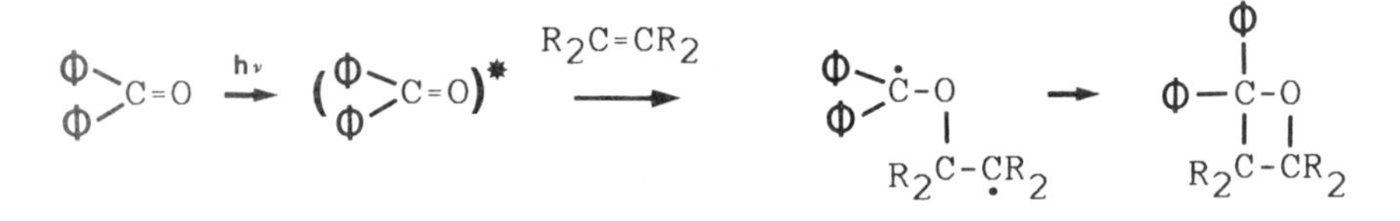

Figure 4.53 *Mechanism of oxetane formation between aromatic carbonyl compounds and olefins, going through a biradical intermediate*

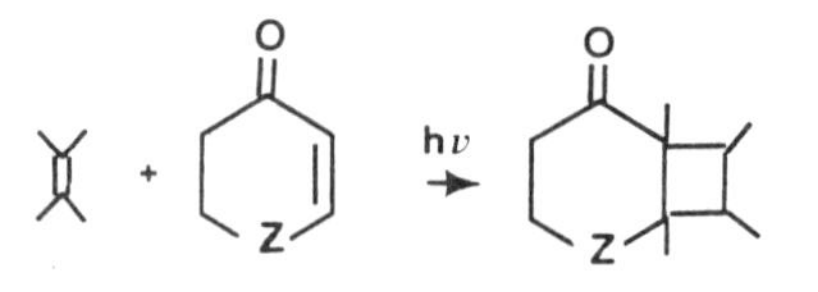

Figure 4.54 *Cycloaddition of olefins to enones and pyridinones. Z=CH (enone) or NR (pyridinone)*

partner molecule is much less likely than in the case of the longer-lived triplet. In spite of this limitation, intermolecular cycloadditions through singlet excited states are very interesting when they are of the concerted type, for in that case the rules of orbital symmetry conservation will apply and the processes can be stereo-specific.

Figure 4.55 shows the different ways in which two molecules can approach to form the new σ bonds. For each molecule the π surface has been drawn in a quasi-cyclic structure, and the terminal orbitals only are considered. A new bond is formed by the interaction of two terminal orbitals of the molecules, and these must give a positive overlap (+ + or − −). From the point of view of either molecule, the positive overlap is obtained either with its own orbital lobes on the same side of its π plane, called the 'supra' approach of the partner, or on opposite sides of this plane, called the 'antara' approach. There are therefore three distinct ways in which the two molecules can come together, known as 'supra–supra', 'supra–antara' and 'antara–antara'. These will lead to different stereochemistries of the products if there are different substituents on the terminal carbon atoms of the reactants.

Photoreduction occurs by hydrogen atom abstraction or by electron transfer. The first process is a common photochemical reaction of carbonyl derivatives and other unsaturated molecules in the presence of suitable hydrogen atom donors (which can be alcohols, paraffins, ethers, *etc.*, that is, almost any molecule with a not-too-strong C—H bond).

In Figure 4.56 the hydrogen donor is noted simply as ZH. The primary photochemical process leads to a pair of radicals which can undergo various secondary processes leading to the final reduction product. The reactivity of the ketone or aldehyde depends greatly on the nature of the lowest excited states S_1 and mainly T_1. The n–π^* states are the most reactive and this can be explained by the fact that a half-filled n orbital on the oxygen atom acts as an electron acceptor; π–π^* states are less reactive since the C═O bond is less weakened when the antibonding orbital is delocalized over the π system, and the reactivity diminishes as the charge-transfer character of the π–π^* state increases with stronger electron donor substituents.

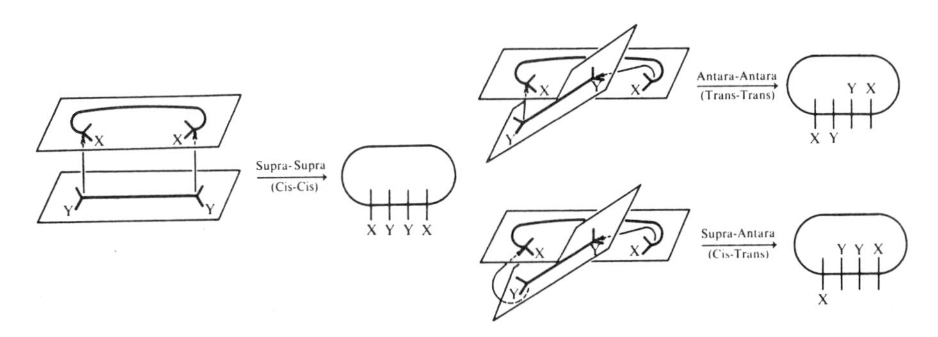

Figure 4.55 *The three distinct orbital interactions in an intermolecular cycloaddition leading to three stereoisomers*

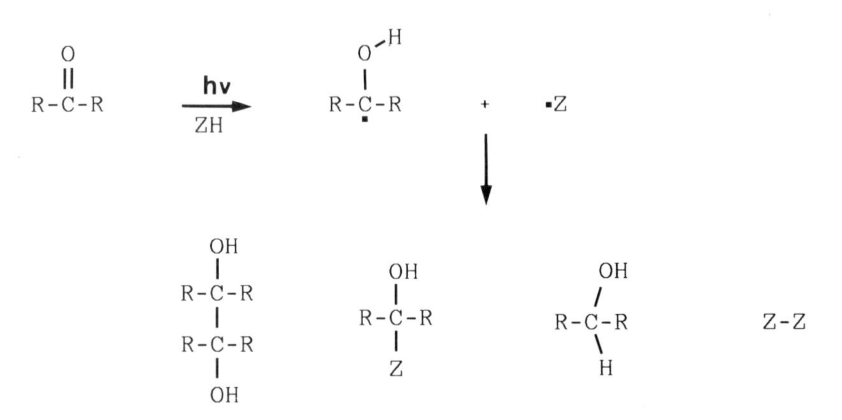

Figure 4.56 *Mechanism of the photoreduction of carbonyl compounds with hydrogen atom donors, ZH. Ketyl radicals and donor radicals are formed in the primary photochemical process*

A molecule with a π–π^* state of very high CT character (such as 4-aminobenzophenone) acquires a large dipole moment in the excited state and the relative energies of n–π^* and π–π^* levels become solvent dependent because of the solvation of the dipolar state. The photochemical reactivity of such molecules often shows a dependence on solvent polarity. Figure 4.57 shows such a crossing of n–π^* and CT (π–π^*) states in the singlet and triplet manifolds.

The photoreduction quantum yields and the triplet yields of 4-aminobenzophenone in several hydrogen donor solvents, ZH, are listed in Table 4.1, with comparison to a benzophenone which has an n–π^* lowest triplet state in all these cases. Although dimethylformamide (DMF) is a far better

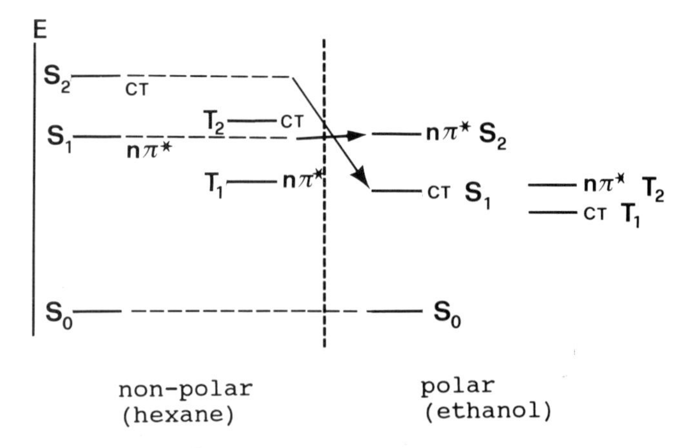

Figure 4.57 *Reversal of n–π^* and CT excited states in 4-aminobenzophenone with change of solvent polarity*

Table 4.1

Solvent		*Benzene*	C_6H_{12}	*Toluene*	*THF**	*DMF*	*EtOH*
4AB*	Φ_T	0.67	0.82	—	—	0.1	$< 10^{-5}$
	Φ_R	2×10^{-2}	0.21	0.1	0.11	$< 10^{-5}$	$< 10^{-5}$
DMBP*	Φ_R	—	1.7	—	—	0.2	0.5

*THF: tetrahydrofuran; 4AB: 4-aminobenzophenone; DMBP: 4,4′-dimethoxybenzophenone
Φ_R is photoreduction yield and Φ_T is triplet yield

hydrogen atom donor than cyclohexane, the CT state is unreactive in spite of its lifetime of several microseconds; this is a remarkable inversion of reactivity linked to the electron distribution of the lowest excited states. In the case of ethanol as H-donor solvent, there is an additional complication due to the very low triplet yield which results from the quenching of the excited singlet state S_1 by protonation (see section 4.4.3).

The ability of various molecules to act as H atom donors in photoreduction varies greatly with the Z–H bond energy and with the electron density on the abstractable H atom. Molecules with electron donor groups (amino, hydroxy, *etc.*) near the H-donor CH group are very efficient. These include some amines, amides, alcohols and ethers; hydrocarbons are substantially poorer H-donors and pure aromatics like benzene show very little reactivity. Acetonitrile H_3C–CN is a very poor H-atom donor because its C–H bonds are made dipolar by the action of the strong electron acceptor CN so that the electron density on H is low. Water itself is an extremely bad H-atom donor in view of both the high OH bond energy and the acidic character of these bonds.

Hydrogen abstraction can lead to the dissociation of some carbonyl derivatives with aliphatic chains. The primary process is the reduction of the carbonyl group, but the biradical formed in the primary photochemical process then rearranges to two closed-shell molecules (Figure 4.58).

In the presence of electron donors such as amines the photoreduction reaction can take place by direct electron transfer rather than by hydrogen abstraction.

In general an exciplex is formed between the reactant molecules (although it may be non-fluorescent when it is formed from the triplet excited state of the carbonyl compound). The primary photoproducts are radical ions rather than neutral radicals, and these may either recombine to restore the original (ground state) molecules or may undergo further reactions leading to the final reduction products (Figure 4.59).

The rate constants of electron transfer with amines are much larger than those of hydrogen atom transfer, *e.g.* in the case of benzophenone, by over three orders of magnitude between triethylamine and 2-propanol. However, hydrogen atom transfer leads in most cases to irreversible reactions, but electron transfer is often reversible through the recombination of the ions

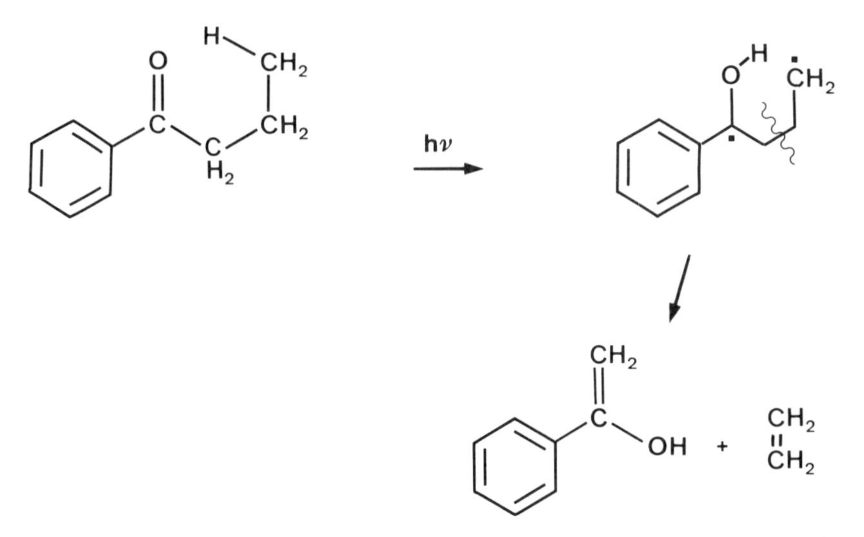

Figure 4.58 *Dissociation of alkylphenyl ketones following intramolecular hydrogen atom abstraction*

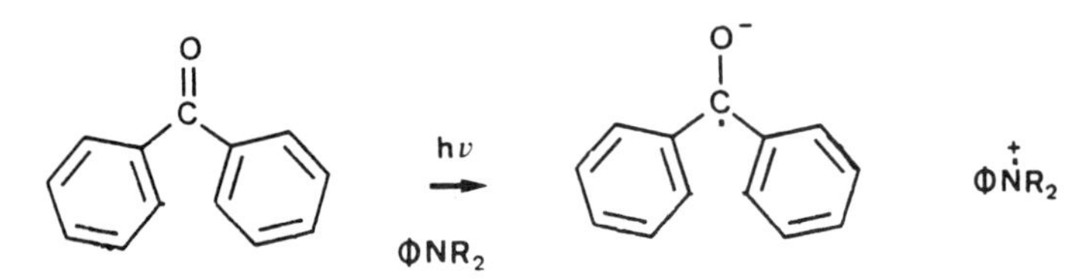

Figure 4.59 *Electron transfer between an aromatic ketone and an amine, leading to the formation of a geminate radical ion pair*

either within the solvation cage prior to separation or through diffusional encounter after separation in polar solvents. In such cases there is apparently no overall chemical reaction, and a sample of benzophenone/1,4-diazabicyclo[2.2.2]octane, DABCO in acetronitrile is photostable; in reality electron transfer does occur with unit quantum yield, the ions separate but recombine at the first encounter. Why should they separate in the first place if they will in any case recombine at the first encounter? In this particular case, there is a simple explanation: benzophenone reacts in its triplet excited state, so that the geminate radical ion pair cannot recombine prior to spin inversion; this takes a long time compared with separation and escape from the solvent cage (Figure 4.60.)

Electron transfer does not follow the simple rule of order of reactivity observed for n–π^*, π–π^* and CT states for hydrogen atom transfer. The π–π^* singlet excited state (S_1) of 9-cyanoanthracene for instance also undergoes electron transfer with amines at a diffusion-controlled rate.

Photo-oxidation and *photo-oxygenation* reactions are the additions of O atoms

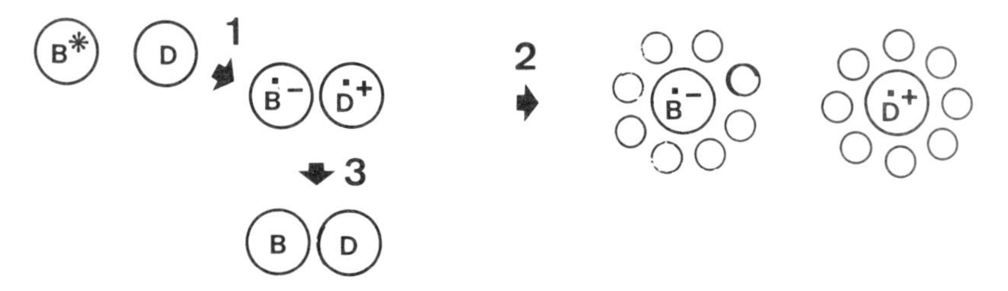

Figure 4.60 *Electron transfer quenching (1), charge recombination (3) and ion separation (2) in polar solvents*

or O_2 molecules to some reactant R. Here the photophysical properties of the oxygen molecule O_2 are the most important, since many of these reactions involve the attack of 'excited singlet' oxygen $^1O_2^*$ on the ground state reactant molecule

$$R + {}^1O_2^* \rightarrow RO_2 \tag{4.36}$$

where the products are usually peroxides or hydroperoxides (*e.g.* Figure 4.61).

The molecule of oxygen O_2 in Figure 4.61 differs from closed-shell organic species in having a *triplet* ground state; it has therefore triplet and singlet excited states, as shown in Figure 4.62, and the first excited state is a singlet $^1O_2^*$ (denoted $^1\Delta_g$ in spectroscopy, but known simply as 'singlet oxygen' in photochemistry). This is a very reactive form of molecular oxygen which is at the basis of most photo-oxidation processes. The energy of this excited state $^1O_2^*$ is very low. The transition $^3O_2 \rightarrow {}^1O_2^*$ corresponds to an absorption in the near IR (NIR) region. Singlet oxygen can therefore be formed through energy transfer from many excited organic molecules according to the following schemes.

$$^1M^* + {}^3O_2 \rightarrow {}^3M^* + {}^1O_2^* \tag{4.37}$$

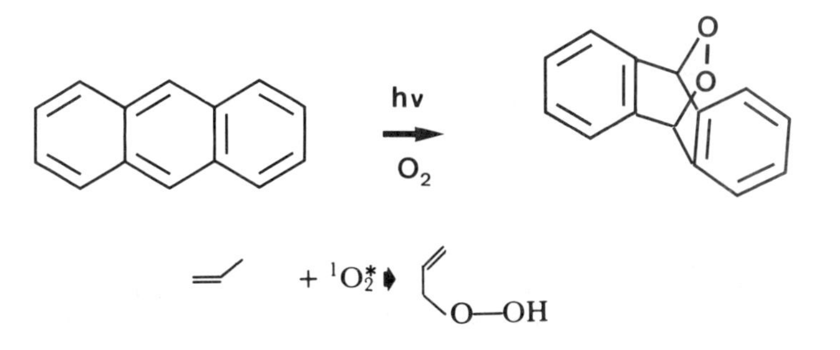

Figure 4.61 *Peroxide and hydroperoxide formation in photo-oxygenation reactions: addition of molecular oxygen to anthracene to form the transannular peroxide, and insertion into an olefin*

$$^3M^* + {}^3O_2 \rightarrow M + {}^1O_2^* \tag{4.38}$$

These processes are both spin-allowed, whereas the deactivation of M* to its ground state, M, would be spin forbidden. The energy requirement for the energy transfer $^1M^* \rightarrow {}^3M^*$ is of course that the singlet–triplet gap should be wider than the excitation energy $^3O_2 \rightarrow {}^1O_2^*$, and this can be a stringent limitation. However, the triplet–triplet energy transfer (eqn. 4.38) is very widespread, and carotenes are some of the very few molecules that can act as quenchers of singlet oxygen $^1O_2^*$.

Figure 4.62 shows the Jablonski diagram of the oxygen molecule, restricted to the first few states relevant to photo-oxidation processes. The phosphorescence S_1–T_0 is very weak and is difficult to detect because it comes in the NIR at 1270 nm. There are however two other emissions at 634 nm and 703 nm which are due to a biphotonic process

$$2\,{}^1O_2^* \xrightarrow{h\nu} 2\,{}^3O_2 \tag{4.39}$$

Since the transition $S_1 \rightarrow T_0$ is strongly spin forbidden in such a light molecule, the gas phase lifetime of $^1O_2^*$ reaches some 45 min (!) and is then still shorter than the natural radiative lifetime. In solution the lifetime decreases as a result of quenching actions to somewhere between a few μs and a few ms, depending on the solvent: water 2 μs; ethanol 5 μs; cyclohexane 15 μs; chloroform 60 μs.

Substitution reactions show in many cases a remarkable difference between photochemical and thermal processes. In this example the orientation effect of electron donor and acceptor substituents in the aromatic ring changes from the classical '*ortho*, *para*-activation' in the ground state to '*meta*-activation' in the excited state (in this case the excited singlet state S_1; Figure 4.63). This can be rationalized from the electron distribution in the ground state S_0 and the excited state S_1. In the case of the S_1 state there is a small but significant charge transfer from the donor substituent to the aromatic ring, specifically to the '*ortho*' (2-) and '*meta*' (3-) positions (Figure 4.64).

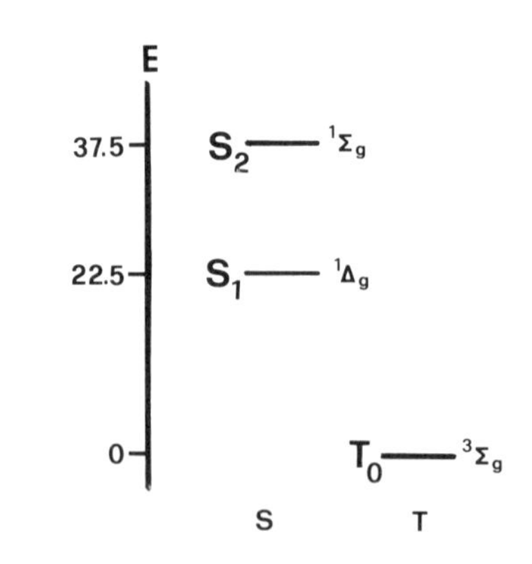

Figure 4.62 *Jablonski diagram of the oxygen molecule. The energy is in kcal mol^{-1}*

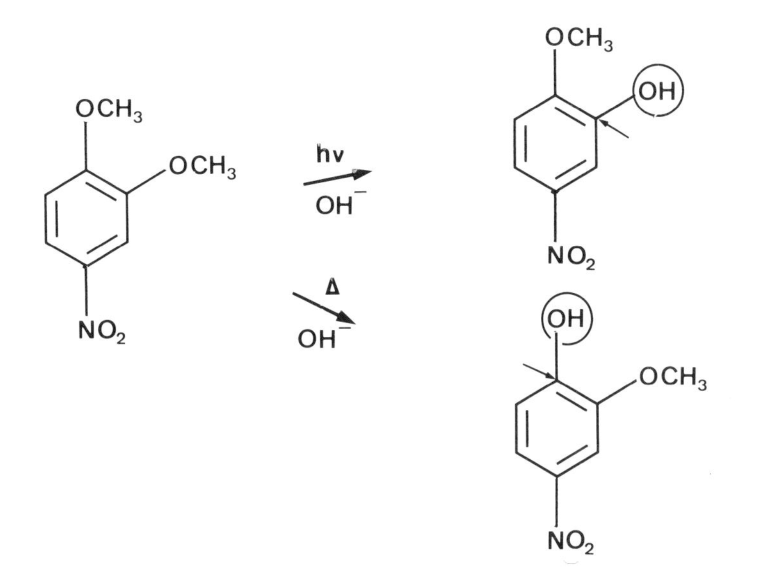

Figure 4.63 *Photochemical and thermal (dark) substitution reactions of 3,4-dimethoxynitrobenzene in alkaline solution*

Photochemical substitution reactions can however follow other pathways than the concerted one which is the rule in the ground state processes. The orientation effects of electron donor and electron acceptor substituents are based on the model of a transition state of σ complex which implies a concerted reaction (Figure 4.65).

Photochemical substitution reactions can proceed through high-energy products such as radical ions, the primary process being a dissociation or an ionization of the excited molecule. Such processes do not have to follow the orientation rules dictated by the charge distribution of the excited molecule, and in many instances the product distribution is still little understood.

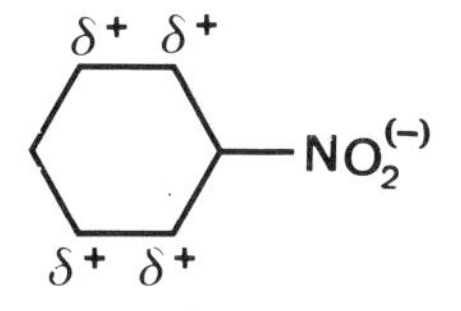

Figure 4.64 *Schematic charge distribution in the ^1L_a state of nitrobenzene*

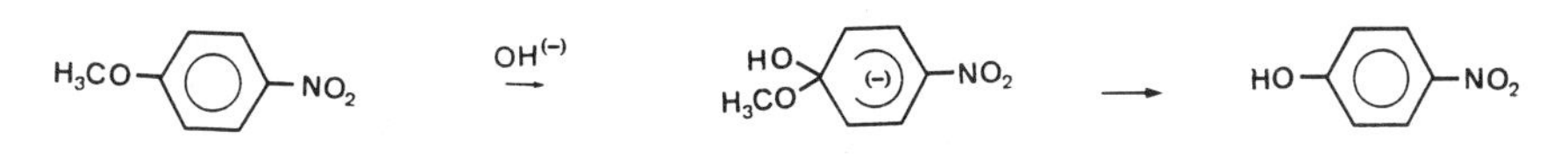

Figure 4.65 *Substitution reaction through a σ-complex transition state*

4.5 PHOTOELECTROCHEMISTRY

Electrochemistry is the study of chemical reactions of molecules at electrodes. Photoelectrochemistry is the study of the reactions of excited molecules or photoproducts at an electrode, or the reactions of ground state molecules at electronically excited electrodes.

An electrochemical reaction in solution is the anodic oxidation or the cathodic reduction

$$\begin{aligned} M &\rightarrow M^{\bullet+} + e^- \\ M + e^- &\rightarrow M^{\bullet-} \end{aligned} \tag{4.40}$$

These reactions have a characteristic free energy which implies the minimal voltage required. As discussed in section 4.1, an excited molecule is at the same time more easily oxidized and reduced than the ground state species. Reactions of excited molecules at electrodes are however practically unknown because their short lifetimes preclude the contact with the electrode when irradiation takes place in the bulk of the liquid. In practice the photoelectrochemical reactions at non-excited electrodes are simply the thermal reactions of photoproducts. We shall give here two examples of such reactions.

(1) Reduction of thionine. The photoelectrode is a solution of thionine (Figure 4.66) which is reduced photochemically in two steps to semithionine and then to leucothionine. The dark electrode is a system Fe^{2+}/Fe^{3+}; in the dark an equilibrium is reached between the cells and there is no flow of current. When the thionine cell is illuminated, the concentrations change and there is no more equilibrium between oxidized and reduced species. A current flows between the cells, its intensity being dependent on the level of irradiation.

(2) Heterolytic dissociation of a cyano derivative. Triphenylcyanomethane dissociates photochemically to form cyanide ions CN^-, the reaction being reversible in the dark.

$$\Phi_3C\text{–}CN \underset{\Delta\ (\text{slow})}{\overset{\lambda<340\ \text{nm}}{\rightleftarrows}} \Phi_3C^+ + CN^- \tag{4.41}$$

thionine

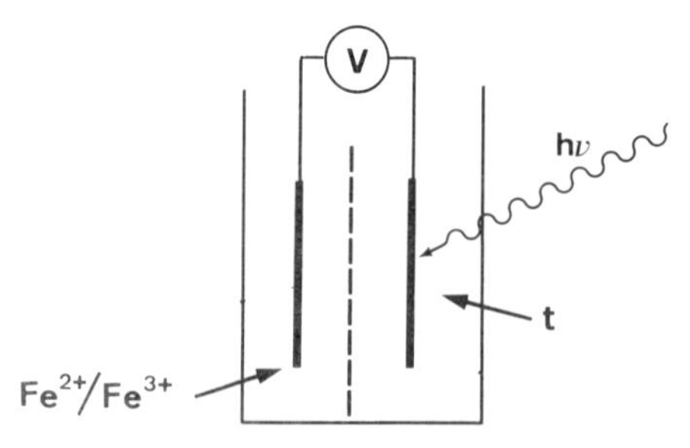

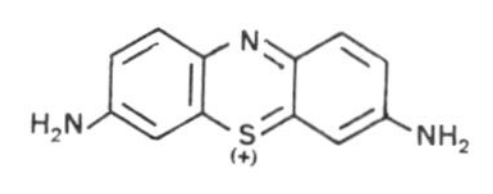

Figure 4.66 *Outline of a photoelectrochemical cell based on the reduction of thionine (formula given on the right). Light is absorbed by thionine (t) in the illuminated half-cell*

This cell uses Ag/AgCN electrodes which are reversible towards CN^-. The potential difference between the electrodes is then given by Nernst's law

$$\Delta E = 2t_+ \frac{RT}{F} \ln \frac{[CN^-]^{dark}}{[CN^-]_{h\nu}} \tag{4.42}$$

These two examples represent photochemical concentration cells. Their common problem is their long-term stability. In practice there are always unwanted side photochemical reactions which slowly destroy the chemicals of the cells.

4.5.1 Reactions at Electronically Excited Semiconductor Electrodes

In section 3.6 we have described the valence (VB) and conduction (CB) bands of semiconductors. When such an electrode is irradiated with light of energy beyond its bandgap, electrons are promoted from VB to CB and positive holes (h^+) are left in the VB. If a molecule in contact with the electrode has an orbital (*e.g.* π or π^*) between the energy levels of VB and CB, oxidation and/or reduction can take place by electron transfer to or from the excited electrode (Figure 4.67).

In an isolated electrode, the VB, the CB and the Fermi level all have constant values throughout the depth of the electrode. If it is immersed in an electrolyte there will be a movement of charge at the interface so that the Fermi level coincides with the redox level of the species contained in the electrolyte. This is somewhat similar to the equalization of the levels of two liquids in contact, the equilibrium condition being that the pressures should be equal at the point of contact. In an energy diagram this is shown as a

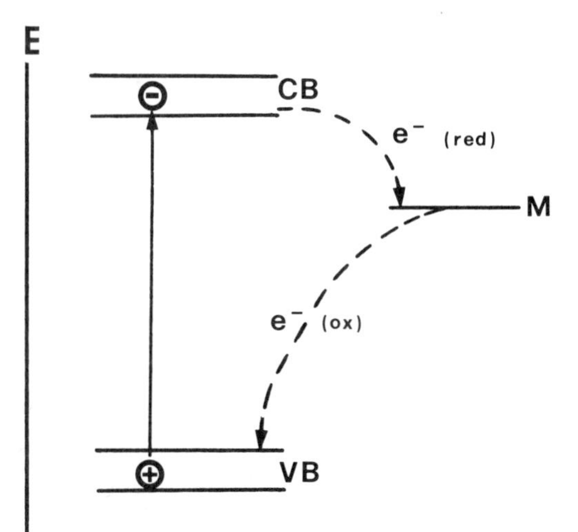

Figure 4.67 *Redox processes at an excited semiconductor electrode. The molecule M must have an orbital of intermediate energy between the valence band and the conduction band of the semiconductor*

bending of the VB and CB levels near the interface; in that region there is an electric field within the semiconductor such that electrons are driven towards the interior while positive holes are pushed towards the electrolyte interface (Figure 4.68). This charge separation in space is essential for any application of photoelectrochemistry, for without it recombination of the electron with the positive hole would be the major energy wasting process.

The chemical reactions of organic molecules at excited semiconductor electrodes are of course reduction and oxidation processes, but these depend on the solvents and other reactants such as water, electrolytes and molecular oxygen. Figure 4.69 gives a few examples of many such reactions which are finding applications in chemical synthesis.

4.6 INORGANIC PHOTOCHEMISTRY

The separation between inorganic and organic chemistry is rather artificial and there are many areas of overlap especially in the field of metal complexes. Within a narrow definition, inorganic photoinduced processes involve a metal which has its own electronically excited states. We consider

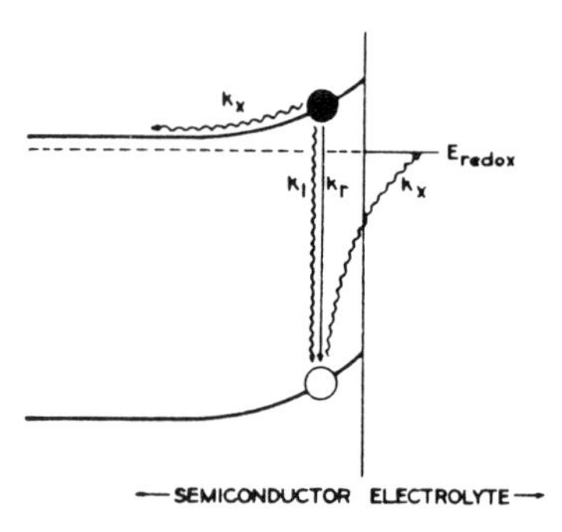

Figure 4.68 *Energy diagram of the valence and conduction bands of a semiconductor–electrolyte interface. The bending of the bands corresponds to the effect of the electric field*

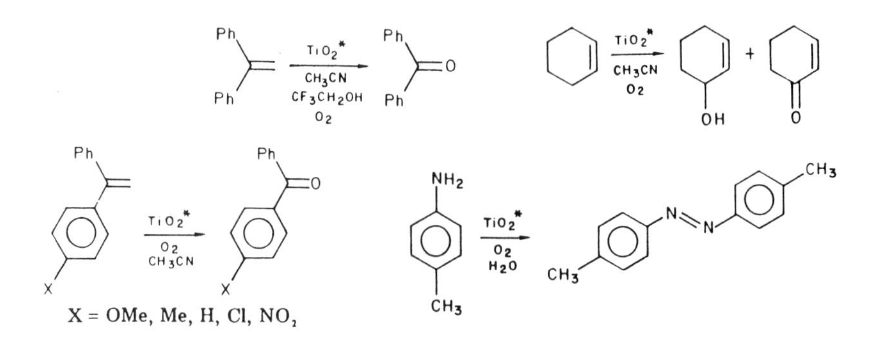

Figure 4.69 *Examples of oxidation reactions of organic molecules at the surface of photoexcited titanium dioxide in polar solvents in the presence of molecular oxygen*

here three different types of reactions: those of excited atoms with organic molecules, the photoreduction of electrolytes and most importantly the photochemistry of metal complexes.

4.6.1 Excited Atom Reactions

The atoms of mercury and of cadmium provide many examples of such reactions, though they are not the only ones. The most important processes of this type involve excited mercury Hg*, this label hiding the fact that there are several different excited states in this atom. (The population of these states depends on the pressure; see Chapter 7 about the Hg arc emissions.) In the gas phase, Hg* will attack organic molecules such as hydrocarbons through a reaction of hydrogen abstraction

$$Hg^* + ZH \rightarrow HgH + Z \tag{4.43}$$

The excited mercury atom Hg* can attack many different molecules, either by a direct chemical mechanism or by energy transfer. It will abstract the halogen atom from molecules like HCl

$$Hg^* + HCl \rightarrow HgCl + H \qquad \text{or} \qquad Hg + Cl + H \tag{4.44}$$

Dissociation of water can be sensitized by excited mercury atoms

$$Hg^* + H_2O \rightarrow (Hg \cdot H_2O)^* \rightarrow HgOH + H(+M) \tag{4.45}$$

This process requires the presence of a 'third body', M, to take up the excess vibrational energy, otherwise the excited complex dissociates with re-emission of luminescence.

Some excited atom-sensitized reactions proceed in fact through energy transfer to the reactive molecule, others by formation of different reactive atoms, *e.g.*

$$Hg^* + N_2O \rightarrow N_2 + O^*(^3P \text{ or } ^3D) + Hg \tag{4.46}$$

followed by

$$>C{=}C< + O^* \rightarrow \underset{\diagdown\, O \,\diagup}{C{-}C} \tag{4.47}$$

4.6.2 Photoinduced Redox Reactions of Ions in Solution

These are photoinduced electron transfer reactions between two ions. The closed-shell ions then form free radicals which can be charged or neutral, these primary photochemical products being very reactive. One example of this process is the electron transfer between a uranyl cation and a nitrate anion

$$(UO_2^{2+})^* + NO_3^- \rightarrow UO_2^{\bullet +} + NO_3^{\bullet} \qquad k_r = 5 \times 10^4\ \text{M}^{-1}\,\text{s}^{-1} \tag{4.48}$$

The uranyl cation (closed shell, UO_2^{2+}) absorbs light in the visible (VIS) region around 430 nm. The open-shell species can be detected by their absorptions at longer wavelengths (*e.g.* $NO_3^{\bullet}$ absorbs in the region 600–700 nm). Here we have an example of a strictly inorganic photochemical reaction between the two ions of an electrolyte solution. Somewhat similar reactions exist in the salts of organic molecules, but in general it is the organic ion which reacts in the excited state

In Figure 4.70, in the overall process, 5-nitroindole is reduced to aminoindole; the chloride ion is present in the form of HCl and the 5-nitroindole is protonated under these acid conditions.

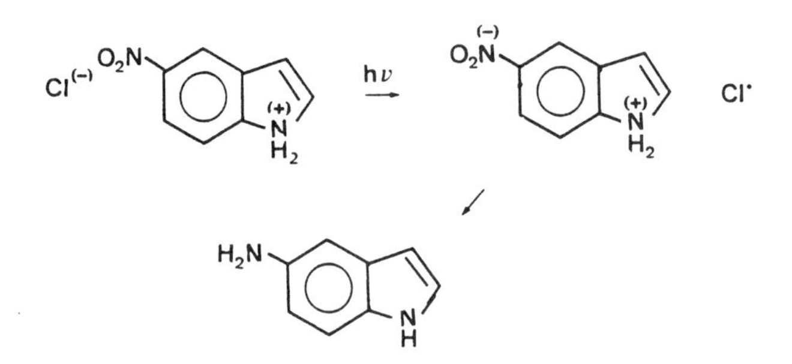

Figure 4.70 *Photochemical reaction of a nitroindole with a chloride ion*

4.6.3 Photophysics and Photochemistry of Metal Complexes

The molecules known as metal complexes have a metal centre linked to several ligands, according to the general formulae shown in Figure 4.71(a).

Many metals form such complexes, in particular the transition metals (Fe, Ni, Co, *etc.*) but there are also many complexes of heavier elements such as Ru or Pt. The ligands range from single atoms or ions (*e.g.* Cl^-) to relatively large organic molecules. A few examples are shown in Figure 4.71(b). There has been growing interest in the photochemistry of these metal complexes over the last 20 years not only from an academic point of view but also for potential applications such as solar energy conversion.

4.6.3.1 Electronic States of Metal Complexes. Figure 4.72 shows the orbital structure of a typical metal complex; for the sake of simplicity it is assumed

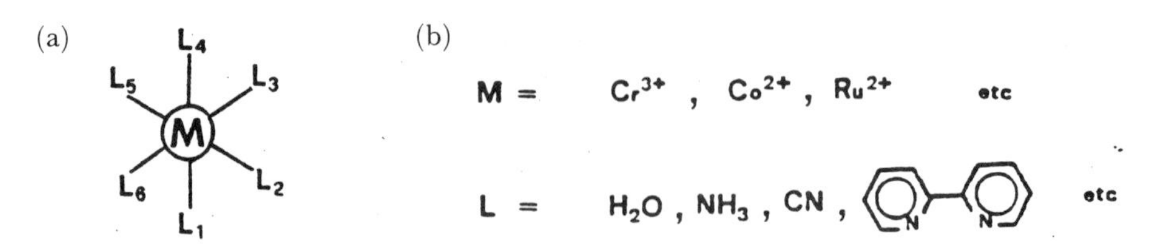

Figure 4.71 *(a) General structure of a metal complex: M is the metal centre (which is in general an ion) and L is a ligand. (b) Some examples of inorganic and organic ligands*

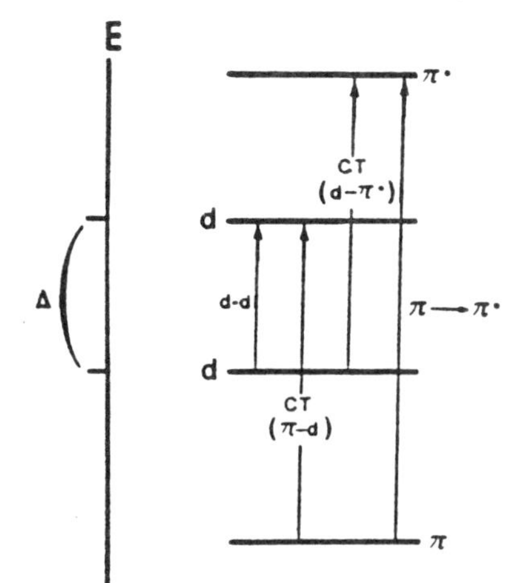

Figure 4.72 *Simplified orbital diagram of an organometallic complex*

that all the ligands are the same, so only one set of ligand orbitals is represented. In this example we take a transition metal centre which has a set of five d orbitals. In the isolated atom these would be degenerate but they are split into two subsets when the metal is bonded to the ligands. The energy difference between the subsets is the 'crystal field' splitting which plays an important role in the photophysical properties of the complex. The ligand orbitals are shown here simply as π and π^* and remain in general very similar to those of the free ligand, at least when the ligand is an organic molecule. From this orbital diagram the four types of excited states of metal complexes can be derived, and their specific properties can be described. The lowest energy excitation is of d–d type, so long as all the d orbitals are not filled. In this respect it should be noted that the metal centre is usually an ion (*e.g.* Fe^{3+}, Ru^{2+}) but it can be neutral if it has 10 d electrons in the five d orbitals.

Next in order of energy come the charge transfer states which can be MLCT (metal to ligand charge transfer), also known as d–π^*, or the ligand to metal charge transfer (LMCT) known as π–d. Finally there remain the electronic transitions of the ligands themselves, called L–L transitions in metal complexes; these would be for instance π–π^* states in a bipyridyl ligand. This is of course only one simple example and there are many variations. The d–d states are not necessarily the lowest excited states, this depends on the magnitude of the crystal field splitting relative to the d–π^* or π–d transitions, so there are complexes in which the order of the d–d and the charge transfer states is reversed.

So far we have considered only an orbital diagram without reference to the spin of the electrons. This is however an important point for the photophysics of metal complexes, just as for organic molecules.

4.6.3.2 Multiplicity (Spin Quantum Numbers) of Metal Complexes. The various metal centres have between 0 and 10 electrons distributed in the five d orbitals in the ground state. When there is a choice between two spin configurations, as in the example of Figure 4.73 this state will be either a low-spin or a high-spin configuration depending on the magnitude of the crystal field splitting. Consider the case of a metal centre with three d electrons. According to Hund's rule these electrons will occupy the three lower d orbitals with parallel spins (this is a quartet state). In a four d-electron metal centre, the fourth electron can be fitted either into one of the half-filled orbitals of low energy, or into one of the vacant orbitals of higher energy. The former leads to a triplet state (low spin), the latter to a quintet state (high spin). The outcome depends on the relative values of the crystal field splitting and of the energy difference of the low- and high-spin configurations.

The Jablonski diagrams of metal complexes differ in general from those of closed-shell organic molecules because the ground state is not necessarily a singlet state. Figure 4.74 gives an example of such a diagram for a metal centre with three d electrons; the electronic states are then doublet (D) and quartet (Q) states. There is a major difference between ground state closed-shell and open-shell species, in that in the latter the excited quartet state Q_1 is not necessarily lower than D_1. In closed-shell molecules (and complexes) the triplet state plays the role of a reservoir of excitation energy because its decay to the S_0 ground state is spin forbidden. This is not the case if the lowest excited state of an open-shell molecule is D_1 and its ground state is D_0; such states then have very short lifetimes. In the example in Figure 4.74 however D_1 would be relatively long-lived since its deactivation is formally spin forbidden.

4.6.3.3 Electron Distribution in Excited States of Metal Complexes. The highly simplified picture of orbitals of Figure 4.75 illustrates the motions of the electronic charges associated with the five types of electronic transitions of importance in metal complexes. Of these two are localized excitations—the dd and LL transitions—which do not involve any actual charge transfer. In the CT transitions shown as dπ* (or MLCT) and πd (or LMCT) there can be a change in the dipole moment of the complex if the ligands are of different types. In the 'charge transfer to solvent' (CTTS) state the metal acquires an additional positive charge, but this is not strictly speaking a transition of the metal complex itself since it involves directly the solvent (usually H_2O).

Figure 4.73 *Low-spin (L) and high-spin (H) electron configurations of the d orbitals of a metal centre; Δ is the crystal field splitting*

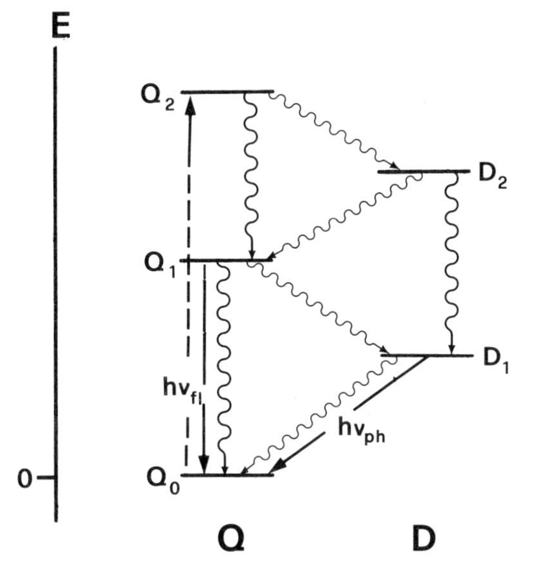

Figure 4.74 *Jablonski diagram of a metal complex with three d electrons. The wavy arrows show non-radiative transitions. Note that intersystem crossings between higher states can be important as a result of the heavy atom effect*

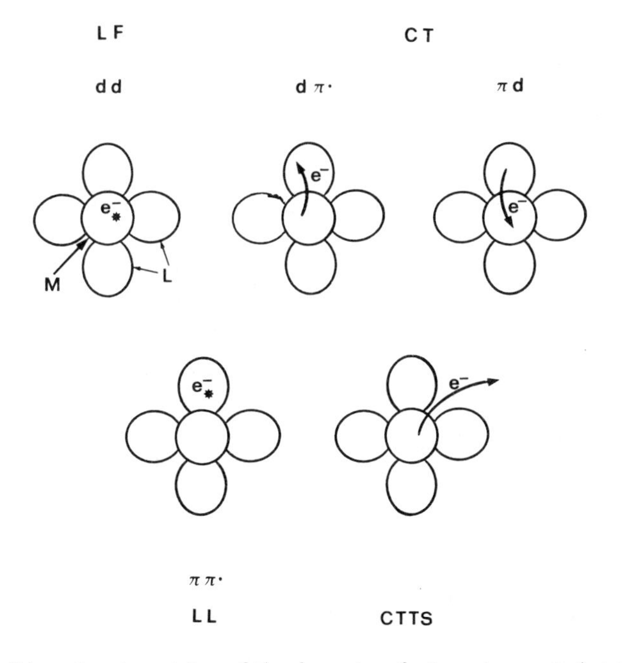

Figure 4.75 *Schematic representation of the charge transfer in various excited states of a metal complex. M is the metal centre and L stands for a ligand. LF is a ligand field transition, CTs are the charge transfer transitions, LL is an intraligand transition, and CTTS is a charge transfer to solvent*

4.6.3.4 Photophysics of Metal Complexes. Figure 4.74 shows a typical absorption spectrum of a transition metal complex. The weak, long wavelength bands correspond to d–d transitions which have small transition moments since they involve only the angular redistribution of electron density around the metal centre. These d–d transitions arise often in the VIS or even in the NIR regions and they give the characteristic colours to solutions of these complexes. At shorter wavelengths the stronger charge transfer bands are observed.

The intraligand (L–L) transitions remain very similar to those of the free ligands and the corresponding absorption bands are therefore also similar both in wavelength and in intensity.

The luminescence and non-radiative transitions are dominated by the heavy atom effect. Few metal complexes show luminescence in solution, which is not surprising in view of the low energy of their lowest excited states; non-radiative deactivation to the ground state competes efficiently with the rather long-lived emissions. The luminescence of metal complexes follows Kasha's rule (it comes only from the lowest excited state of a multiplicity manifold), but the lifetimes fall between those of fluorescence and phosphorescence or organic molecules. The complex $Ru^{2+}(bpy)_3$ for example has a luminescence lifetime of about 1 μs, and this should be termed 'phosphorescence' according to a simple model of orbitals. The metal centre acts however as a heavy atom and states of nominally different multiplicity become strongly mixed through spin-orbit coupling. In the case of complexes of the heavier metals this mixing is so strong that the spin labels become meaningless and are dropped from the notation of state properties.

The heavy atom effect affects also the non-radiative transitions, as shown above in Figure 4.76. Transitions between higher excited states of different nominal multiplicities are efficient, so that the luminescence quantum yields depend on the excitation wavelength. (Vavilov's rule breaks down in such cases.)

The excited state lifetimes and luminescence properties of metal complexes are related to the relative positions of the potential energy wells shown in Figure 4.77. On the left we have a lowest excited state which resembles geometrically the ground state (the internuclear distances, *r*, are similar). The crossing between these states requires a high activation barrier *E* (in a classical picture) and the excited state lifetime is therefore relatively long. The Stoke's shift between the absorption band (a) and the emission band (e)

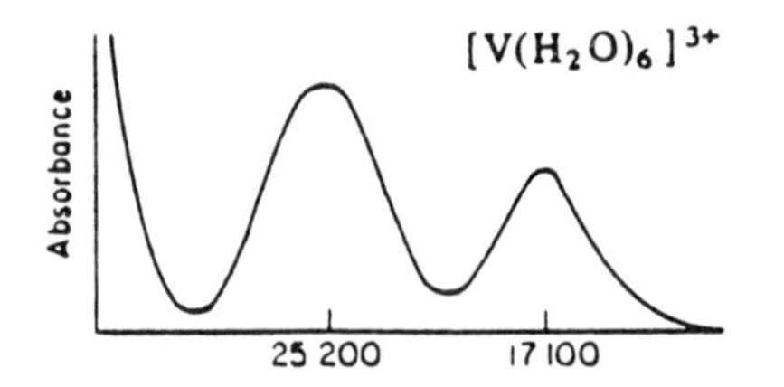

Figure 4.76 *Absorption spectrum of a metal complex. (Abscissa in cm^{-1}.)*

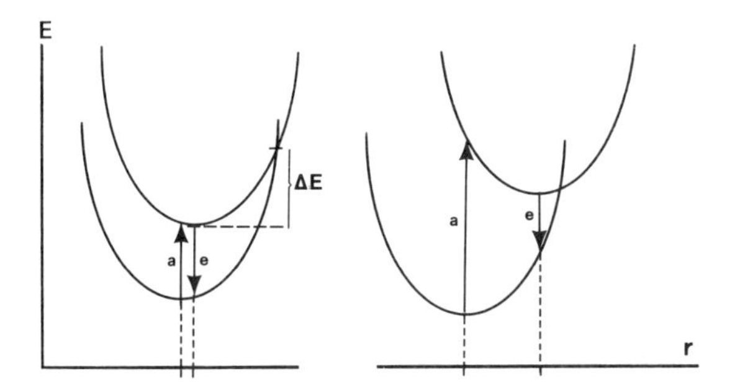

Figure 4.77 *Potential energy diagrams of weak and strong couplings between the ground and excited states of a metal complex. (r is the distance between the metal centre and a ligand, a are the absorption and e are the emission spectra.)*

is therefore small; this is a case of 'weak coupling'. The opposite situation of 'strong coupling' is characterized by efficient non-radiative crossing from the excited state to the ground state with a very small activation barrier; the Stoke's shift between the absorption and emission spectra is large, but the luminescence is very weak because of the short excited state lifetime.

4.6.3.5 Photochemical Reactions of Metal Complexes. The major photoinduced reactions of metal complexes are dissociation, ligand exchange and reduction/oxidation processes. The quantum yields of these reactions often depend on the wavelength of the irradiating light, since different excited states are populated. This is seldom the case with organic molecules in which reactions take place almost exclusively from the lowest states of each multiplicity S_1 and T_1.

It can be expected that d–d states of metal complexes should lead to dissociation, since this is a metal-centred excitation in which an electron leaves a d orbital which binds a ligand. However, CT states d–π^* and π–d will be involved in redox reactions of the metal centre; such reactions often lead to the oxidative or reductive dissociation of the complex.

In addition to these reactions which in effect destroy the original complex, there are important isomerization processes which can be stereo, or valence, isomerizations. Some of these proceed in fact through primary dissociation followed by recombination of the fragments with a different geometry.

Ligand exchange reactions. A molecule of solvent S is exchanged for the ligand L, according to the general scheme

$$ML_n \xrightarrow[S]{h\nu} ML_{n-1}S \tag{4.49}$$

Photoaquation is a common reaction of this type, *e.g.*

$$Cr^{3+}L_6 + H_2O \xrightarrow{h\nu} Cr^{3+}L_5(H_2O) \tag{4.50}$$

The primary photochemical process is the dissociation

$$Cr^{3+}L_6 \xrightarrow{h\nu} Cr^{3+}L_5 + L \tag{4.51}$$

The quantum yield of this reaction is independent of the irradiation wavelength when all the ligands, L, are identical. In mixed ligand complexes two new observations require attention:
(1) which of the ligands will be replaced?
(2) what is the effect of the irradiation wavelength?

In the following example a mixed ligand complex of Cr^{3+} can undergo two different photoaquation reactions:

$$[Cr(NH_3)_5(SCN)]^{2+} \xrightarrow[H_2O]{h\nu} \begin{cases} [Cr(NH_3)_5(H_2O)]^{3+} + SCN^- & (1) \\ [Cr(NH_3)_4(H_2O)(SCN)]^{2+} + NH_3 & (2) \end{cases} \quad (4.52)$$

The relative quantum yields of the two processes depend on the irradiation wavelength (Table 4.2).

The absolute quantum yields tend to increase with the photon energy within an electronic transition. In addition, these reactions can also take place in the dark, but their ratio is different and there is a large difference in the activation barriers; the effect of temperature on the reaction rate constants shows that this activation energy is very small in the photochemical reactions, of the order of 1–2 $kcal\,mol^{-1}$. However, the dark reaction proceeds with a large activation barrier of some 25 $kcal\,mol^{-1}$.

Reduction and oxidation reactions. These are electron transfer reactions which can be either inter- or intramolecular, as in the following example.

$$Br{-}Co(NH_3)_5 \xrightarrow[H^+]{h\nu} Co^{2+} + 2NH_4^+ + 3NH_3 + Br^- \quad (4.53)$$

Electron transfer to the metal centre results in oxidation of the halogen and the complex then dissociates. The relative quantum yields are high (0.1 to 1) when irradiation is made in the CT band; in complexes with Br or I this reaction is still observed with irradiation in the d–d bands, but the quantum yields are then lower.

There are also many intermolecular electron transfer reactions such as

$$Fe(CN)_6^{4-} \xrightarrow[H_2O]{h\nu} Fe(CN)_6^{3-} + e^- \quad (4.54)$$

This is an electron transfer to the solvent, in which a hydrated electron is formed. In this complex there are well defined d–d transitions in the VIS and near UV (NUV), and CT bands in the UV regions. Irradiation in the d–d bands leads to ligand exchange, for instance to photoaquation in water. Irradiation in the CT bands results in electron transfer to the solvent. This provides a very good example of the dependence of the nature of photochemical reactions on irradiation wavelength in metal complexes.

Table 4.2

Wavelength/nm	*Relative quantum yield*
652	(2)/(1) = 8
492	22
373	15

Isomerization reactions. Some metal complexes have distinct stereochemical forms and can undergo reactions similar to the *cis–trans* isomerizations of organic molecules. The square planar Pt complex shown in Figure 4.78 is of this type; here the NO ligand is glycine.

Cis–trans isomerization can take place either photochemically or in the dark, but the reaction pathways are quite different. In the light-induced process the reaction goes through a tetrahedral intermediate formed from the triplet excited state, whereas the dark reaction involves a dissociation of the complex, followed by recombination. In the latter case the presence of free glycine is demonstrated by the use of radioactive tracers; no free glycine appears in the photochemical reaction.

In valence isomerization reactions the most common case is the rearrangement of a ligand, as in the example in Figure 4.79 in which a nitro group, NO_2, isomerizes to a nitrito group, O—N═O. The primary photochemical process is in fact a dissociation of the Co–N bond, with formation of free radicals. The nitro radical isomerized to its nitrito form, then binds again to make the new complex.

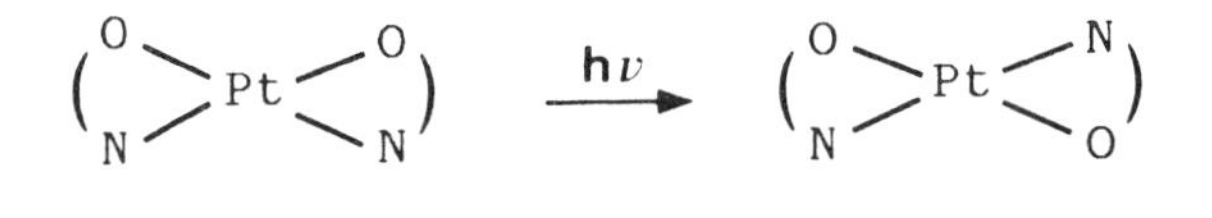

cis trans

Figure 4.78 *Mechanisms of photochemical and thermal (dark) stereo-isomerizations of a square planar Pt complex*

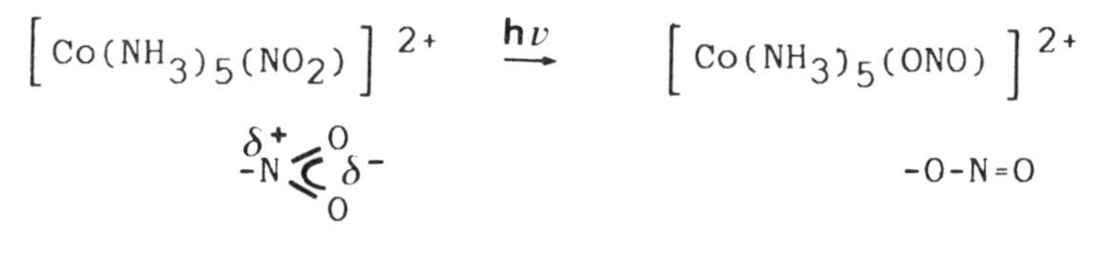

Figure 4.79 *Example of valence isomerization of a ligand*

4.7 PHOTOCHEMISTRY IN SOLIDS AND ORGANIZED ASSEMBLIES

Most of the photophysical and photochemical processes discussed so far concern isolated molecules in the gas phase, or molecules in homogeneous liquid solutions. There are important environmental effects in some cases, due to solvent polarity in particular. When we consider the photochemical properties of molecules in solids some new features appear, and these are related not only to the viscosity of the medium. Taking a step further and we reach the realm of more or less organized assemblies of molecules, from the simple micelles to the monomolecular layers known as Langmuir–Bloggett films. This last subject is left to the final chapter, since here we reach indeed the frontiers of current research in photochemistry.

4.7.1 Types of Solids

The photochemical properties of molecules that are part of a solid or are embedded in a solid matrix depend greatly on the nature and organization (if any) of the solid lattice. In order of increasing organizational structure, solids can be classified for our purpose as *glasses*, *polymers* and *crystals*.

(1) A 'glass' is formed when certain liquids are cooled below a temperature known as the glass transition temperature. The molecules retain the structure of the liquid state at that temperature, but they are frozen in a motionless state. There is however no special ordering of the solvent molecules around a solute.

(2) There are many different 'polymers' which are long chains of molecules linked by covalent bonds. The chains can be either intertwined in a loose assembly, or they can be cross-linked by covalent bonds to form a very strong lattice. Chromophoric molecules can be included in polymers in two different ways: they can be dispersed at random, rather like in a glassy matrix, or they can be part of the polymer chains themselves.

(3) A 'crystal' is an ordered molecular structure, with precise symmetry properties. Some chromophoric molecules form pure crystals in which they are forced to stay relatively close and in fixed orientations. In other cases such molecules can be dispersed as 'defects' in the lattice of some other molecules.

4.7.2 Photochemical Reactions in Glasses

Diffusion of molecules in rigid glasses is negligible within the lifetimes of excited molecules, so that the main reactions are unimolecular dissociations and isomerization. These are rather similar to liquid state reactions, but the fragments cannot separate through diffusion and often recombine to restore the reactants. There are exceptions when the photoproducts are in fact more stable than the reactants, as in the case of photoeliminations.

$$CH_3N_3 \xrightarrow[\text{Ar matrix}]{h\nu} H_2C{=}NH \quad (+N_2) \tag{4.55}$$

$$H_2C{=}NH \xrightarrow{h\nu} HCN + H_2$$

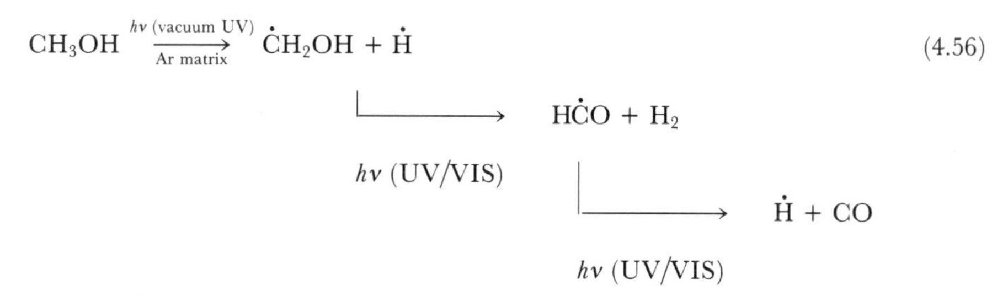

Irreversible photochemical reactions can take place when one of the fragments of a photodissociation reaction reacts with a molecule of the glass matrix itself; hydrogen atom abstraction is a common secondary reaction of this type.

$$CH_3{-}\overset{\overset{\large O}{\|}}{C}{-}CH_3 \xrightarrow{h\nu} \cdot CH_3 + \cdot\overset{\overset{\large O}{\|}}{C}{-}CH_3 \searrow \dot{C}H_3 + CO \nearrow^{ZH} CH_4 + \dot{Z}\ldots \qquad (4.57)$$

4.7.3 Excitons in Polymers and Crystals

When two or several similar molecules are in close contact the excitation energy of any one molecule can be delocalized; in a simplified picture this can be thought of as a hopping of electronic excitation by isoenergetic energy transfer from one molecule to another. This exciton interaction is at the basis of the stability of excimers, but it is most significant in large arrays of similar molecules in polymers and especially in crystals. The exciton is then a 'pseudo-particle' which travels very fast throughout the lattice; the excited molecule thus appears to travel at enormous speeds, well beyond the diffusional limit in liquid solvents. The exciton formed by the initial excitation of a single molecule in a polymer or in a crystal can be trapped very quickly by a molecule of lower excitation energy, even when it is present in very low concentration. Such molecules can be either foreign species or lattice molecules at defects within the crystal structure. Exciton trapping is a significant process in the photochemical damage of nucleic acids (section 5.5).

4.7.4 Bimolecular Photochemical Reactions in Solids

Although actual diffusion in solids is not significant within the lifetimes of excited molecules, bimolecular reactions can take place when molecules are kept in close contact in a polymer or crystal lattice. In some crystals the molecules are ideally spaced for cycloaddition, as in the example of cinnamic acid (Figure 4.80). The geometrical requirements are quite stringent and the reaction cannot proceed if the interplane separation of the molecules exceeds about 4Å.

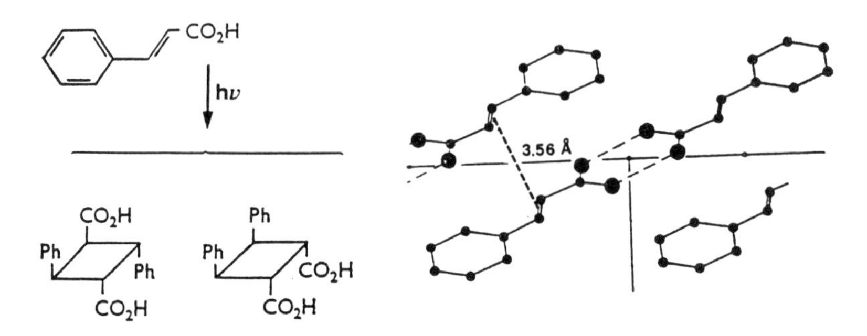

Figure 4.80 *Photochemical cyclodimerization of cinnamic acid in a crystal*

The photodimerization of cinnamic acid and similar molecules is observed in crystals, but reactions of the same type occur in some polymers as well. Polymers such as polystyrene are made of long, saturated hydrocarbon chains with pendant groups in close contact dangling from the chain; these chromophores can then interact in bimolecular photoaddition reactions. Polyvinylcarbazole and its derivatives are important examples of polymers which lead to such bimolecular interactions (*e.g.* exciplex formation).

4.7.5 Photochemistry in Micelles

A micelle is an assembly of 'amphiphilic' molecules dispersed in water. Such molecules are made of two parts, a polar 'head' group and a non-polar 'tail'. The polar head is for example a carboxylic acid which can dissociate into ions ($-COO^-$ and H^+); the non-polar tail is a saturated hydrocarbon chain. Since the non-polar parts are insoluble in a polar solvent, these molecules aggregate in water to form micelles which are microscopic droplets with a non-polar interior and polar groups at the water interface. This picture of micelles is probably an oversimplification, because water penetrates to some extent between the molecules; it is however sufficient for an understanding of the special properties of micellar suspensions in photochemistry.

4.7.5.1 Micellar Catalysis. Non-polar molecules such as aromatic hydrocarbons are practically insoluble in water. In a micellar suspension they concentrate in the non-polar interior of the micelles where they can reach relatively high concentrations, even though their overall concentration may be very low. The quantum yields of bimolecular reactions like photoadditions are therefore greatly increased in micellar suspensions.

4.7.5.2 Orientational Effects. Solute molecules which have distinct polar and non-polar parts take up specific orientations in micelles, such that their non-polar end stays in the non-polar interior of the micelle, the polar group residing at the water interface. Photocycloaddition of these molecules will therefore lead preferentially to the head-to-head dimer, even if the head-to-tail dimer is the major product in non-polar solutions, because of steric effects (Figure 4.81).

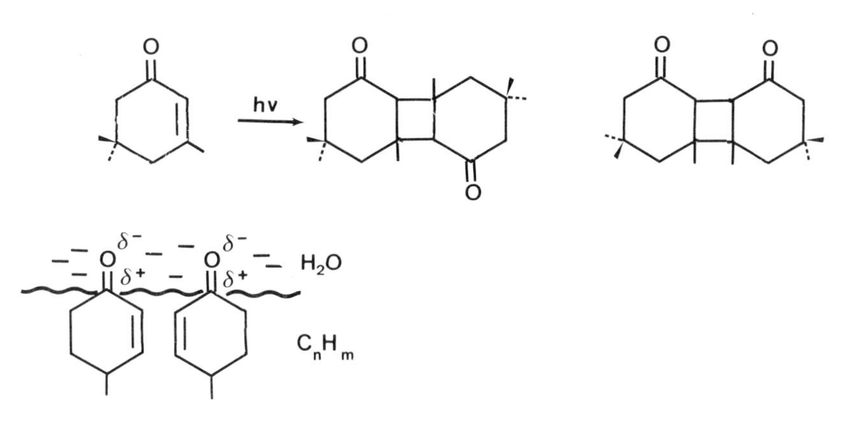

Figure 4.81 *Example of head-to-tail and head-to-head photodimerization of an enone in solution and at the interface of a micelle*

4.8 CHEMILUMINESCENCE

In a photophysical or photochemical process light is absorbed by a chemical sample to produce for instance luminescence or photochemical reactions. There are some chemical processes which have the opposite result, that is the production of light from a thermal reaction. This process of 'chemiluminescence' is widespread in the natural world, where it is described as 'bioluminescence'; the best known example being the emission of VIS light by fireflies and glow-worms.

Chemiluminescence can occur when a thermal (dark) reaction is so exothermic that its energy exceeds that of the electronically excited state of one of the product molecules. The major pathway for these reactions is the decomposition of cyclic peroxides, and this is at the basis of most bioluminescence processes. There are some other physico-chemical processes which can lead to the formation of excited states and thereby to the emission of light; these are based on the bimolecular recombination of high-energy species such as free radicals and radical ions.

4.8.1 Electroluminescence

When two radical ions formed by electrolysis recombine to restore the closed-shell neutral species, there is in principle enough energy liberated to produce one excited molecule (and one ground state molecule).

$$M^{\bullet+} + M^{\bullet-} \rightarrow M^* + M \qquad (4.58)$$

$$M^* \longrightarrow M + h\nu$$

Figure 4.82 shows that in a simple picture of orbitals the energy of the excited state is exactly equal to the energy spent in the formation of the ions; this is then true for all positive and negative ions of the same chemical species. This is however a very simple picture of isolated molecules which takes no account of the solvation energies or of the electrostatic interactions. It is clear that electroluminescence requires in the first place the formation of high-energy chemical species, their thermal reaction of recombination leading to electronically excited products.

Thermoluminescence is a related process of light emission when some solids are heated to temperatures close to their melting point. If ions have been trapped in the solid so that they cannot move within the time of observation in the rigid matrix, they will recombine when diffusion can take place. The energetics of this process is similar to that of electroluminescence.

4.8.2 Chemiluminescence Sensitized by the Decomposition of Cyclic Peroxides

The photochemical oxidation reactions of unsaturated organic molecules lead to the formation of peroxides which are characterized by the weak bond O–O. In terms of bond energies this is one of the weakest covalent bonds; weak enough to be split in thermal processes. When a cyclic peroxide dissociates into carbonyl compounds, one of these may be formed in an electronically excited state (Figure 4.83). The energy difference results from the loss of the weak O–O bond and the formation of two C=O π bonds; over 60 kcal mol^{-1} is then available as free energy. There are also reasons of

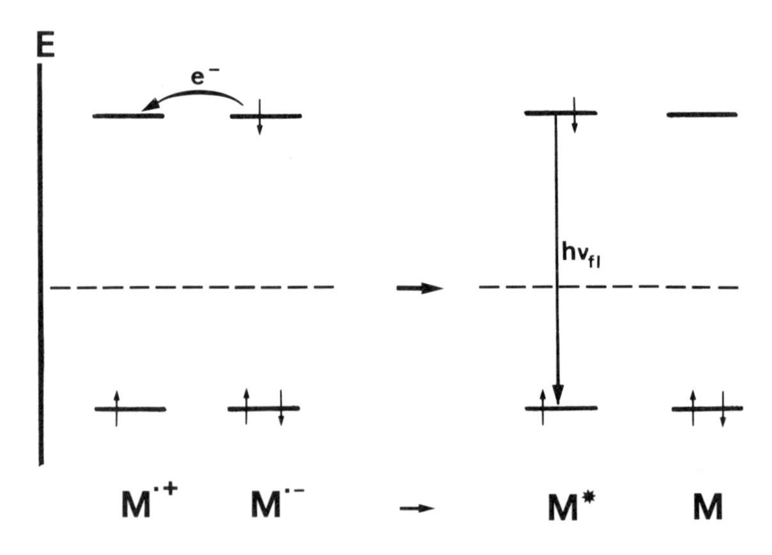

Figure 4.82 *Orbital diagrams of electroluminescence (or recombination luminescence) of positive and negative radical ions*

Figure 4.83 *Thermal decomposition of cyclic peroxides producing an electronically excited state of a carbonyl group*

conservation or orbital symmetry that favour the production of one of the carbonyl groups in an electronically excited state.

The luminescence arises from dye molecules which are sensitized by energy transfer from the excited carbonyl group.

The chemiluminescence spectrum is therefore the fluorescence spectrum of the dye. Many different colours of 'cold light' can be obtained in chemiluminescent systems according to the nature of fluorescing species.

4.8.3 Chemiluminescence of Free Radical Reactions

$$NO + O \rightarrow (NO_2)^* \rightarrow NO_2 + h\nu \qquad (4.59)$$

The recombination of free radicals can be highly exergonic, since in general a new bond is formed by the formerly non-bonding electrons. In the example of eqn. (4.59) the closed-shell molecule, NO_2, is then formed in an excited state which can decay through luminescence emission. Some organic free radicals show similar reactions but it must be stressed that in most cases the emission yields are low.

4.9 REACTIONS OF FREE RADICALS

Free radicals and radical ions are some of the most important primary photochemical products. These are 'open-shell' species, since they have one unpaired electron so that their total spin quantum number is $\pm(\frac{1}{2})$. They can disappear finally only through reactions with other open-shell molecules; such reactions involve in many cases the addition or the disproportionation of two radicals (*e.g.* Figure 4.84).

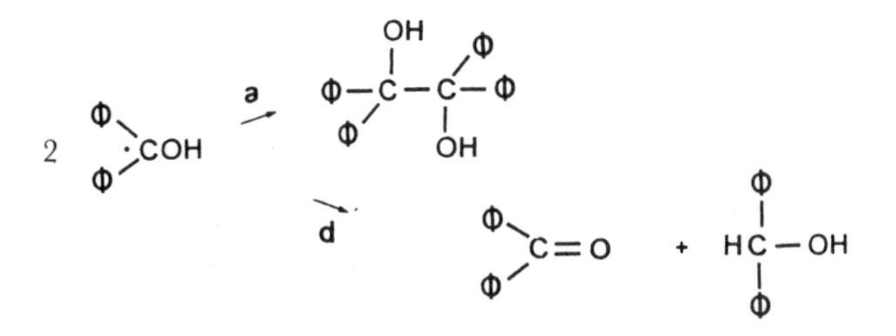

Figure 4.84 *Radical addition (a) and disproportionation (d) reaction of ketyl radicals*

Two important properties of free radicals are their *stability* and their *reactivity*. Many free radicals are intrinsically unstable, so that they isomerize to a new, more stable radical, or they dissociate into a smaller radical and a closed-shell species (Figure 4.85).

The reactivity of a free radical can be defined by the rate constants of its reactions with other molecules and other free radicals. In general this reactivity depends on the extent of localization of the unpaired electron. When it is highly localized on a single atom, as in the methyl radical for example, this site is highly reactive. However, delocalization of the unpaired electron over aromatic rings reduces the reactivity to the point where some free radicals can be kept virtually for ever in the form of stable, unreactive samples. The triphenylmethyl radical is the best known example.

4.9.1 Resonance Stabilization of Free Radicals

In the case of the triphenylmethyl radical shown in Figure 4.86, it is possible to write many different resonance structures; but in a small free radical such as the methyl radical there is only one possible structure. The reactivity of the radical decreases as the unpaired spin density at each site decreases, and the radical also becomes more stable because of the resonance energy. This resonance stabilization is zero for the phenyl radical, since the unpaired electron resides in an orbital which is orthogonal to the π system. By contrast, the methylphenyl radical has a resonance stabilization energy of some 10 kcal mol^{-1}, and the larger methylnaphthyl radical is stabilized by about 15 kcal mol^{-1}. These resonance stabilizations can have important consequences for the energy balance of photochemical reactions (see *e.g.* sections 4.4.2 and 4.4.4).

$$\begin{array}{c}\cdot CO \\ | \\ CH_3\end{array} \rightarrow CO + \cdot CH_3$$

Figure 4.85 *Example of a radical dissociation reaction yielding a smaller free radical and a closed-shell fragment*

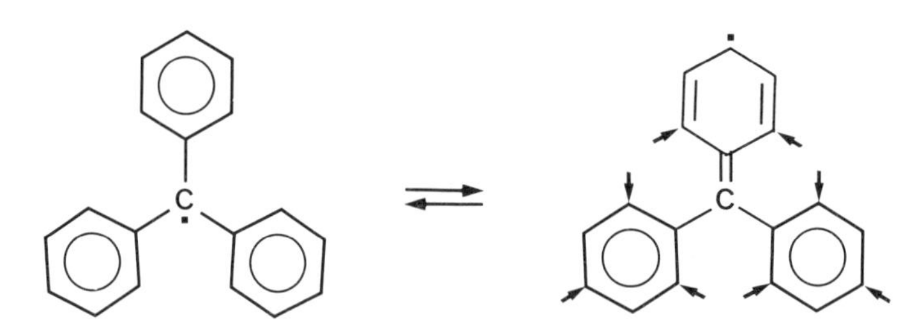

Figure 4.86 *Resonance structures of the triphenylmethyl free radical. The arrows on the right show alternative positions of the unpaired electron*

4.9.2 The Reactions of Free Radicals

A summary of the major chemical reactions of free radicals is given in Table 4.3. Broadly speaking these can be classified as unimolecular reactions of dissociations and isomerizations, and bimolecular reactions of additions, disproportionations, substitutions, *etc.* The complexity of many photochemical reactions stems in fact from these free radical reactions, for a single species formed in a simple primary process can lead to a variety of final products.

4.9.3 Magnetic Field Effects in Free Radical Reactions

The photochemical dissociation of a molecule AB often leads to the formation of a pair of radicals $A^{\bullet} + B^{\bullet}$, *e.g.* as in Figure 4.33. If the reaction takes place from the lowest triplet excited state of AB, the radicals will have parallel spins and cannot recombine unless a spin flip takes place to bring them to the singlet state of the geminate radical pair.

The triplet state of the geminate radical pair has three distinct levels ($\alpha\alpha$, $\beta\beta$, $\alpha\beta + \beta\alpha$) which are isoenergetic in the absence of any magnetic field. They interconvert very rapidly with the isoenergetic singlet state, so that the spin flip is fast and recombination is efficient.

In the presence of an external magnetic field the T^{+} and T^{-} levels become separated, while the T^{0} level remains unchanged. Isoenergetic crossing to the singlet state S can take place only from the level T^{0}, so that the recombination of the radicals is much slower. When these radicals are formed in a restricted space such as a micelle, their probability of escape prior to recombination is much greater (Figure 4.87).

This magnetic field effect can be used to control to some extent the molecular weight of some polymers formed by free radical initiators. Such initiators undergo homolytic dissociation to form a radical pair $A^{\bullet}B^{\bullet}$ which attacks the ends of growing polymer chains to add further monomeric units (section 6.2); polymerization ends when the free radicals $A^{\bullet}$ and $B^{\bullet}$ (or $A^{\bullet}$ and $A^{\bullet}$, $B^{\bullet}$ and $B^{\bullet}$) recombine. Assume there are only two radicals ($A^{\bullet}$ and $B^{\bullet}$) in a micelle; if one of these escapes from the micelle there can be no

Table 4.3 *Reactions of free radicals*

Unimolecular	(1) Isomerization $R^{\bullet} \rightarrow R'^{\bullet}$
	(2) Dissociation $R^{\bullet} \rightarrow R'^{\bullet\prime} + X$
Bimolecular	(1) Addition $R^{\bullet} + R'^{\bullet} \rightarrow RR'$
	(2) Disproportionation $R^{\bullet} + R'H^{\bullet} \rightarrow RH^{\bullet} + R'$
	(3) Transfer $R^{\bullet} + R'X \rightarrow RX + R'^{\bullet}$
	(4) Addition $R^{\bullet} + X{=}Y \rightarrow RXY$
	(5) Electron transfer $R^{\bullet} + R'^{\bullet +} \rightarrow R^{\bullet +} + R'^{\bullet}$
	$R^{\bullet} + R'^{\bullet -} \rightarrow R^{\bullet -} + R'^{\bullet}$
	$R^{\bullet +} + R'^{\bullet -} \rightarrow R^{\bullet -} + R'^{\bullet +}$
	(6) Substitution $R^{\bullet} + ArY \rightarrow RAr + Y^{\bullet}$
	(7) Displacement $R^{\bullet} + R'Y \rightarrow RR' + Y^{\bullet}$

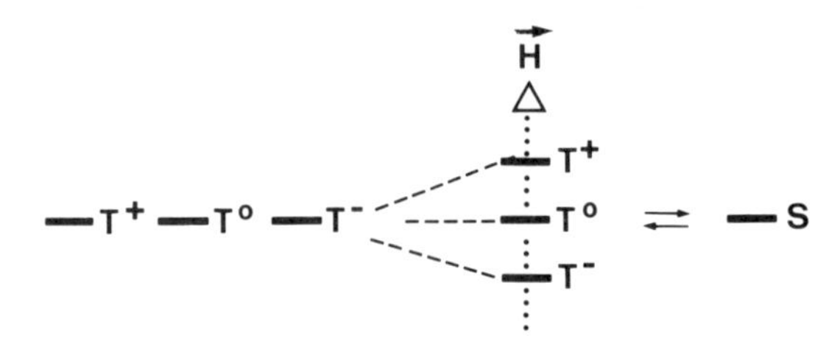

Figure 4.87 *Splitting of the sub-levels of a triplet state in a magnetic field; isoenergetic crossing to the singlet state S is then restricted to the sub-level* T^0

radical recombination, so that the polymer chain would grow indefinitely. Thus in the presence of a magnetic field polymers of higher molecular weight are obtained (Figure 4.88).

4.9.4 Excited States of Free Radicals

Free radicals and radical ions usually have excited states of low energy, and this can be understood from a simple orbital picture (Figure 4.89). In a closed-shell organic molecule the energy spacing between the HOMO and LUMO is quite large, but in the open-shell species the lowest excitations involve transitions of an electron between closely spaced orbitals. Many free radicals and radical ions of organic molecules absorb in the VIS or NIR, while the closed-shell precursors absorb only in the UV.

Luminescence is seldom observed from free radicals and radical ions because of the low energy of the lowest excited states of open-shell species, the benzophenone ketyl radical being however a noteworthy exception. There are few reports of actual photochemical reactions of free radicals, but the situation is different with biradicals such as carbenes. These have two unpaired electrons and can exist in singlet or triplet states and they take part in addition and insertion reactions (Figure 4.90).

4.9.5 Photochemical Reactions of Free Radicals

Although we think of free radicals mostly as highly reactive transient species which are intermediates in the chemical transformations of closed-shell molecules, there are some stable free radicals of which the triphenylmethyl

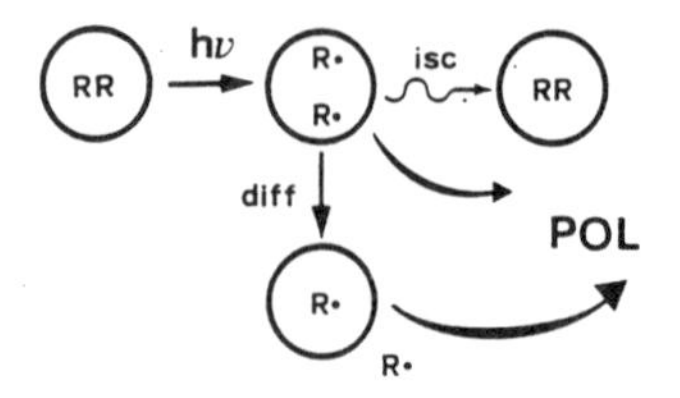

Figure 4.88 *Effect of a magnetic field on the escape-to-recombination ratio of free radicals in a micelle. The concentration of these radicals determines the extent of polymerization*

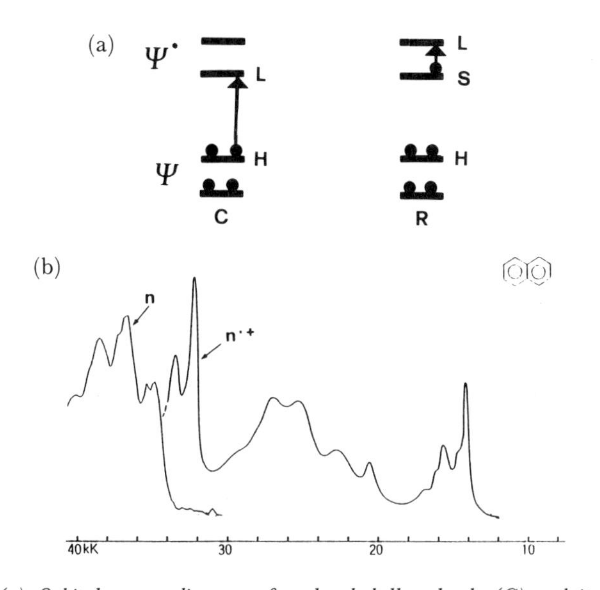

Figure 4.89 *(a) Orbital energy diagram of a closed-shell molecule (C) and its radical anion, illustrating the much smaller first transition in the latter (R). (b) Absorption spectra of naphthalene and of its radical cation; H = HOMO, L = LUMO, S = SOMO (singly occupied molecular orbital)*

$$-\overset{|}{\underset{|}{C}}-H \quad + \quad :CR_2 \xrightarrow{h\nu} -\overset{|}{\underset{|}{C}}-CR_2H$$

Figure 4.90 *Photoinduced insertion reaction of a closed-shell molecule into a carbene biradical*

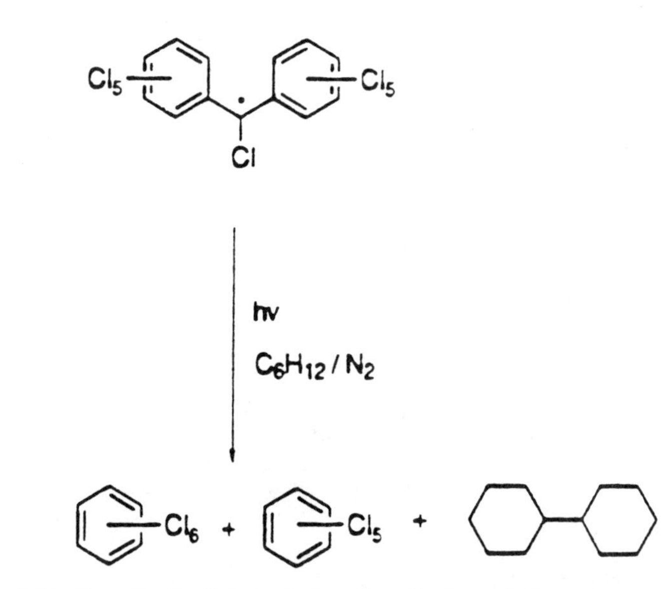

Figure 4.91 *Example of a photochemical reaction of a free radical*

radical is a well known example. Such stable free radicals can of course be irradiated just like any closed-shell species, and their photochemical reactions are similar: unimolecular dissociations and rearrangements. There are also some examples of bimolecular reactions, but the efficiency of these is limited by the short lifetime of the excited states (Figure 4.91).

CHAPTER 5

Light and Life

The presence of sunlight on the surface of the Earth has played an absolutely essential role in the development of life. It should be remembered in this respect that the 'quality' of this sunlight may have varied greatly in the course of time, since it is determined largely by the presence of light-absorbing species in the atmosphere.

Figure 5.1 shows the spectrum of sunlight in the upper atmosphere, having travelled through empty space, and as it reaches the surface of the Earth, after absorption and scattering by the atmosphere. The incident light above the atmosphere is referred to as 'AMO', which stands for 'air mass zero'. If it arrives perpendicular to the surface of the Earth, it will be defined as AM1 (air mass 1) at sea level, and for an angle α, the pathlength through the

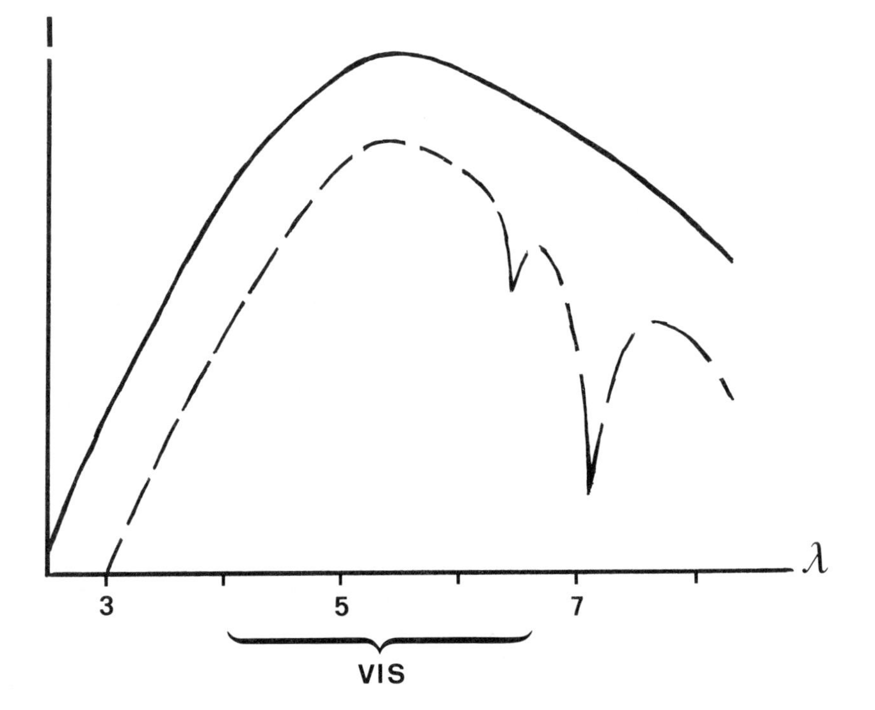

Figure 5.1 *Spectrum of sunlight before (solid line) and after (broken line) its passage through the atmosphere. Intensity I in relative units, wavelength λ in nm/100*

atmosphere is of course longer, so that it will become AM1.5, AM2, *etc.* (Figure 5.2).

It is a matter of conjecture as to whether sunlight was involved in the buildup of early organic molecules which eventually formed DNA, RNA and proteins. Photosynthetic bacteria may well have been the very first independent life forms, and from the time of the growth of green plants the atmosphere must have become gradually richer in oxygen, since it appears as a byproduct in the major process of photosynthesis. It is generally accepted that early life developed in the oceans, and it may be surmised that its eventual development on land was made possible by the formation of the protective ozone layer in the upper atmosphere.

Unfiltered sunlight (as it shines on the moon, for example) contains an important amount of far UV light which is potentially damaging to living matter. Nowadays these short wavelength components are largely cut out of the spectrum of sunlight by the UV-absorbing ozone layer (containing the molecule O_3). The composition of the Earth's atmosphere and consequently the spectral distribution of sunlight at the surface of our planet has very probably changed greatly in the course of time, from the very beginnings of life through its evolution over many millions of years.

5.1 ULTRAVIOLET LIGHT IN BIOLOGY AND MEDICINE

It is usual to divide UV radiation in three broad groups, according to its biological effects. These are:

UVA, which covers the wavelength region below 290 nm. This radiation is also described as 'abiotic', that is detrimental to life. It will be seen that this wavelength region corresponds to the direct absorption of UV light by nucleic acids and proteins. The low-pressure mercury arc lamp which emits at 254 nm is known as a bactericidal lamp.

UVB, between 290 and 330 nm, is the deeper region of 'biotic', that is biologically beneficial, UV radiation. Light of this wavelength region is used for the biosynthesis of vitamin D, and it produces pigmentation of the skin.

UVC, from 330 nm to the onset of the visible (VIS) region at around 390 nm. This is the near UV part of biotic UV radiations, absorbed essentially by the blood cells.

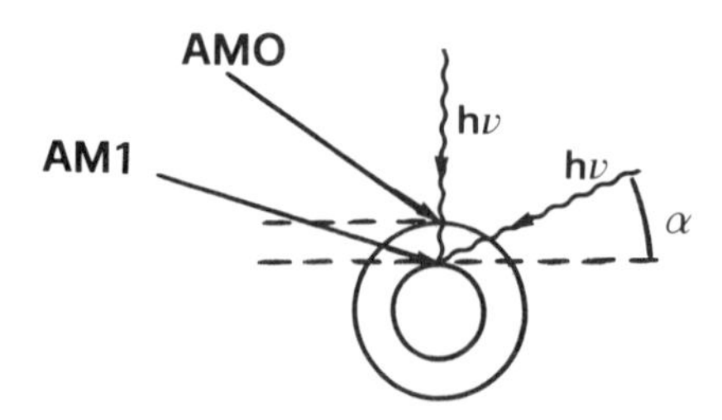

Figure 5.2 *Pictorial definitions of 'air mass' (AM) units of absorption by the atmosphere*

Natural light (sunlight) at the surface of the Earth contains very little UVA, as this is filtered out by the ozone layer in the upper atmosphere.

5.2 PHOTOSYNTHESIS

The overall chemical reaction of photosynthesis (of green plants for instance) is the combination of water and carbon dioxide to form saccharides, or polysaccharides

$$nCO_2 + nH_2O \xrightarrow{h\nu} C_nH_{2n}O_n + nO_2 \tag{5.1}$$

These products are the most important components of the living matter of plants, *e.g.* cellulose. In these simple terms photosynthesis is the photochemical oxidation of water, and reduction of carbon dioxide, by means of a photoactivated catalyst which in green plants is the molecule *chlorophyll*.

There are in nature many different 'chlorophylls', all based on the same general structure shown in Figure 5.3. These molecules absorb strongly in the wavelength regions 600 and 400 nm, hence their green colour. Their absorption spectrum makes them efficient absorbers for much of the spectrum of sunlight, but the excited state energy of the chlorophyll molecule is then relatively low (around 2 eV for the singlet state S_1).

In green plants the process of photosynthesis is localized in small regions of the cells, these regions corresponding to 'photosynthetic units'. One such unit includes many molecules of chlorophyll (up to several thousands) as well as other molecules which absorb in other regions of the VIS spectrum and which are called 'auxiliary pigments'. These auxiliary pigments absorb the sunlight which the chlorophylls would let through, and the electronic excitation energy is then handed over to the chlorophylls by processes of energy transfer.

If a scale of electrochemical redox potentials is considered, it is obvious that water is not an easily oxidizable species; nor is carbon dioxide easily reduced. The energy requirement of photosynthesis in green plants can be met only by the cooperation of *two* excited chlorophyll molecules. The first one gives a part of its excitation energy to the second one (Figure 5.4).

This coupling of two 'photosystems' is an important feature of the photosynthesis of green plants. Water is oxidized in photosystem 2, carbon dioxide is reduced in photosystem 1, but neither of these photosystems could have the energy needed for both processes. Their cooperation is essential, and the entire process stops if one or the other photosystem is unable to function.

The primary process of photosynthesis (in both photosystems) is an electron transfer reaction from the electronically excited chlorophyll molecule to an electron acceptor, which is in most cases a quinone. This primary electron acceptor can then hand over its extra electron to other, lower energy, acceptors in 'electron transport chains' which can be used to build up other molecules needed by the organism (in particular adenosine triphosphate; ATP). The complete process of photosynthesis is therefore much

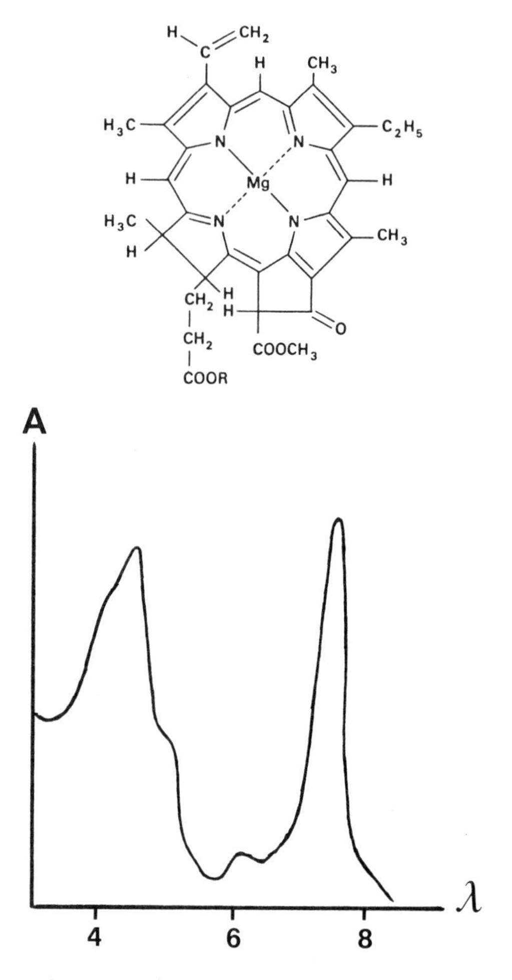

Figure 5.3 *The chemical structure of a chlorophyll and its absorption spectrum*

more complex than the mere oxidation of water and reduction of carbon dioxide to produce polysaccharides.

The presence of two linked photosystems is clearly shown by a comparison of the absorption spectrum of the plant and the 'action spectrum' for oxygen evolution. This action spectrum is the relative quantum yield for O_2 production as a function of wavelength. It follows the absorption spectrum (as expected) up to around 680 nm, then it drops sharply at longer wavelengths. However, additional illumination with long-wavelength light (around 700 nm) restores the expected quantum yield. The conclusion is that the two photosystems have slightly different absorption spectra, such that the absorption peaks are at 683 nm for photosystem 1 and 672 nm for photosystem 2 (Figure 5.5).

In green plants there are two different types of chlorophyll which are known as chlorophyll *a* and *b*. The former is the active substance, since it is

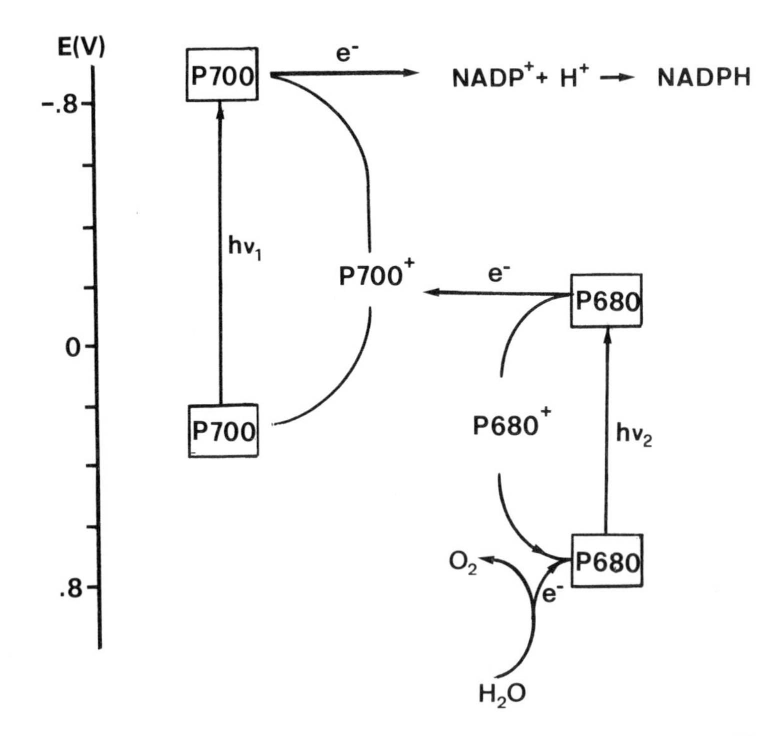

Figure 5.4 *Energy diagram of the two linked photosynthetic systems of green plants. The vertical scale is the energy (in eV) of an electron at the various stages of electron transport. P700 (pigment 700) and P680 (pigment 680) are chlorophylls with absorption maxima at 700 nm and 680 nm, respectively*

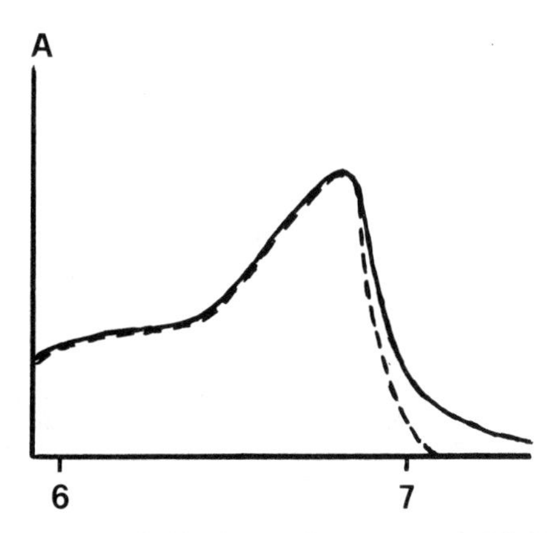

Figure 5.5 *Absorption spectrum of the photosynthetic system (solid line) and action spectrum (broken line) for oxygen evolution. Wavelength λ in nm/100*

the final energy trap for the excitation energy which is harvested by chlorophyll *b* and other auxiliary pigments. The slight difference in the absorption spectra of the chlorophylls in the two photosystems results from their different environments.

It is established that the primary electron acceptor in photosystem 1 is the molecule ferredoxin, while in photosystem 2 it is a quinone. The identity of the primary electron donor in photosystem 2 is still unknown; the oxidation of water must take place by electron transfer to this primary donor, X.

Chlorophyll itself shows a short-lived red fluorescence in solution. The green plants show a delayed fluorescence of approximately the same spectrum under specific conditions in which the electron transport chain is blocked. This delayed fluorescence results from the recombination of the charges (a process well known in electroluminescence), and its kinetics are complex and the decay quite long (several seconds).

In the overall scheme of the photosynthesis of green plants the electron transport cycle starts with the excitation of chlorophyll *a* in photosystem 2. The excited electron then follows a downward electron acceptor chain which eventually reaches the chlorophyll *a* of photosystem 1 (P700) in which it can fill the positive hole left by electronic excitation. The energy released in the electron transport chain which links photosystems 2 and 1 is used for other biochemical processes which are thereby related to photosynthesis. One of these is the process of 'photophosphorylation' which is the production of molecules with phosphate chains used as energy transfer agents in many biochemical reactions.

Photosynthetic bacteria provide the simplest examples of photosynthesis, probably dating back to earlier times in evolution. The process in photosynthetic bacteria is fundamentally different from that in the higher plants in that one single photosystem is used instead of the two coupled photosystems of the plants. Another important difference is that in most cases the photosynthetic pigment is a *bacteriochlorophyll*, a molecule which absorbs at much longer wavelengths than the chlorophyll of green plants (in the region 800 to 1000 nm); as a result the usable excitation energy is lower in the photosynthetic bacteria.

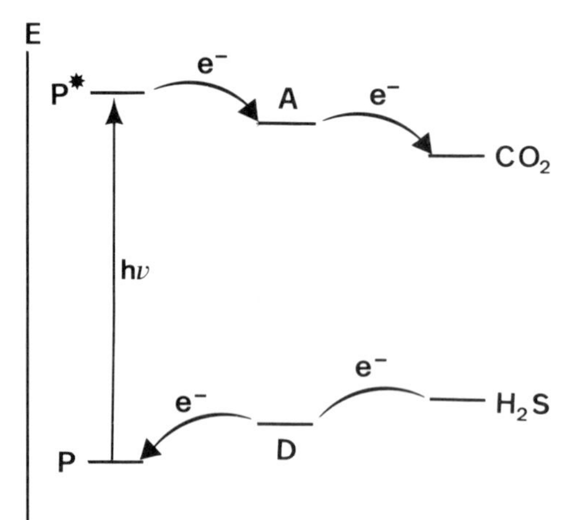

Figure 5.6 *Electron transport chain in photosynthetic bacteria. P = pigment; A = electron acceptor; D = electron donor*

These bacteria cannot in general oxidize water and must live on more readily oxidizable substrates such as hydrogen sulfide. The reaction centre for photosynthesis is a vesicle of some 600 Å diameter, called the 'chromatophore'. This vesicle contains a protein of molecular weight around 70 kDa, four molecules of bacteriochlorophyll and two molecules of bacteriopheophytin (replacing the central Mg^{2+} atom by two H^{+} atoms), an atom Fe^{2+} in the form of ferrocytochrome, plus two quinones as electron acceptors, one of which may also be associated with an Fe^{2+}. Two of the bacteriochlorophylls form a dimer which acts as the energy trap (this is similar to excimer formation). A molecule of bacteriopheophytin acts as the primary electron acceptor, then the electron is handed over in turn to the two quinones while the positive hole migrates to the ferrocytochrome, as shown in Figure 5.7. The detailed description of this simple photosynthetic system by means of X-ray diffraction has been a landmark in this field in recent years.

This electron transfer chain, shown in Figure 5.7, may well provide an important clue to the way in which nature has managed to achieve an

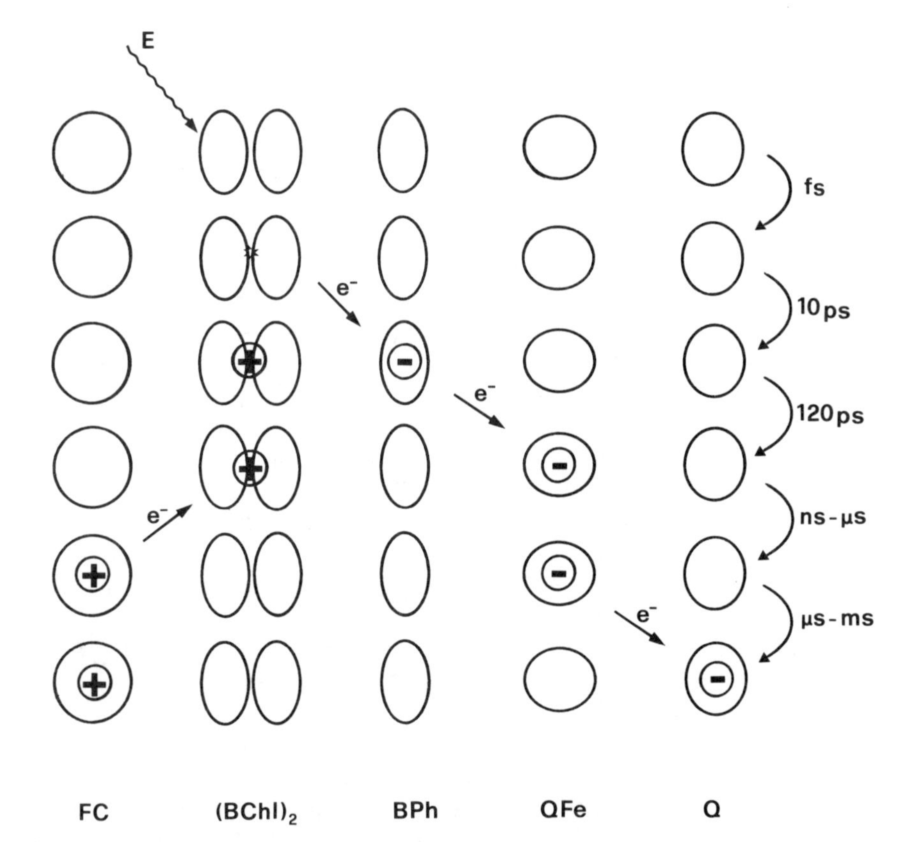

Figure 5.7 *Electron transfer processes in the first stages of photosynthesis. The energy of light E absorbed by the antenna chlorophylls is transferred to the special pair* $(BChl)_2$. *FC is the ferrocytochrome, BPh the bacteriopheophytin and QFe, Q are quinones*

efficient separation of the positive and negative charges formed in the primary photochemical process. At the present time much effort is being devoted to this very problem; for it is essential to avoid the rapid recombination of the excited electron with the positive hole if any useful oxidation and reduction processes are to be obtained.

In the excited state no actual charge separation has yet taken place, although the two half-filled orbitals are already potential oxidation and reduction centres. Chemical reactions will take place so long as the excited electron does not recombine with the positive hole left in the starting orbital, and this charge recombination can take several forms. In the first place the electron could go back to the radical cation it has just left, a process which is simply the 'reverse reaction' of charge separation. This can be avoided only if the energy level of the electron decreases substantially on going to the second electron acceptor; then it will decrease still further on going to the third acceptor so that *thermal* repopulation of the previous electron acceptor becomes very slow.

The second process that can reduce the yield of useful ions is charge recombination between any reduced electron acceptor and the original positive hole left in the initially excited molecule (shown as ⊕ and ⊖ in Figure 5.8). This can be prevented only by the *spatial* separation of the positive and negative charges. The reduction potentials of the successive electron acceptors decrease as their spatial separation increases, so that both the reverse electron transfer and the charge recombination become very slow.

It is obvious that such a situation can be obtained only in a highly organized system; the molecules must be held fairly rigidly in their geometrical situation, and certainly no diffusion can be allowed. The energy level of the excited electron must follow the spacing of the electron acceptors, but the energy difference between the various acceptors must not be so great as to imply a drastic loss of efficiency in the conversion of the primary excitation ($h\nu$) into the usable redox potentials.

How the process of evolution, spread out over many millions of years, has managed to produce such an organized electron transport system is a tale which is yet to be told. Of more immediate concern is the possibility of making artificially some chemical systems which might have similar properties. The choice of the various electron acceptor molecules is one thing; to place them, and to keep them in the required geometrical arrangement is quite another.

This last problem is perhaps not strictly within the realm of photochemistry. It is however so important potentially that it cannot be overlooked on grounds of arbitrary separations between different branches of scientific research (perhaps the expression of 'interdisciplinary' approach would best describe it). A few pages will therefore be devoted to the science and technology of artificially organized molecular systems such as monomolecular and multimolecular layers, micelles and spatially restricted environments like zeolites; and since we reach here another of the frontiers of photochemistry, section 8.4 in the final chapter is devoted to these systems.

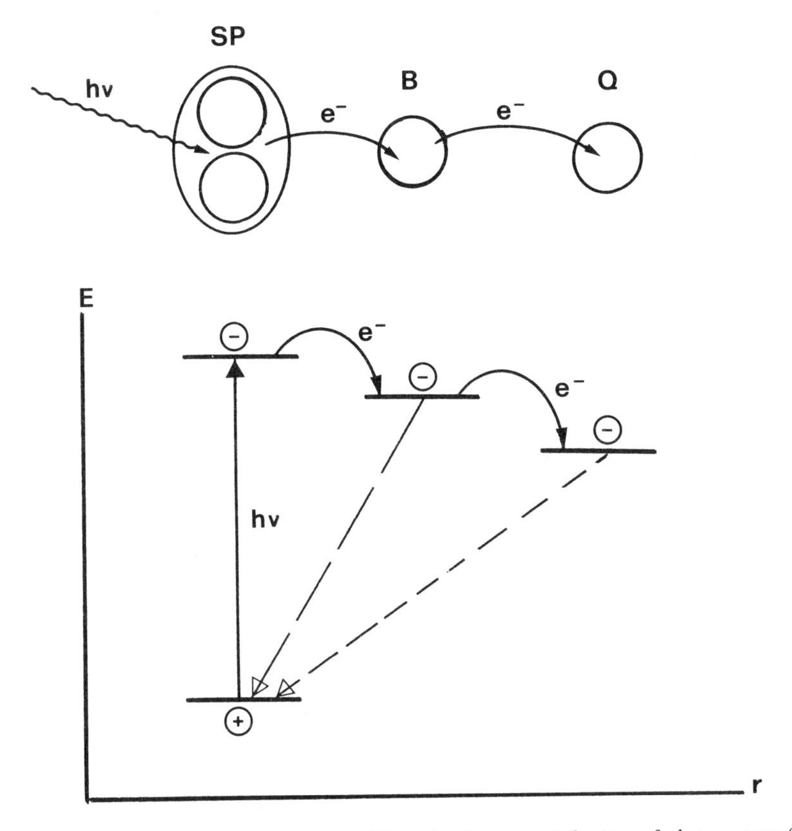

Figure 5.8 *Increasing spatial separation of ions in the sequential steps of electron transfer in photosynthesis. SP = special pair*

5.3 VISION AND PHOTOTAXIS

In human terms 'vision' is a very complex process of information about the surroundings of a living organism; it is information conveyed by light. The 'information' itself can take various forms, from the simplest, restricted to a general intensity of light, to the most complex, which can deal with shapes, colours and sizes of surrounding objects. The overall process of 'vision' involves therefore the interpretation of many nerve impulses which originate from a fundamental photochemical reaction. This fundamental reaction only can be discussed in terms of photochemistry. The processes of formation and eventual decoding of the nerve impulses is well beyond the realm of simple physical and chemical processes.

Before considering the photochemistry of vision in higher animals, it will be interesting to look briefly at the process known as 'phototaxis' which is sometimes considered to be a very early form of 'vision'. It has been observed that some photosynthetic bacteria are able to swim selectively towards illuminated areas, and to avoid dark places; this light-controlled motion has been named 'phototaxis' (Figure 5.9). This typc of behaviour is

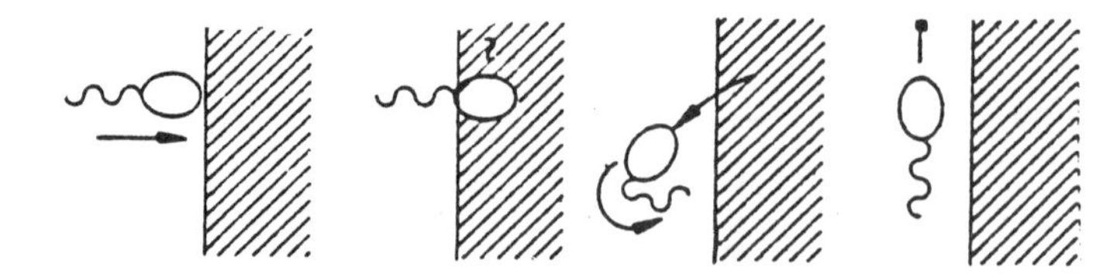

Figure 5.9 *Phototaxis: some simple organisms can move only in the presence of light*

observed with the photosynthetic bacterium '*Chromatium*'. Although the detailed mechanism of this phototaxis is not known, it may be that light simply provides the energy for motion through the production of ATP in the photosynthetic process. When the bacterium enters a dark area the production of ATP would stop, the motion of the flagella then coming to a halt. As soon as the organism finds a lighted area again, its motion is resumed.

In higher life forms various types of eye have evolved in the course of time, the simplest one being that of insects; these are arrays of elementary 'eyes' (forming the compound eye) each of which respond to light of a particular direction. Insects therefore do not perceive *shapes* in the human sense, but rather distributions of light intensity in various directions (Figure 5.10).

The vision of the shapes of objects requires the formation of an image on a photosensitive surface. In higher animals the optics of the eye make use of a lens (rather like a photographic camera), but there are some lower species which use a simple pinhole aperture (as was used, incidentally, in the early days of photography).

In the human eye the photosensitive surface is the *retina*, and this consists essentially of a carpet of photosensitive cells called the *rods*. Each rod is an elongated cell which includes at one end the photosensitive microstructures (thylakoids), at the other end the synapsis which leads to the nerve fibre, and in between a ciliated structure, mitochondria and the nucleus (Figure 5.11). The thylakoids are arranged in parallel arrays so that the probability of light absorption is greatest for light which comes along the long axis of the rod. It may be surmised that a 'coincidence detection' method is used by the rod, such that a nerve impulse requires the near simultaneous absorption of several photons in different thylakoids. The visual pigment contained in the thylakoids is called *rhodopsin*; it is a combination of a chromophoric molecule of *retinal* with a protein called *opsin*. The absorption spectrum of rhodopsin

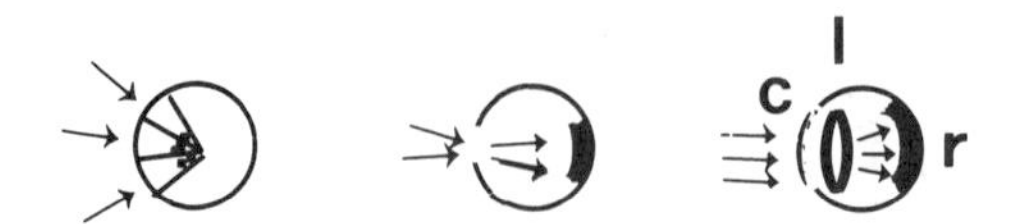

Figure 5.10 *Three types of eye in evolution (from left to right): the compound eye of insects, the pinhole eye and the eye of higher animals. The photosensitive surface is shown as a thick black line; c is the cornea, l the lens and r the retina*

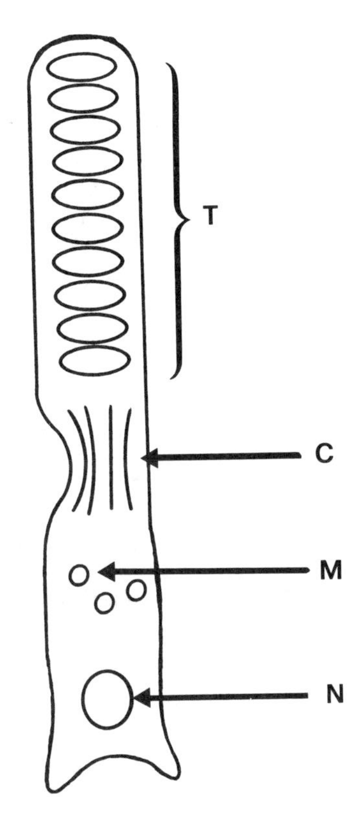

Figure 5.11 *Main components of a rod cell: thylakoids (T), ciliate structure (C), mitochondria (M) and nucleus (N)*

follows of course the spectrum of VIS light, but in the UV the eye has no sensitivity only because of the absorption by the lens (some animals can in fact see near UV light) (Figures 5.12 and 5.13).

The molecule of retinal is a conjugated hydrocarbon chain with a terminal carbonyl–hydrogen group; its reduced form is an alcohol called vitamin A. Such conjugated chains can exist in several stereoisomeric forms, each bond giving rise on 180° twisting to a *cis* or *trans* isomer. The important forms of retinal in the process of vision are the 11-*cis* isomer which is the biologically stable form in the dark, and the all-*trans* isomer which is formed by photochemical *cis–trans* isomerization. The 11-*cis* retinal is bound to opsin (in the dark), but after photochemical isomerization the all-*trans* retinal cannot fit into the protein and is expelled in the form of free retinal. The free retinal will later be reduced to vitamin A, re-isomerized in a dark, enzyme-catalysed reaction to its 11-*cis* form, and then returned to the rod to be re-oxidized to 11-*cis* retinal which recombines with the free opsin (Figure 5.14).

The primary photochemical process of vision is therefore the *cis-trans* isomerization of retinal in rhodopsin. The free protein opsin then leads to the production of the nerve impulse through a secondary biochemical process

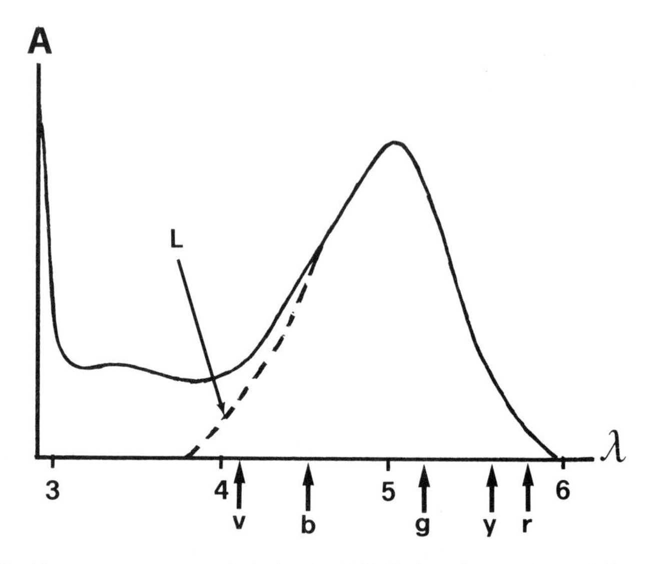

Figure 5.12 *Absorption spectrum of rhodopsin (full line) and sensitivity of the eye. The drop shown by the broken line is the absorption of the lens (L). Wavelength λ in nm/100; approximate colours shown as v (violet), b (blue), g (green), y (yellow) and r (red)*

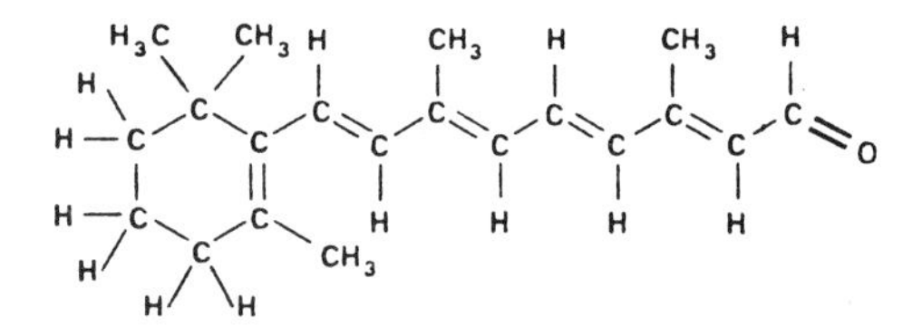

Figure 5.13 *Structure of retinal*

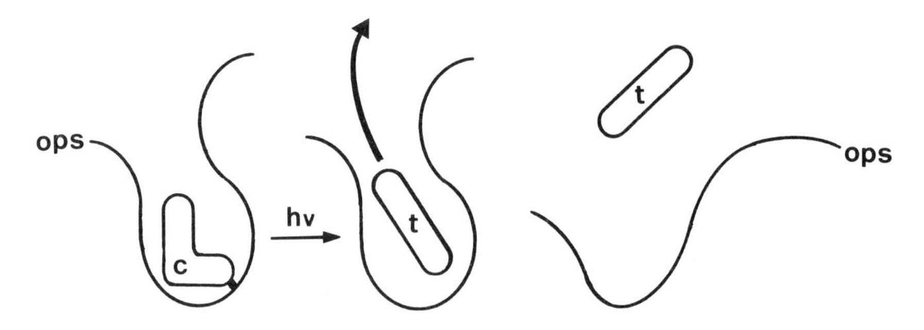

Figure 5.14 *Hypothetical events in the process of vision. Following the structural change of retinal, its link to opsin (ops) is broken and the protein separates from the chromophore (C)*

which is not yet fully understood. A nerve impulse results from the depolarization of the cell membrane which is normally kept positively polarized inside by a difference of concentration of ions.

The effect of absorption of a quantum of light by rhodopsin is to cause an influx of some 10^5 Na^+ ions. It can be surmised that the rhodopsin molecule normally blocks ion transport across the cell membrane, but its change of shape as free opsin leaves a 'pore' which allows ion conduction.

Colour vision is quite separate from the 'black and white' vision considered so far. It requires specialized cells called *cones*, which contain slightly different visual pigments which have absorption spectra with wavelength maxima near 450, 540 and 580 nm, respectively (Figure 5.15). These pigments may not be fundamentally different from the rhodopsin of the rods, the spectral changes coming probably from different environments. Any one wavelength (colour) of light then gives a specific relative stimulation of the cells which contain these pigments, and this relative stimulation is somehow decoded by the brain in the form of colour.

The eye of higher animals is remarkably sensitive to low light intensities, but saturation can occur relatively easily at high light intensities and the recovery is then slow. After the absorption of light rhodopsin (which has a red colour) is bleached when the retinal leaves the opsin, and recovery goes through a sequence of enzyme-catalysed reactions which takes several seconds.

5.4 PHOTORESPONSE MECHANISMS IN PLANTS AND ANIMALS

Although photosynthesis and vision are the most obvious uses of light by living organisms from simple bacteria to the highest forms of life, there are

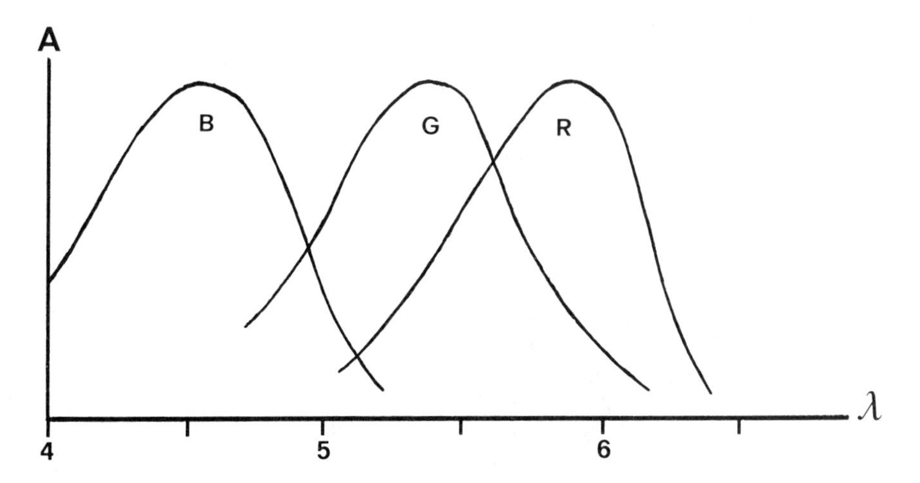

Figure 5.15 *Colour vision relies on three pigments of different absorption spectra, B, G and R. Wavelength λ in nm/100*

many other biological processes controlled by light. Few of these can be described at this time in terms of well established photochemical reactions, but the influence of light (specifically of the intensity of sunlight) is apparent in a variety of ways.

(1) The growth of plants is regulated by light (among other factors).

(2) Many plants are able to follow the direction of sunlight; this is the process of 'phototropism'.

(3) Daily cyclic processes (such as the waking and sleeping times) are probably controlled in part by the variations of sunlight. Although the organism has an 'internal clock' which regulates these daily cycles, it is observed that the timing of this clock goes slowly wrong when the organism is kept in darkness, and it may even assume (in time) a cycle imposed by the variations of artificial lighting.

(4) It is a matter of conjecture as to what extent the much longer cycles of biological processes, over 1 year in particular, may be determined by the seasonal variations of the intensity of sunlight. In many green plants chlorophyll is degraded slowly in the autumn, and in its place quite different pigments are formed (anthocyanins).

It is not surprising that the detailed mechanism of such processes is very difficult to establish. It is known however that the molecule *phytochrome* is involved in many cases; this molecule can exist in two different forms which are called 'Pr' (red phytochrome) and 'Pfr' (far-red phytochrome) which are interconvertible by the action of light

$$\mathrm{Pr} \underset{\Delta}{\overset{h\nu}{\rightleftarrows}} \mathrm{Pfr} \qquad (5.2)$$

Depending on the wavelength distribution of light, a photostationary state can be established between Pr and Pfr.

Phytochrome is a 'chromoprotein', that is a molecule which consists of a relatively small chromophore unit linked to a large protein. The structure of the chromophore is now well established and the action of light leads to an isomerization; the association of the isomer Pfr with the protein chain is different from that of the Pr isomer (Figure 5.16).

The Pfr isomer formed photochemically is the biologically active molecule, its mode(s) of action being complex dark reactions which are not fully understood at the present time.

There is some structural similarity between the phytochrome chromophore and that of rhodopsin, and in both cases the light-induced process is a simple isomerization. The mode of action at the molecular level may well be similar, the phytochrome acting as a 'plug' in the cell membrane. The action of light is then essentially to remove (or to open) the plug, thus allowing diffusion of biologically active molecules through the membrane (Figure 5.17).

5.5 PHOTOCHEMICAL DAMAGE IN LIVING SYSTEMS

So far some beneficial actions of light on living systems have been considered (photosynthesis, vision, photoresponse mechanisms) but it is important to

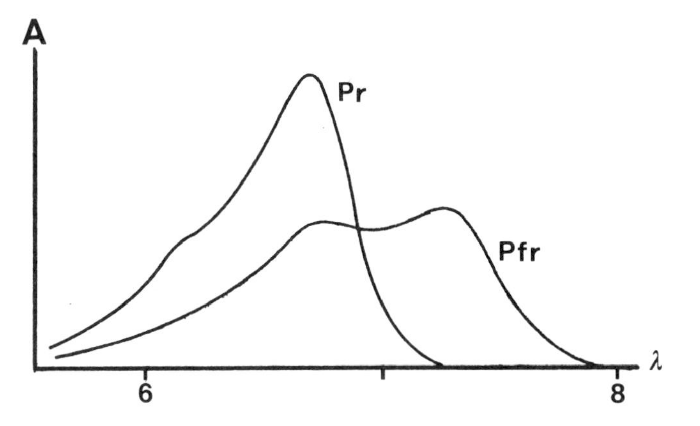

Figure 5.16 *Absorption spectra of 'red' phytochrome (Pr) and 'far red' phytochrome (Pfr). Wavelength λ in nm/100*

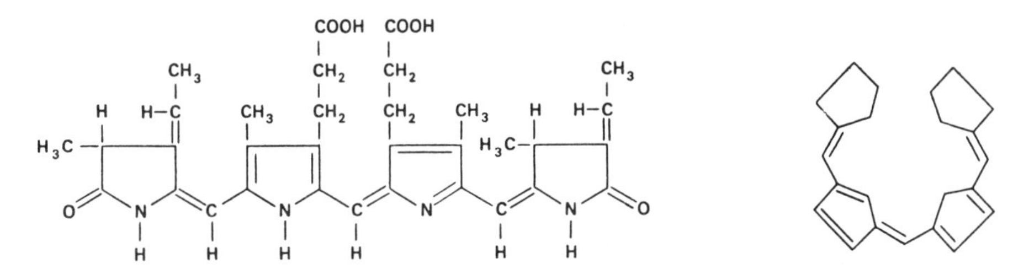

Figure 5.17 *Chemical structure of phytochrome and hypothetical folded conformation*

realize that photochemical reactions can have damaging, indeed lethal, results on living organisms. In this respect we shall consider specifically two of the fundamental molecules which make up living matter, namely the *nucleic acids* which are the molecular basis of genetic information (of heredity), and the *proteins* which are the basic constituents of animal matter and its most important chemical reactants or catalysts (enzymes).

5.5.1 Photochemistry of Nucleic Acids

A nucleic acid (so called because it is found in the nucleus of the cell) is a large 'heteropolymer' molecule built up of four basic constituent groups. Each of these 'repeating units' consists of a *sugar*, linked to a *phosphate* on one side and to a *base* on another, there being four different bases which are abbreviated as U (uracil), A (adenine), G (guanine), C (cytosine) and T (thymine) (Figure 5.18). The phosphates and sugars form a backbone for the long chain which carries the bases in a sequence and is the carrier of the genetic information of the cell. The important feature of the complete molecule of DNA is that it is a 'dimer' of two such chains or 'strands' which form a double helix held together by hydrogen bonds between specific pairs of bases (Figure 5.19).

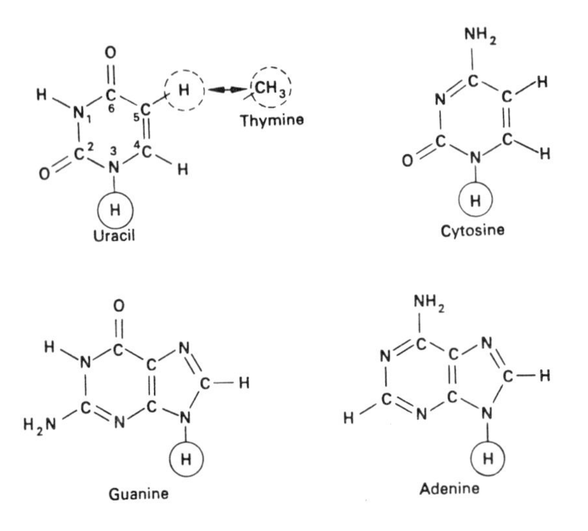

Figure 5.18 *Chemical structures of nucleic acid bases: uracil (U), cytosine (C), thymine (T), guanine (G) and adenine (A)*

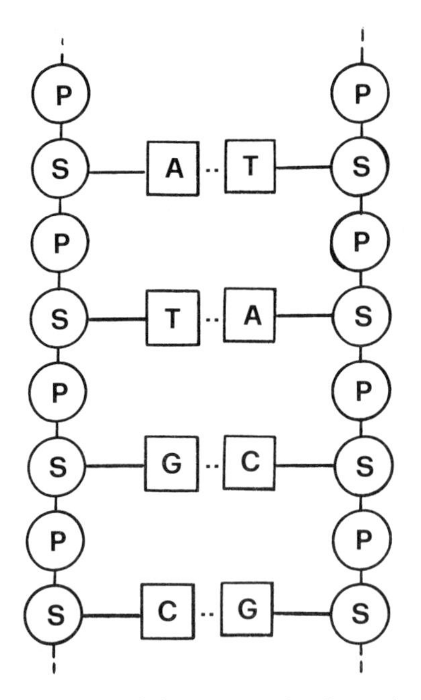

Figure 5.19 *Schematic representation of the two strands of a nucleic acid. P = phosphate and S = sugar; bases are as in Figure 5.18. The two strands of the double helix are held together through specific hydrogen bonds (base pairing)*

The sugar and phosphate groups are essentially saturated molecules which would absorb only in the far UV. The bases are the chromophores of the DNA molecule, with an absorption maximum around 260 nm.

The free base molecules can undergo various photochemical reactions, in particular cytosine will add to water in a process which is however reversible in the dark (Figure 5.20).

The most important photochemical reaction of the bases of DNA is the dimerization of thymine, because this can take place not only between free molecules but also within the DNA chain when two such bases happen to be close together. Since the bases in a DNA strand form a continuous array of relatively close-packed molecules, energy transfer is efficient throughout the chain and any excitation energy will eventually find a 'trap' such as a thymine dimer (or excimer) where the dimerization reaction can take place.

The dimerization reaction of thymine is 'photoreversible'. At the shorter wavelengths where the dimer absorbs there is a dissociation of the dimer to reform the two monomers. In this way photostationary states can be obtained according to the irradiation wavelength (Figure 5.21).

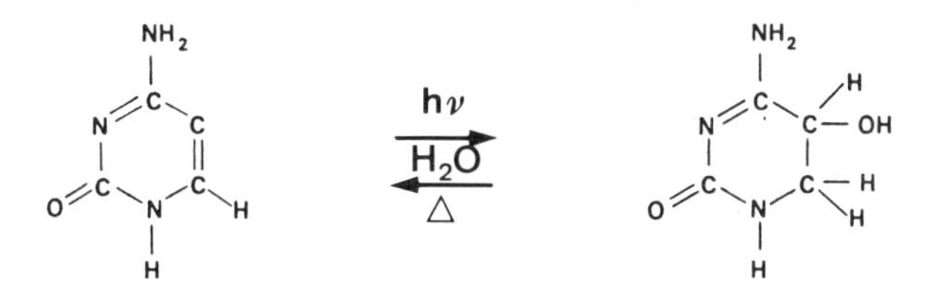

Figure 5.20 *Reversible photochemical hydration of cytosine*

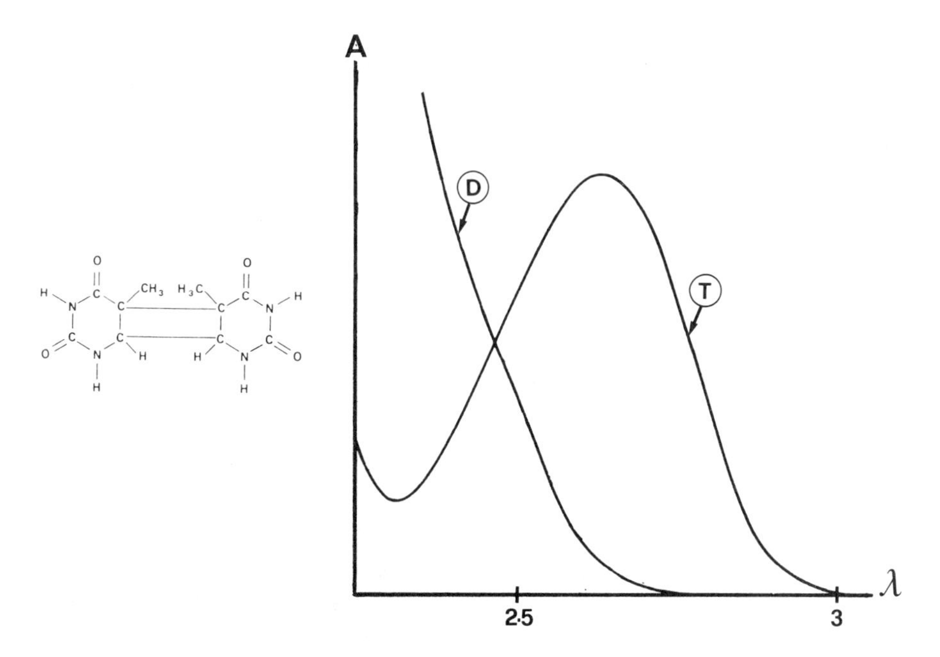

Figure 5.21 *Chemical structure of the thymine dimer (left) and absorption spectra (right) of thymine (T) and its dimer (D). Wavelength λ in nm/100*

The reactive excited state in the photodimerization of thymine is probably the triplet state T_1; the dimerization process can be quenched by triplet acceptor molecules such as ketones and quinones.

Since the photodissociation of the thymine dimers requires light of very short wavelengths (below 250 nm) it is not an important process in nature. The nucleic acids themselves absorb below 300 nm, with a peak of absorption near 260 nm which corresponds to the individual chromophore bases. These short wavelengths do not occur in sunlight (at the surface of the Earth) as they are filtered out by the ozone layer in the upper atmosphere. However, low pressure mercury arc lamps essentially produce light at 254 nm, close to the absorption maximum of DNA, and these lamps are termed 'bactericidal' since they are very effective in destroying microorganisms which are irradiated with this light. It should be borne in mind that bacteria are not the only living organisms which are sensitive to these short wavelength rays. Skin damage, and in particular irritation or actual damage to the eye, can also occur in human beings exposed to this type of radiation.

5.5.2 Photochemistry of Proteins

A protein is a long chain heteropolymer of amino acids. The absorption maximum of proteins is found at around 285 nm, and of the naturally occurring amino acids two are particularly important in photochemistry: the aromatic molecule *tryptophan* and the sulfur derivative *cysteine* (Figure 5.22).

Tryptophan itself can undergo several photochemical reactions such as deamination (the NH_2 group is then replaced by H or OH) and decarboxylation (loss of COOH, replaced by H or OH), ring opening, *etc*. The role of singlet (excited) oxygen is however equally important, and this can occur even with relatively long wavelength light (*e.g.* in the VIS region). The molecule $^1O_2{}^*$ can be formed by sensitization by many chemical species such as quinones, flavines, nucleotides, *etc*. It will then attack proteins and other essential constituents of the cell. It is noteworthy that the excitation energy of oxygen (to form $^1O_2{}^*$) is so low that even the chlorophyll of green plants can act as a sensitizer. $^1O_2{}^*$ is also very difficult to quench because of this low excitation energy, some carotenoids being the only readily available quenchers in living systems. Such long-chain carotenoids (with more than nine double bonds) are indeed often found in association with the chlorophyll

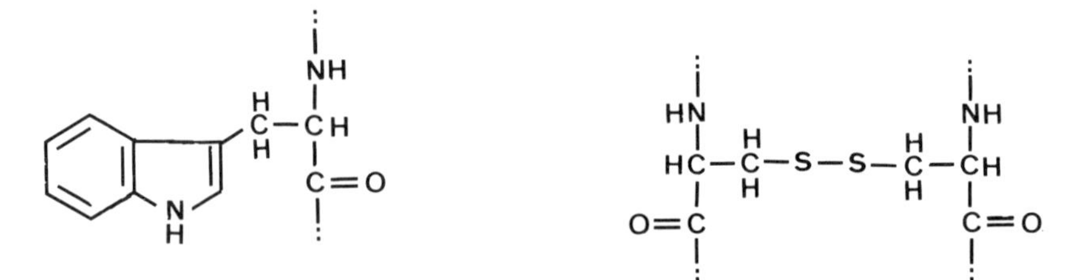

Figure 5.22 *Chemical structures of tryptophan and cysteine as subunits of a protein chain*

in photosynthetic units, and it may be surmised that they have a protecting role against the action of singlet oxygen besides their role of auxiliary pigments.

The amino acid cysteine readily undergoes a photodissociation through the weak disulfide linkage, with a high quantum yield (of the order of 0.1). The sensitivity of proteins to UV irradiation follows roughly their cysteine concentration. This molecule is clearly the main target in the UV damage of proteins (though not in the VIS damage which is the attack by $^1O_2^*$). Dissociation of the S–S bond leads to the formation of free radicals which then react further in secondary reactions.

5.6 PHOTOCHEMISTRY IN BIOSYNTHESIS

The action of light through photochemical reactions is used in living organisms for synthetic processes other than photosynthesis itself. The full range of these reactions may yet have to be established. We shall consider here only the synthesis of vitamin D.

Vitamin D deficiency leads to defects in the structure of bones, a disease known as 'rickets', common in miners who lived and worked underground during the daytime. They did not have the normal share of sunlight, but it took some time to realize that the lack of vitamin D was related to a photochemical reaction. This reaction is shown in Figure 5.23 in a simplified form, in order to point out the important electrocyclic process.

5.7 PHOTOMEDICINE

Under this heading we shall consider the genuine therapeutic actions of light through photophysical and photochemical processes. The former rely on the

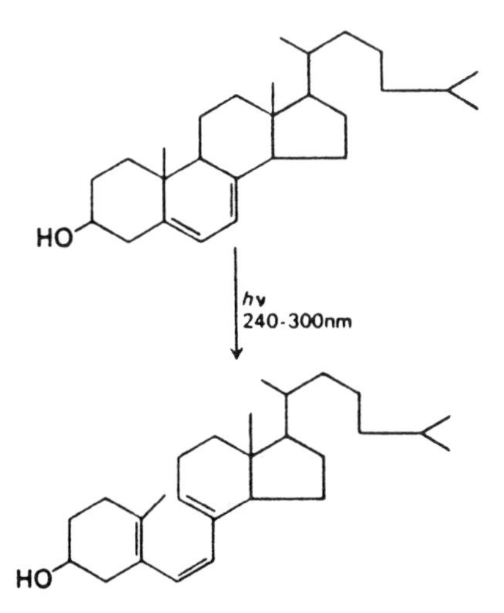

Figure 5.23 *Chemical structures of 7-dehydrocholesterol and vitamin D. The photochemical reaction is an electrocyclic ring opening*

heating effects which follow the absorption of light. Such effects could also be obtained in principle by contact with a hot body, but light absorption has the great advantages of spatial and spectral selectivity.

5.7.1 The Repair of Detached Retinas by Laser Welding

The retina is the light-sensitive layer which forms a sort of carpet on the inside of the eye. It sometimes happens—especially in old age—that this carpet becomes detached from the eyeball and floats in the surrounding liquid. In the areas of detached retina vision is of course lost.

The modern technique of repair of the detached retina uses the properties of pulsed laser light. This light is highly monochromatic, so that the lens of the eye can focus it to a very sharp spot of the retina, near to the fault line where detachment has occurred. The heat generated by this light then acts like spot welding, and with a number of these 'molten' contact points the retina is cemented back on the inner surface of the eye. Although vision is lost in the welded spots because the intense heating has destroyed much of the organic matter, these spots are very small and hardly impair the visual process over the much larger surface of the retina. In those unirradiated parts which are again forced into contact vision is restored.

Before the development of this laser welding technique, the repair of a detached retina required major surgery; the photophysical process is now the technique of choice.

5.7.2 Photochemical Effects of VIS and UV Light

It has long been known empirically that sunlight is essential for a healthy life, quite apart from its role in vision. The essential role of a photochemical reaction has been mentioned already with respect to the biosynthesis of vitamin D (section 5.6). The lack of sunlight can indeed lead to serious diseases, such as have been known in the mining communities when workers had to spend very long times in the darkness of underground tunnels. In this case light is certainly a life supporting agent, but not exactly a 'therapeutic' agent like a drug in chemical therapy. There are also some genuine therapeutic uses of light, of which two are singled out here, and a potential application which is still at an experimental stage will be described briefly.

5.7.2.1 The Treatment of the Jaundice of New-born Infants. This disease results from the accumulation of bilirubin in the blood, when the liver is not yet able to function with its full efficiency. It takes its name from the yellowish colour of the skin, due to the relatively high concentration of bilirubin which absorbs in the violet and blue regions of VIS light.

When the red blood cells die after their average lifetime of some 120 days, their haemoglobin is degraded to bilirubin which is then filtered by the liver. In new-born infants this process cannot take place often for several days, while the liver becomes fully functional. The accumulation of bilirubin in the blood then causes a jaundice which can, if untreated, lead to serious consequences. The molecule of bilirubin (Figure 5.24) is highly insoluble in water because of its internal hydrogen bonds. When it is exposed to light (blue, violet or near UV) it undergoes a *cis–trans* isomerization which breaks these hydrogen bonds. Now the new molecules (photobilirubins) can hydrogen bond to water and so become soluble. Unlike bilirubin itself, the water-soluble photobilirubins can be excreted directly into the urine, even without the action of the liver. The treatment of this jaundice of new-born infants is very simple: exposure of the whole body to sunlight or to a suitable lamp is all that is required. This is now a standard application of photomedicine.

5.7.2.2 The Photochemotherapy of Psoriasis. A treatment of this skin disease relies on the light-induced cycloaddition of a molecule of psoralene to a pyrimidine base of DNA (Figure 5.25). Although the detailed mechanism has not been established, it is thought that the psoralene molecule eventually builds a bridge between two strands of the DNA, thereby rendering it inactive in the process of replication. Psoralene contains two double bonds which can in principle take part in cycloadditions, the second one being conjugated to the carbonyl group. (This is similar to enones, section 4.4.)

In practice, 8-methoxypsoralene is used as the photoactive agent. It can be taken orally by the patient who will then be exposed to near UV radiation some 2 hours later. The treatment has however side-effects which make it relatively dangerous.

5.7.2.3 The Phototherapy of Cancer: a Mode of Treatment Still at the Research Stage. Cancer is the uncontrolled proliferation of cells which build up into tumours. Treatments rely on the elimination of diseased cells, through surgery, through the effects of ionizing radiation or through chemicals.

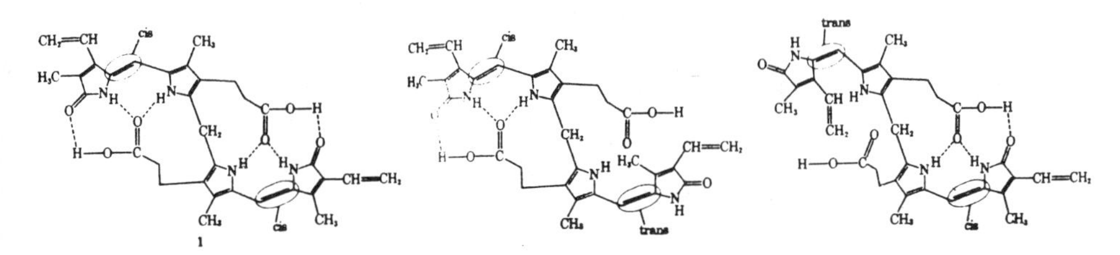

Figure 5.24 *Chemical structures of various forms of bilirubin showing the intramolecular hydrogen bonds*

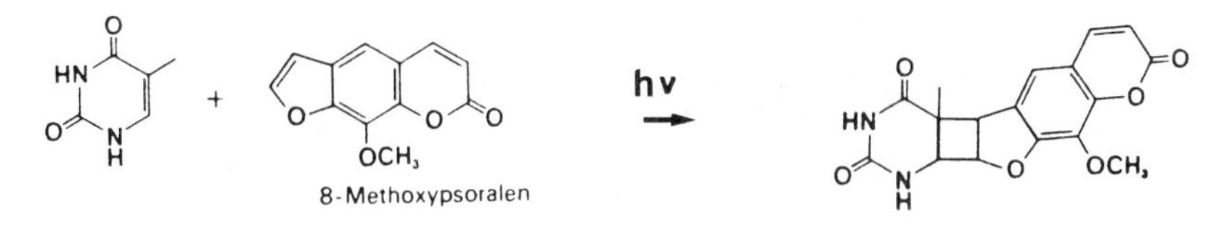

Figure 5.25 *The photochemical reaction between 8-methoxypsoralene and a pyrimidine base*

Cancer cells differ from normal cells in several ways, and these differences must be exploited in chemical treatments. In particular, it is found that certain chemicals concentrate in cancer cells in preference to normal cells, and phototherapy is based on a sensitized photo-oxidation process which destroys the essential components of the cell.

A porphyrin dye is used as sensitizer, chosen to concentrate in cancer cells. Irradiation with VIS light then produces excited singlet oxygen $^1O_2{}^*$ which leads to oxidation reactions. However, the difference in concentration of the sensitizer in cancer cells and normal cells is not large, and irradiation must therefore be kept localized. In some cases a light-guide is used to reach inner tumours, for instance in the oesophagus. At the time of writing (1993), this technique of phototherapy is undergoing clinical trials which will lead to a realistic assessment of its potential.

5.8 BIOLUMINESCENCE

The emission of VIS 'cold' light by animals and plants is widespread in the living world. The 'cold' nature of this light is particularly fascinating since in common experience the source of most VIS light is associated with heat; the heat of flames released in the combustion of organic matter for instance. We must therefore restrict the definition of 'bioluminescence' to the light emitted by *living* organisms in their normal environment. This excludes for example the recombination luminescence of chlorophyll in leaves in which the photosynthetic process has been stopped artificially (for instance through addition of carbon dioxide).

The best known example of bioluminescence is that of the glow worms, but many other examples are found among molluscs, fishes and mushrooms. Bioluminescence is fairly common among marine life, especially in the deep waters where sunlight cannot penetrate.

There are many different light-emitting dyes used for bioluminescence, usually called 'luciferins'. The overall chemical reaction is rather complex and involves at least one enzyme-catalysed step in the formation of the excited dye molecule, but the energy is always derived from the decomposition of a cyclic peroxide. In this respect it is quite similar to the process of chemiluminescence (Figure 5.26).

5.8.1 The Role of Bioluminescence: Its Origins in Evolution

Many animals use bioluminescence as a signal to attract potential mating partners over large distances. One particular species will then use a specific signal, for instance a burst of flashes of a specific frequency that can be recognized by other individuals of the same species. This is not a totally safe way to communicate, for in some cases there are 'pirate' species which will use similar light signals to attract their prey.

This is indeed the second important biological role of bioluminescence, common among deep sea fishes. Many small animals are attracted by the

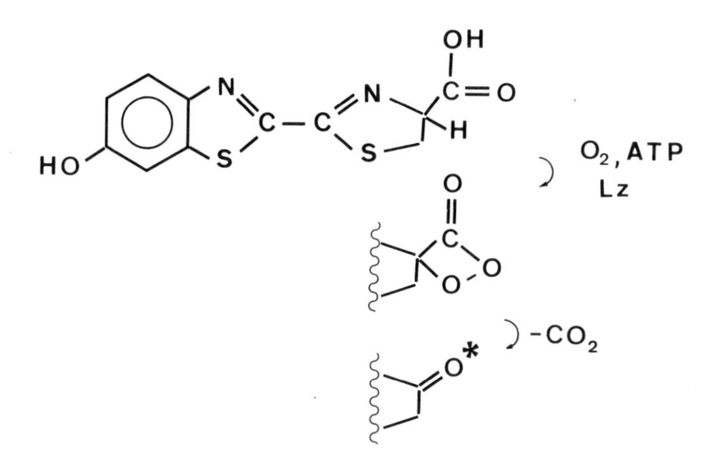

Figure 5.26 *Mechanism of a bioluminescence reaction. A luciferin is oxidized in the presence of an enzyme (luciferase; Lz) and adenosine triphosphate, ATP*

faint light of such fishes in the darkness of the deep oceans and find their way to the mouth of their predator.

There are however instances in which the biological usefulness of light emission is not at all clear. Why should some mushrooms for example show bioluminescence? This brings us to some speculations concerning the origins of bioluminescence in evolution.

It is thought that the primitive atmosphere of the Earth was devoid of oxygen, so that the very early forms of life had no defence mechanisms against oxidation processes which—if uncontrolled—are equivalent to a slow combustion. As the concentration of oxygen increased with the advent of photosynthesis, so the need arose to deal with oxidation products through some harmless chemical reaction. The most important of these products are the peroxides, and here bioluminescence may have evolved to release the energy from the cell in the form of light instead of keeping it as a destructive local heat.

CHAPTER 6

Light in Industry

Although the systematic development of photochemistry is comparatively recent (from the end of the Second World War), some direct and indirect 'industrial' applications based on empirical discoveries date back to much earlier times. Probably the earliest problem of photochemistry that faced man was the light-induced degradation of dyestuffs which were used in antiquity; but this is only an indirect application of photochemistry, though an important one, and as we shall see it is still today a matter of great concern to industry. The first truly 'useful' application of the effect of light on matter was the invention of *photography* in the nineteenth century. This was due to purely empirical observations, well before anything was known about the excited states of molecules or any of the laws of photophysics and photochemistry.

Throughout the history of technology there has been an interplay between empirical discoveries and rational scientific theories. In almost every case a new science must start from an empirical base of simple observations. From a number of systematic observations some laws may emerge, then these laws can be illustrated by means of models, and eventually the models can be moulded into a more general theory. These theories in turn may suggest new experimental approaches, and so the circle (or rather, spiral) of knowledge goes on.

Applied photochemistry is a very good example of this interplay of observation and theory. As in photobiology, it is important to recognize that light can have both beneficial and detrimental actions. In industrial applications, the former are processes such as light-induced polymer formation or polymer solubilization, imaging processes, photochemical syntheses, and the conversion of light (in particular, of sunlight) into other forms of energy. Protection of matter against the detrimental action of light is the 'indirect' application of photochemistry, and this takes many forms: prevention of the photodegradation of polymers, dyestuffs, agrochemicals such as weed killers and insecticides, *etc.* In this chapter it will be seen how the basic knowledge of the laws of photophysics and photochemistry are applied to these varied industrial problems.

6.1 PHOTOGRAPHIC PROCESSES

The conventional process of photography relies on the light-induced reduction of silver halides to metallic silver. The silver halides (AgCl, AgBr) form nearly colourless crystals which are insoluble in water. The absorption of light results in an electron transfer from the halide anion to the silver cation, producing a neutral halogen and silver atoms.

$$Ag^{+}Br^{-} \rightarrow Ag^{\bullet} + Br^{\bullet} \tag{6.1}$$

The colloidal metallic silver formed in this way is opaque (it would appear black in reflected or transmitted light). The halogen atoms combine with an organic substrate mixed with the silver halide crystals to produce bromides.

Although the primary photochemical process of photography is quite simple [it is an electron transfer from X^- ($X =$ halogen) to Ag^+ within a microcrystal of AgX], the technology of photographic emulsions and of their treatments after irradiation is complex.

The photographic emulsion is essentially a dispersion of microcrystals of AgBr in an organic substrate called 'gelatin'. The emulsion therefore consists of microscopic grains which have a specific size distribution. This distribution is important in determining the *sensitivity* and the *resolution* of the emulsion.

In conventional photography the sequence of operations is:
(1) *exposure* (or irradiation);
(2) *development* (a sort of amplification of the image formed during irradiation);
(3) *fixation* of the final image so that it will not be affected by further exposure to light.

During irradiation of the emulsion (exposure) those silver halide grains which have absorbed some light have one or a few silver cations reduced to metallic silver. These form a 'latent image' on the emulsion, so called because it cannot be seen by the human eye on account of the very low concentration of metallic silver atoms. At this stage the photochemical process itself is over, the next steps in the processing of the exposed emulsion being dark (thermal) chemical reactions.

The irradiated emulsion is then treated with a reducing agent (*e.g.* thiosulfate) which attacks those grains which already contain metallic silver. This is somewhat similar to the growth of crystals from pre-existing crystalline centres in glasses, the pre-existing metallic silver of the latent image acting as a centre from which the reduction eventually spreads to all the silver halide of the grain. However, unirradiated grains are not attacked by the reducing agent, as they have no pre-existing metallic silver centres. (Actually such unirradiated grains would also be attacked eventually if the contact with the reducing agent or *developer* was far too long, or if its concentration was too high; thus the process of development must be carefully controlled for optimal results).

In the correct conditions each irradiated silver halide grain is completely reduced to metallic silver, while each unirradiated grain is totally unaffected and still contains only the silver halide. The process of development of the exposed emulsion is therefore equivalent to an amplification of the action of light. A few photons are used to produce a few silver atoms in the grain, but after development the fully reduced grain contains many millions of metallic silver atoms and is a grain of colloidal silver embedded in the gelatin (Figure 6.1).

At this stage the final (amplified) image is formed by the development of the latent image. The emulsion still contains grains of unirradiated silver halide and it is therefore still photosensitive. In the final step the non-reduced silver halide is dissolved in water, in the form of a soluble salt (silver nitrate). This is called *fixation* (of the image). When all the remaining silver halide has been washed away, the colloidal metallic silver remains as the final image and the emulsion is then photostable and can be handled in daylight.

The resolution of a photographic emulsion can be defined by the smallest distance between two spots such that they will form two separate images after development. This is clearly related to the average grain size; the smaller the grains the higher the resolution (Figure 6.2). However, the sensitivity of the emulsion follows the opposite dependence on grain size, since the 'amplification' depends on the number of silver atoms which can be reduced in a grain which has been exposed to light. Sensitivity and resolution are to that extent contradictory properties, and in practice there is a wide choice of photographic emulsions for different applications.

6.1.1 Spectral Sensitization

If the photographic process required the absorption of light by the silver halide in all cases, it would be of limited use because it could 'see' only UV light; AgBr for example has practically no absorption in the visible (VIS) region. A photographic emulsion in which AgBr acts as the light absorber is used in 'unsensitized' films, and these are used in some special professional and industrial applications. In practice, most commercial photographic films are based on emulsions which contain a dye, M, which is the primary light absorber. The reduction of silver ions to metallic silver then relies on an electron transfer process, as depicted in Figure 6.3. The electronically excited

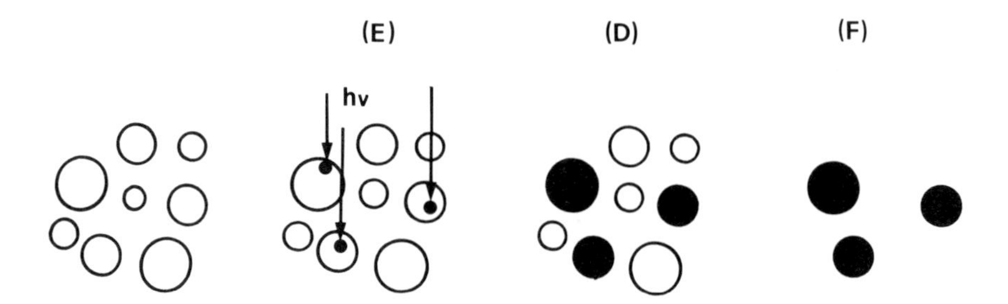

Figure 6.1 *The sequence of processes in the silver halide photographic process. In the exposure phase (E) a few silver atoms are formed in each irradiated grain. These grains are reduced to metallic silver in the development phase (D), and finally the unirradiated silver halide is removed in the fixation phase (F)*

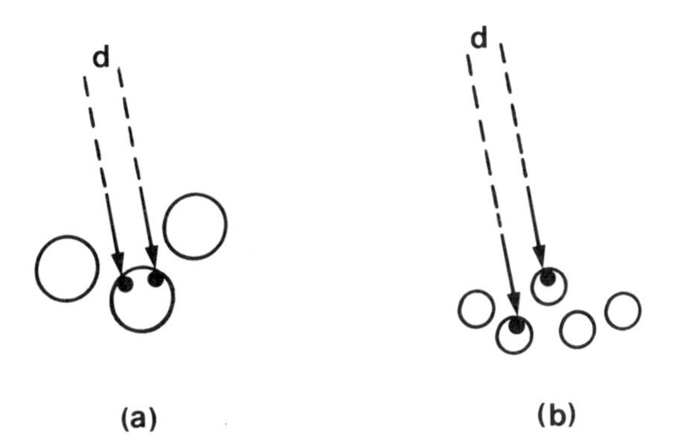

Figure 6.2 *The resolution of a photographic emulsion depends on the size of the grains. d = distance between two beams of light*

dye molecule, M*, is here the electron donor and AgBr (or rather, Ag^+) is the electron acceptor. Over the years, a very wide range of dyes has been developed by the photographic industry, many of them based on the cyanine series of which a few examples are shown in Figure 8.6, p. 261. As the conjugated chain gets longer, so the absorption spectrum moves towards

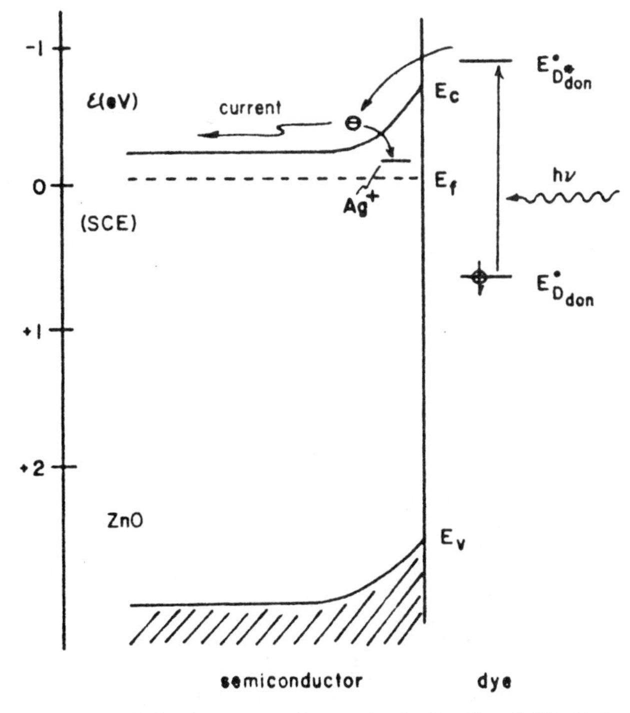

Figure 6.3 *Sensitization of the electron transfer reaction in the silver halide photographic process. Light is absorbed by a dye which acts as an electron donor to Ag^+. SCE = silver–camomel electrode*

longer wavelengths and so the spectral sensitivity of photographic emulsions can be extended to the near IR region.

6.1.2 Colour Photography

In the black and white photographic process, the dark areas result from the presence of reduced metallic silver, Ag. In the colour process metallic Ag is used to synthesize a dye of the appropriate colour, and all the silver is removed from the film. The photochemical reaction is therefore the same as in the black and white process, namely photoinduced electron transfer within the AgBr grains.

The colour film is made up of several layers which are sensitive to three distinct spectra, as shown in Figure 6.4. The first layer contains unsensitized AgBr which absorbs only blue light. The lower layers consist of AgBr associated with green and red absorbing sensitizer dyes, G and R respectively. These are separated by a yellow filter to absorb any remaining blue light.

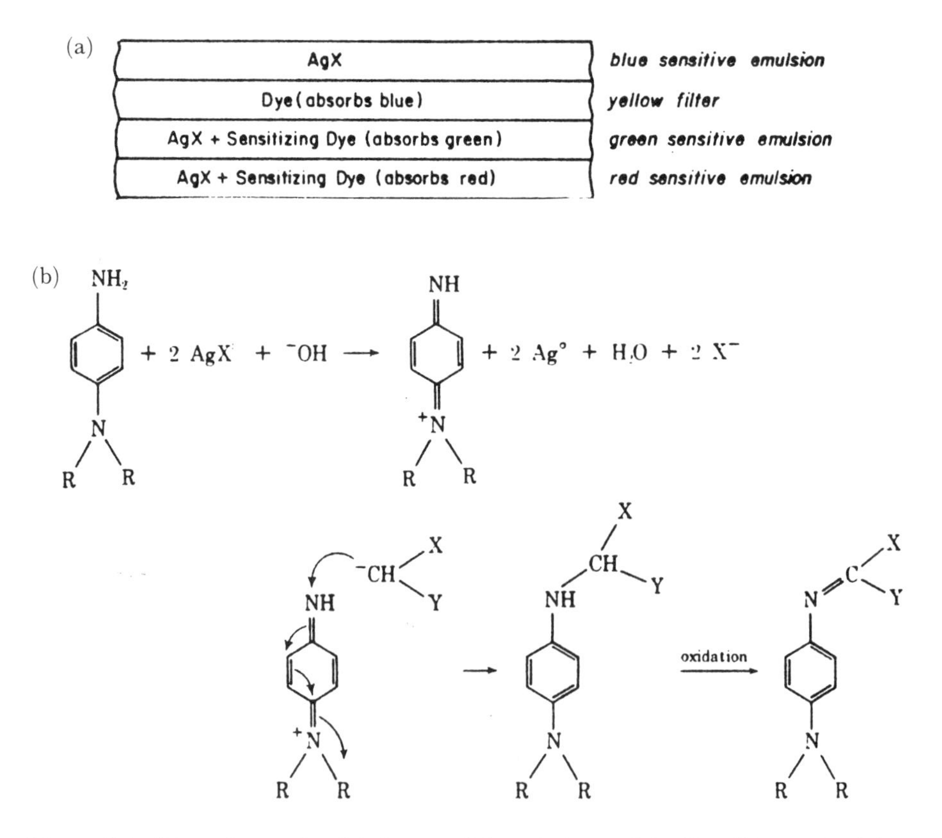

Figure 6.4 *Colour photography: (a) diagram of the main layers of the film; (b) example of dye formation with a p-phenylene diamine developer and a coupler*

The three light sensitive layers contain some organic molecules which can be transformed into dyes of the desired colour by the action of metallic Ag. While the development stage relies on hydroquinones in black and white photography, derivatives of *p*-phenylene diamine are used in the colour process. These are oxidized by Ag present in the 'activated' grains (those which contain some metallic Ag) and these oxidized species react with a 'coupler' to form the dye (Figure 6.4b). Depending on the colour of the dyes, a positive or a negative picture is obtained.

6.1.3 'Instant' Photography

Instant photography (*e.g.* of the 'Polaroid' type) also relies indirectly on the silver halide process; but here the reduced Ag (metallic silver) acts as a reducing agent to modify the solubility properties of a dye. The dye is linked covalently to a molecule such as a quinone (Figure 6.5), which is insoluble in water. When this is reduced to hydroquinone, it becomes soluble in alkaline solutions. The principle of this photographic process is the following:

(1) Before irradiation, all the silver is in the ionic form Ag^+, and all the dye is linked to non-ionic hydroquinones.

(2) At the development stage the hydroquinones in contact with irradiated AgBr grains are oxidized to insoluble quinones.

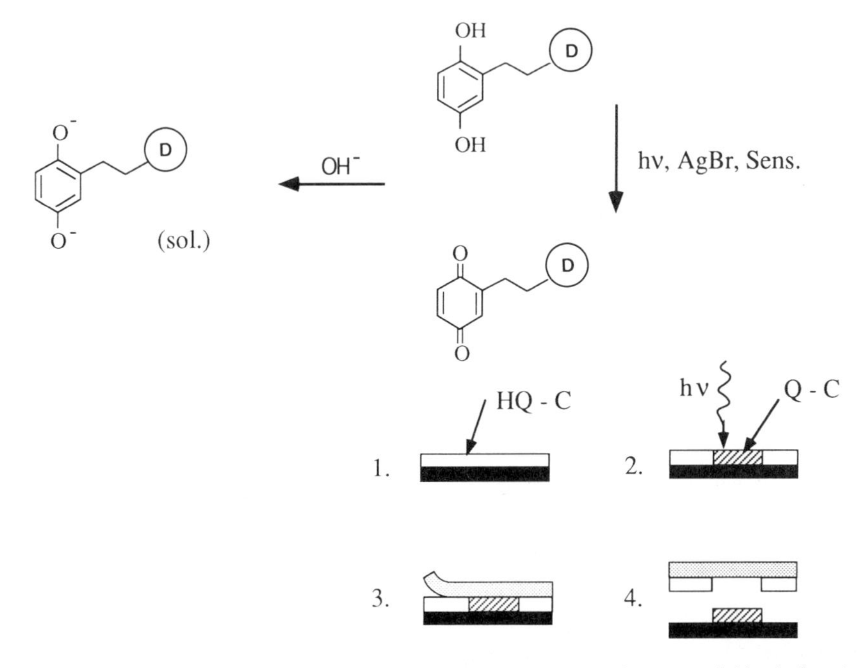

Figure 6.5 *Outline of the process of 'instant' photography. A water-soluble hydroquinone (HQ-C) linked to a dye (D) is oxidized to an insoluble quinone (Q-C) by metallic silver. The soluble and insoluble parts of the emulsion are separated through contact with an alkaline paste*

(3) A sheet coated with an alkaline paste is then applied to the exposed photosensitive film. In the irradiated areas the dye linked to hydroquinone is removed from the film and deposited on the alkaline sheet.
(4) The film is separated from the alkaline sheet, and these form a positive and a negative of the picture.

6.1.4 Electrophotography; The Photocopying Process

This is the process used in most photocopying machines. It is largely a mechanical process which uses one key photophysical step based on photo-induced electron transfer of a rather special kind.

In the actual photocopying machine all the different parts that will be described form a cylindrical rotary drum. For the sake of clarity the various layers of the drum are in a planar configuration (Figure 6.6).

The metal (M; usually Al) is in fact the centre of the drum. This is coated with a thin layer of semiconductor (Sc) which is an insulator in the dark but becomes conductive when exposed to light. The whole photocopying process takes place in several stages.
(1) Electrical charges are produced on the metal–semiconductor interface and on the outer surface of the semiconductor, by means of a corona discharge (ionization of the air by high voltage).
(2) The semiconductor surface is exposed to light, following the image which is projected onto it. In the areas exposed to light electrons are promoted to the conduction band where they travel freely to neutralize the opposite charges of the metal layer; in dark areas the charge separation is retained. At this stage a latent image has been formed. This latent image is in fact a map of electrically charged and neutral areas.
(3) The latent electrical image is then revealed by spraying electrically charged ink particles which adhere to the oppositely charged areas of the latent image. At this stage the drum is covered with ink according to the electrical latent image.
(4) The sheet of paper is applied to the semiconductor surface so that the ink is transferred to it from the charged areas.

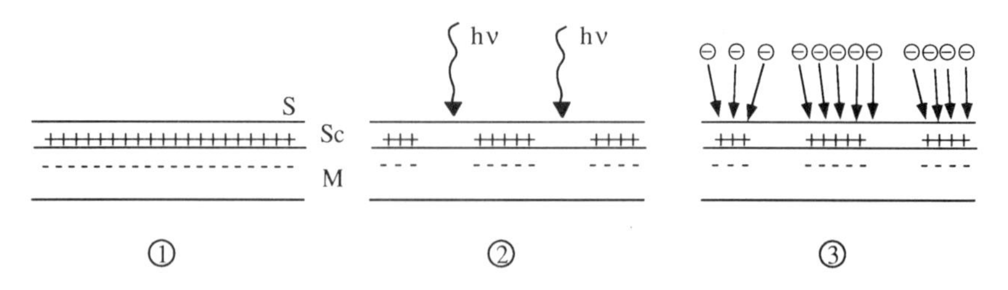

Figure 6.6 *The principle of electrophotography. The heart of the machine consists of a thin semiconductor layer (Sc) on a metal drum (M). Opposite electrical charges are formed by a corona discharge (1), then these charges recombine through photoconduction in the irradiated areas (2). The remaining electrically charged areas are sprinkled with electrically charged ink (3)*

(5) The paper is then heated in order to melt and stabilize the ink; the final image is then obtained, and the copy can be released.
(6) The drum must be cleared of any remaining ink to prepare it for the next cycle. This is a simple mechanical process which uses a brush to wipe the ink particles from the drum.

In this process of electrophotography the image is reproduced directly as a positive, since the areas exposed to light become neutral and do not attract the charged ink particles. This is an important advantage over the silver halide photographic processes which first produce a negative, thereby requiring two complete sequential processings.

There are two drawbacks in this process. The first is that it is largely mechanical as it follows the rotation of the drum. It is therefore rather slow. In addition, it does not have the high sensitivity of the silver halide process because there is no stage of 'amplification' such as the development step in conventional photography. Nevertheless, it has become established as the major process of photocopying because it is readily automated. It offers the important advantage of using ordinary paper instead of the expensive and light sensitive silver bromide films.

6.2 PHOTOPOLYMERIZATION AND PHOTOCHEMICAL DEGRADATION OF POLYMERS

A polymer is a long-chain molecule which consists of many smaller molecules (monomers) linked together by covalent bonds. In a 'homopolymer' all the monomer repeating units are the same, whereas in a 'heteropolymer' there can be two or more different repeating units, according to ordered or random patterns (Figure 6.7).

In a 'plastic' the polymer chains can be either intertwined simply at random to form a three-dimensional pattern, or they can be linked together by chemical bonds. This latter type is called a 'crosslinked' polymer, and it

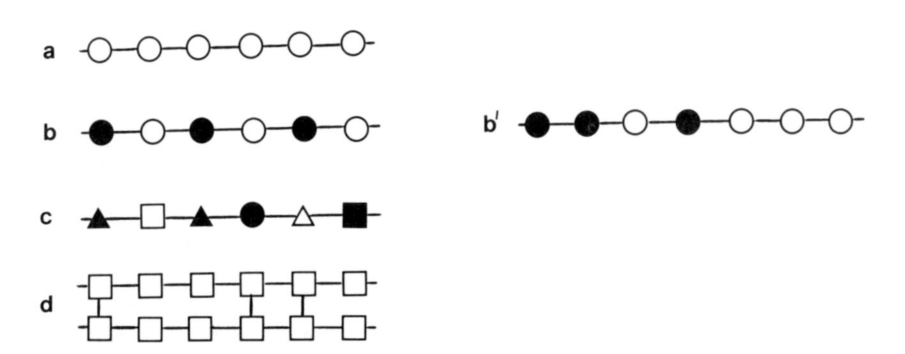

Figure 6.7 *Schematic diagrams of different polymer types: a homopolymer (a), a sequential (ordered) heteropolymer (b), and random heteropolymers (b' with two monomeric units and c with several monomeric units). (d) is a crosslinked polymer*

has a much higher three-dimensional mechanical stability than the former type. In particular, crosslinked polymers are highly insoluble in organic solvents, whereas the non-crosslinked types swell in such solvents and can form liquid solutions in which the polymer chains are solvated separately.

A polymer is formed by the chemical treatment of the liquid monomer, such treatment being either a dark, thermal reaction or a photochemical reaction; the latter only will be considered here. The importance of photo-induced polymerization processes stems from the fact that a polymer can be formed (or can be destroyed) according to an optical image. Before considering the photochemistry of these polymerization and degradation processes, the various industrial applications will be briefly reviewed. In this respect a distinction can be made between the use of light to produce simply bulk polymers, without any particular pattern, and the use of light in 'imaging' processes. The former can in principle be replaced by dark, thermal processes, whereas the latter make full use of the optical properties of light to form (or to remove) polymers selectively, according to a pattern which can be drawn on a 'mask'. Such patterns can be magnified or reduced by means of optical devices, hence the photopolymerization processes are used for the preparation of printed circuits and semiconductor chips in the electronics industry.

Photo-resist technology is widely used for imaging processes in such applications in electronics. If it is wished to produce a metallic pattern of connections between many electronic components (resistors, capacitors, integrated circuits, *etc.*), this can be done by the selective etching of a thin copper plate deposited on an insulating base. The copper layer is protected by a 'resist' which is a polymer, deposited in such a way that it prevents the attack of the metal by an etching solution which will solubilize only the unprotected, exposed copper (Figure 6.8).

A distinction can be made between a 'positive' and a 'negative' resist, according to the action of light. In a positive working resist the monomer is deposited on the copper surface in the form of a viscous liquid. It is then irradiated through a mask (this is simply a drawing of the required pattern on a transparent sheet) and polymerization takes place only at the exposed places. The unirradiated liquid monomer is then washed away in a suitable solvent, and the exposed copper can be dissolved in an etching bath. Finally the protective polymer layer is removed by chemical or mechanical means, and the printed circuit is ready.

In a negative working resist the monomer is first polymerized over the entire copper surface, then the protective polymer layer is irradiated through the mask. In the irradiated areas the polymer is degraded into smaller units and becomes soluble; it can then be removed by treatment with a suitable solvent and the etching bath will attack this exposed copper.

A variation of this technique can be used for the preparation of integrated circuits with a resolution of the order of the μm. In this case a silicon wafer is protected by the resist which must be deposited with an accurately controlled thickness. Exposure to light is done through a mask (with an optical reduction of the size of the image).

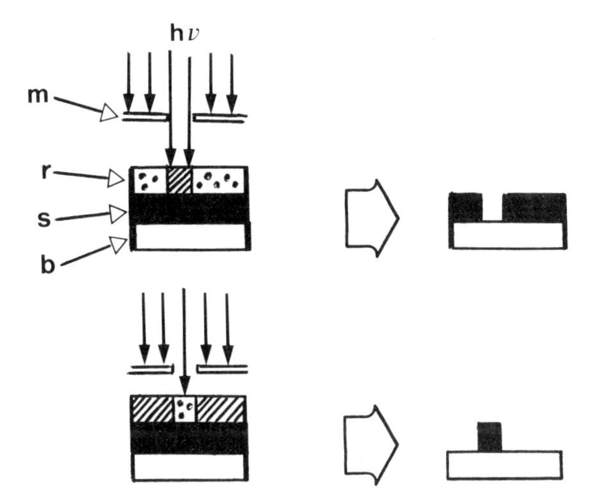

Figure 6.8 *Photo-resist technology. A metal substrate (s) deposited on an insulator base (b) is protected by the resist (r). Irradiation of the resist through a mask (m) results in photoinduced polymerization or polymer destruction*

6.2.1 Photochemical Curing of Surface Coatings

It is often necessary to protect a metal surface against corrosion by covering it with a thin coat of an organic material. This is deposited in the form of a more or less viscous liquid which must then be solidified. This is known as curing. It can be done simply by evaporation of the solvent when a viscous solution is used, but in many cases this solidification process can be initiated by the action of UV light.

Photochemical curing has found many industrial applications, not only for the protection of metal surfaces but also for printing and decoration on a variety of materials. It should be added that optical and mechanical properties of surfaces can also be modified by the coating, for example a porous material can be made impermeable by the thin coat of solid organic substance.

The photochemical curing of surface coatings relies on light-induced polymerization reactions. The organic material is applied in the form of a liquid (which can be quite viscous), then irradiation results in polymerization or in crosslinking of a polymer to form a hard, solid material. In actual industrial applications there are of course many factors which must be taken into account quite apart from the photochemical process itself, such as the adhesion of the coating on the coated surface and the costs of both materials and curing processes. A great advantage of photochemical curing is that it is localized to the irradiated areas, with the possibility of forming images.

6.2.2 Chemical Structures of Polymers

In the few examples selected in Figure 6.9, the structural formulae of some important polymers and of the constituent monomers illustrate the general

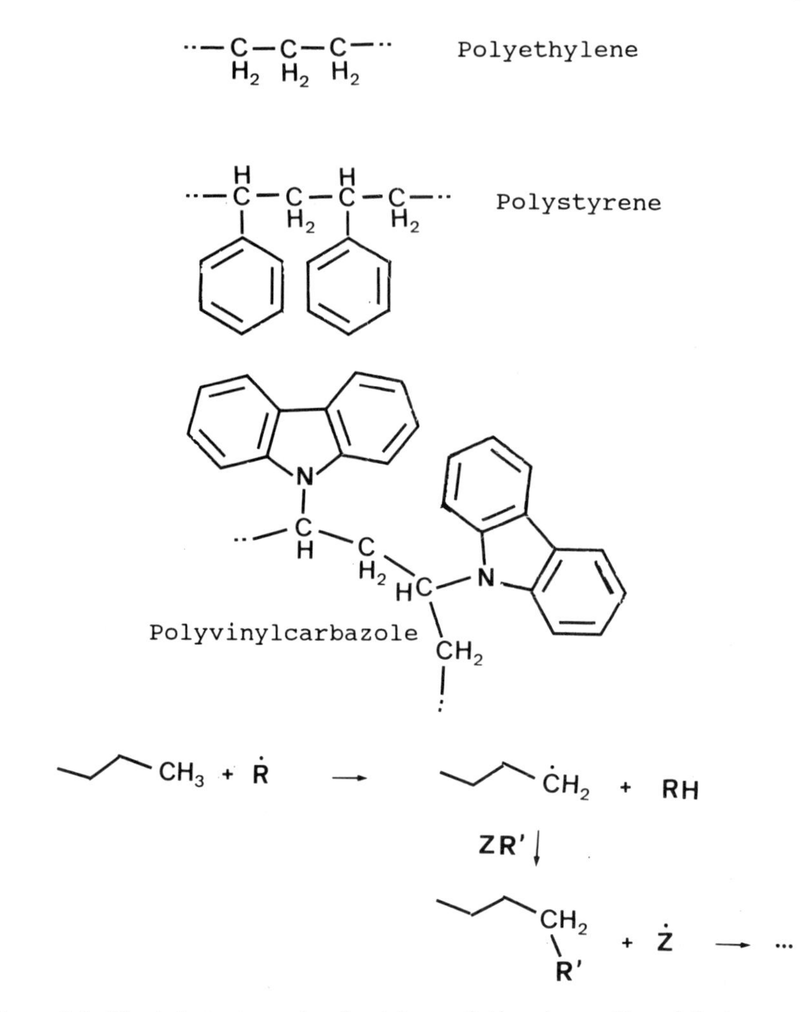

Figure 6.9 *Chemical structures of a few polymers (top) and an outline of the process of free radical-initiated polymerization (bottom). The free radical R′ attacks the end of a polymer chain through hydrogen atom abstraction. The broken polymer chain can then add another monomer unit R′, and so on*

principle of polymerization through free radical reactions. The monomer is characterized by an unsaturated linkage—a double bond—which can add to a free radical to form a larger radical. This new radical can in turn react with another monomer molecule to form a longer chain. The polymer chain is the array of saturated (single) bonds which results from the bonding together of many monomer units, with 'pendant' groups attached at alternate sites in polystyrene and polyvinylcarbazole for example.

The importance of the free radical mechanism of polymerization shown here should not let us forget other mechanisms such as the elimination of water in some thermal processes.

6.2.3 Photoinitiation of Polymerization

Photoinitiation of polymerization can be obtained through a variety of photochemical reactions which produce reactive free radicals. These radicals then lead to the formation of the polymer chains through the addition of further monomer units to the end of a chain in a sequence of radical addition reactions (Figure 6.10). A photoinitiator of polymerization is therefore a molecule which produces free radicals under the action of light. Benzophenone and other aromatic ketones can be used as photoinitiators, since a pair of free radicals is formed in the hydrogen abstraction reaction. Some quinones behave similarly, for example anthraquinone in the presence of hydrogen donor substrates such as tetrahydrofuran.

Benzoic ethers, dioxolane, and sulfur derivatives are photoinitiators which undergo photochemical dissociation reactions into pairs of radicals. The polymerization of vinyl (monomer) into the polymers of the polyvinyl series can be photoinitiated by an ammonium salt of a benzoin ether derivative. In this case the active species is the benzoyl radical (Figure 6.11).

These photoinitiation processes which depend on the formation of free radicals in some photochemical reaction lead to chain reactions, since each molecule of initiator can promote the addition of many monomer units to a polymer chain. The quantum yield of monomer addition can therefore be much larger than unity, but it cannot be controlled since the growth of a polymer chain is then limited by termination reactions in which two free radicals react to produce closed-shell molecules.

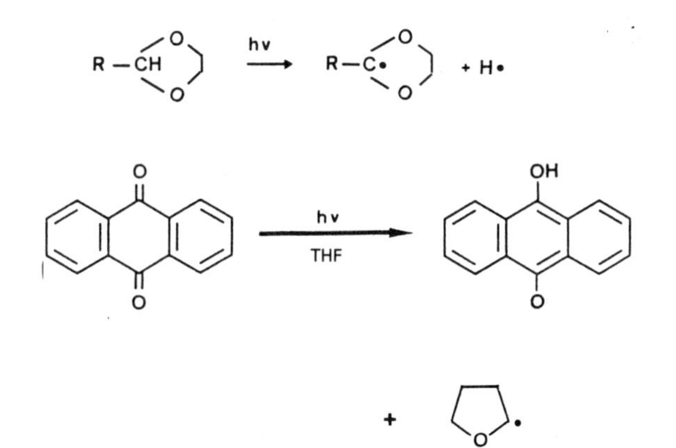

Figure 6.10 *Two examples of free radical photoinitiation reactions: dissociation of dioxolane and hydrogen atom abstraction of anthraquinone in tetrahydrofuran*

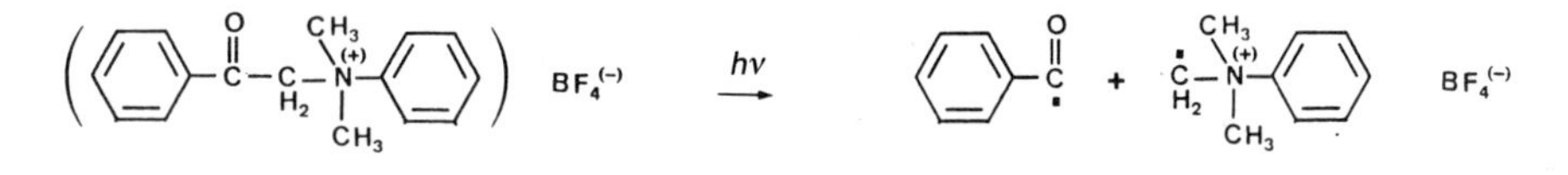

Figure 6.11 *Photochemical reaction of a cationic initiator used in the form of a salt*

A very different process of photopolymerization relies on the reaction of *photocondensation* which is an addition of two molecules to form a longer adduct. There is no radical intermediate in this case and one photon (at least) is required for each step in the polymerization process. Some maleimide derivatives can be polymerized in this way, to form an insoluble crystalline polymer. It should be noted that the monomer molecules must have two reactive groups, one at each end, so that the polymer chains can extend in principle indefinitely (Figure 6.12).

Photo-crosslinking and the reverse process of photodissociation of pre-existing crosslinks relies on a cycloaddition reaction (and on the reverse dissociation of the cyclic adduct). For example, derivatives of vinylcinnamic acid can form crosslinks which are dissociated by irradiation with short wavelength light (*e.g.* 254 nm produced by low-pressure mercury arcs). In this process the polymer chains become separated, and the polymer itself is then soluble in organic solvents.

An example of photo-crosslinking is the reaction which follows the photodissociation of diazofluorenes, with the loss of molecular nitrogen and formation of a carbene. Two carbenes will then react to form an adduct between two polymer chains (the crosslink). In this case irradiation can be made in the near UV, for instance at 365 nm with a medium-pressure mercury arc lamp (Figure 6.13).

Photoinduced polymerization can also be obtained through the dissociation of organic salts such as sulfonium, diazonium and similar salts. The photodissociation leads to several species, a radical cation, a neutral free radical and a closed-shell anion for example. The radical cation can then react further, *e.g.* through hydrogen abstraction from a substrate, ZH, to form another free radical $Z^{\bullet}$.

6.2.4 Photodegradation of Polymers and Protection Methods

In all these cases the important reaction is the formation of reactive free radicals which induce the addition of further monomer molecules to the end

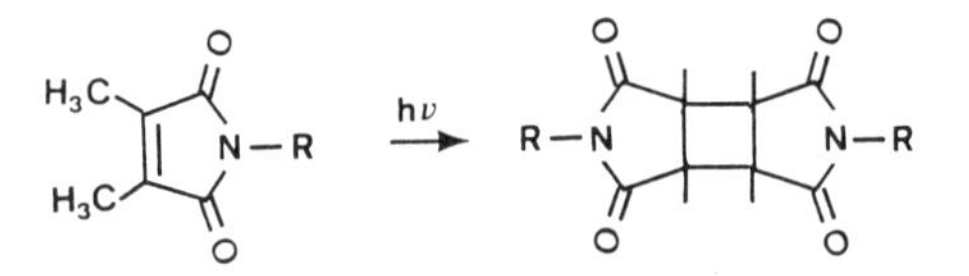

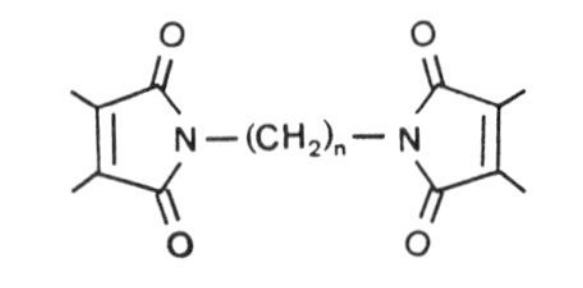

Figure 6.12 *The principle of a photocondensation polymerization reaction. With bichromophoric monomers long chains of polymers can be formed*

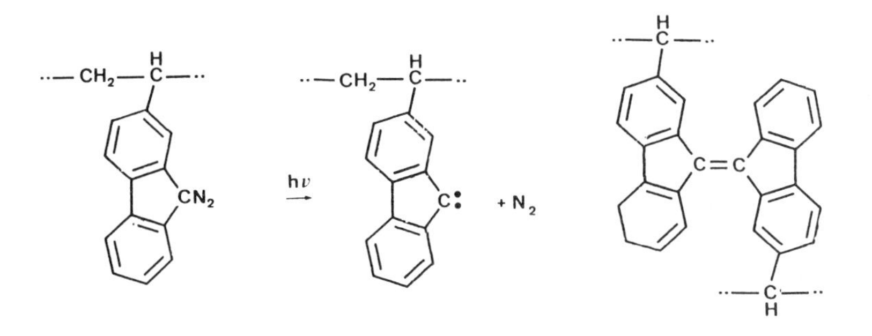

Figure 6.13 *A crosslinking reaction proceeding through biradical addition following nitrogen photoelimination*

of the polymer chain. The photoinitiator is therefore a molecule which forms free radicals in a photochemical reaction, and as we have seen there is a wide variety of such molecules that can be used with different polymers and in different conditions of irradiation. It should be noted that with some polymers a photoinitiator can promote the *degradation* of the polymer rather than its buildup. Negative working resists use this type of system of course, the photochemical reaction leading to the breakdown of polymer chains to form shorter chains of increased solubility in organic solvents (Figure 6.14). In this example we have seen that a polymer/photoinitiator system can be designed deliberately to lead to photodegradation (so that it can be used as a negative working resist). The problem of photodegradation of polymers is however quite general, and the prevention of such photoinduced damage is of great concern in many industrial applications. It must be realized that what is called a 'polymer', prepared on an industrial scale, is not a pure compound according to the chemical definition, but contains inevitably impurities and 'impurity sites' which greatly affect its photochemical properties.

The major impurities which are found in any polymer are the unreacted monomer itself, unreacted initiator (peroxides and all types of photoinitiators) and catalysts used in the polymerization process, as well as traces of the solvent and of water. Within the polymer chain itself there will be some defects or impurity sites which result essentially from oxidation reactions during the making of the polymer. The polymerization process on an industrial scale cannot be performed in the absence of atmospheric oxygen, and this will attack the growing polymer chain at random points to produce

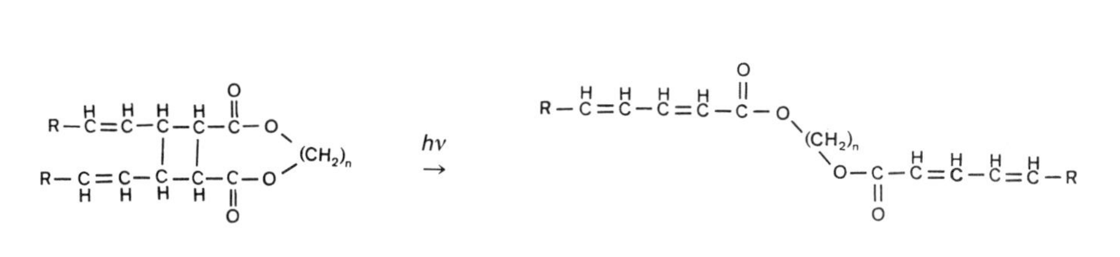

Figure 6.14 *The breaking of pre-existing crosslinks by short wavelength irradiation*

peroxides, hydroperoxides and carbonyl groups; these impurity sites can act as chromophores (light absorbers) when the pure polymer itself should be perfectly clear and transparent down to the far UV; pure polyethylene should not absorb above 220 nm.

Since it is impossible to avoid the presence of impurities in polymers, the protection against photodegradation must rely on additives which are called 'photostabilizers'.

In most cases two different additives are used, one acting as an internal filter or excited state quencher and the other one being a scavenger of free radicals. These two additives have a 'synergic' effect, that is they are much more effective together than either of them separately.

The first additive is a molecule of high extinction coefficient in the wavelength region below 360 nm where damage by sunlight is most important. These molecules are derivatives of 2-hydroxybenzophenone, benzotriazole, oxanilides, *etc.* They have in common a pattern of intramolecular hydrogen bonding between a carbonyl group and a nearby OH or NH group. Intramolecular proton transfer in the singlet excited state then leads to deactivation, according to the process of quenching considered in section 4.4. Figure 6.15 shows the general structures of these molecules and examples of their absorption spectra are given in Figure 6.16. This illustrates their characteristic absorption in the near UV spectrum.

It is still not clear whether these additives act as quenchers of the excited states produced by light absorption in the polymer impurities or impurity sites. Such quenching action is not likely to be very important, since in most cases the direct contact of the excited molecule and the quencher molecule is necessary, and this requires the diffusional encounter of the molecules within the lifetime of the excited state. This cannot take place in polymers (which are solids or extremely viscous liquids) unless improbably high concentrations of additives are used. The photostabilizers certainly act as internal filters, *i.e.* they absorb the light before it has had a chance to reach the photoreactive sites within the polymer, and they degrade, very rapidly, the electronic excitation energy through a non-radiative deactivation process. This is thought to be the reversible proton transfer which is complete in a time of some ps.

The second additive acts most probably as a scavenger of free radicals. A class of compounds known as 'hindered piperidines' is one of the most widely used for this purpose. Their chemical structure is shown in Figure 6.17.

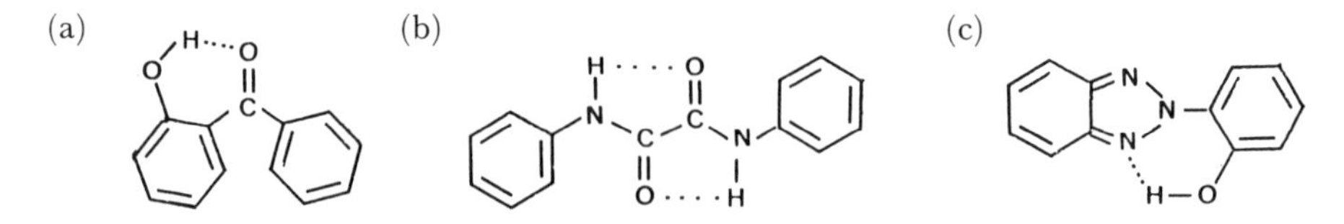

Figure 6.15 *Chemical structures of some polymer photostabilizers: (a) 2-hydroxybenzophenone, (b) hydroxybenzotriazole, (c) oxanilide*

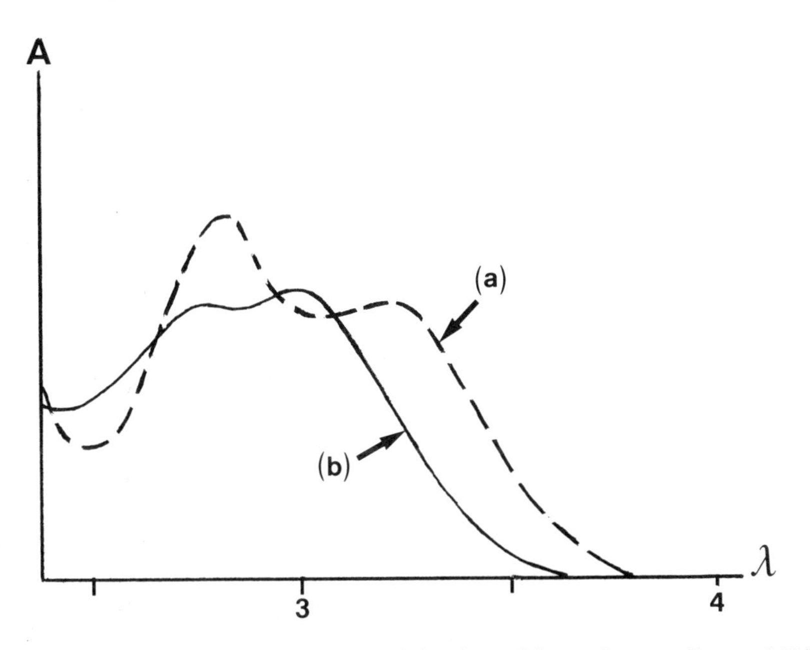

Figure 6.16 *Examples of absorption spectra of the photostabilizers shown in Figures 6.15(a) and (b). Wavelength λ in nm/100*

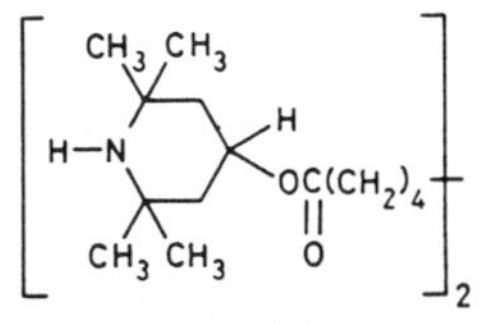

Figure 6.17 *Chemical structure of a hindered piperidine*

These molecules have no absorption in the near UV, so they cannot act as internal filters. There is no evidence that they could act as quenchers of the excited impurity chromophores in common polymers, and it has been mentioned already that such quenching action would be, in any case, unlikely to be important in relatively rigid systems such as polymers. This remark does not apply to free radical scavengers, because a free radical has an unlimited lifetime since it can disappear only through a chemical reaction with another open-shell molecule.

6.3 PHOTOCHEMISTRY IN SYNTHESIS

Photochemical reactions cover a most extensive range of chemical processes, some of which are totally beyond the realm of dark reactions. Industrial applications of photochemistry to the large scale synthesis of chemicals are however relatively unimportant, and this can be traced to the high cost of light as a source of energy. There are some attempts at using sunlight as a 'free' source of energy, but this is not likely to lead to major synthetic processes because sunlight by its nature is an unstable and largely unpredictable light source; not to mention its geographical distribution.

There are however a few important synthetic applications of photochemistry, and we shall consider here three representative examples.

6.3.1 Photochlorination of Polymers

In terms of tons of chemicals per year, this is by far the most important industrial process in which a light-induced reaction plays a key role. The photochemical reaction itself is extremely simple, being the homolytic dissociation of molecular chlorine, Cl_2, into $Cl^{\bullet}$ atoms,

$$Cl_2 \xrightarrow{h\nu} 2\,Cl^{\bullet} \tag{6.2}$$

These $Cl^{\bullet}$ atoms are in fact free radicals and react very efficiently with hydrocarbons through substitution and insertion processes.

$$>\!=\!< + Cl^{\bullet} \rightarrow {}^{\bullet}\!>\!-\!<\!Cl \xrightarrow{Cl_2} Cl\!>\!-\!<\!Cl + Cl^{\bullet} \tag{6.3}$$

$$RH + Cl^{\bullet} \rightarrow R^{\bullet} + HCl \xrightarrow{Cl_2} RCl + Cl^{\bullet} \tag{6.4}$$

Some of these processes lead to chain reactions in which a new $Cl^{\bullet}$ atom is formed in a (dark) free radical reaction. A single $Cl^{\bullet}$ atom formed in the photochemical reaction can therefore lead to overall quantum yields of several thousands.

The purpose of photochlorination of polymers is to improve their stability with respect to fire hazards. Polymers made of C and H atoms only are quite dangerous in this respect, since they undergo combustion reactions in the presence of oxygen to form CO_2 and H_2O. Replacement of H by Cl reduces the flammability of such polymers.

Why should the photochemical formation of chlorine be preferred to the similar, thermal (dark) dissociation? There are two distinct reasons for this.

(1) Although molecular chlorine can be dissociated into atoms at high temperatures, many organic molecules will undergo unwanted chemical reactions as well. When heat is applied to a chemical sample, you cannot pick and choose the molecules which will be thermally 'excited', whereas excitation through the absorption of light is highly specific. Photochlorination does not require heat energy for the dissociation of Cl_2.

(2) The fact that photochlorination can be a chain reaction reduces the cost of light energy to acceptable levels. This would not be the case if one photon would be needed for each Cl_2 molecule.

6.3.2 The Synthesis of Caprolactam

This compound is a key intermediate in the manufacture of nylon 6, an important industrial polymer. The photochemical process may take two distinct pathways (Figure 6.18). The quantum yield of the photochemical

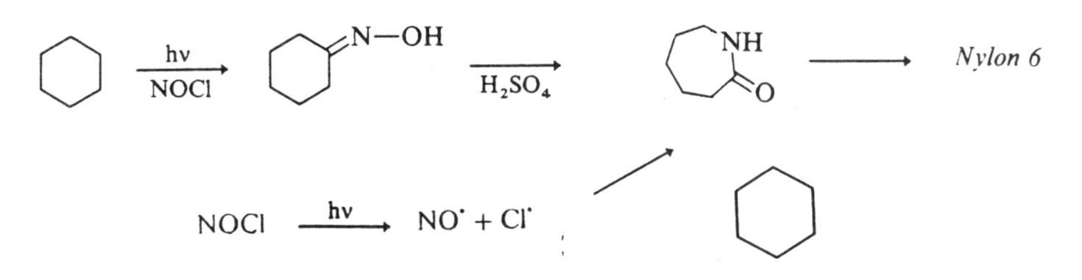

Figure 6.18 *Reaction sequence in the synthesis of caprolactam*

reaction is smaller than unity, there being no chain reaction like in the photochlorination process. The cost of the energy of light is however justified in this case because there is no efficient alternative in dark reactions that would lead to a product of comparable purity.

6.3.3 The Syntheses of Vitamin D and of 'Rose Oxide'

Since light is a costly form of energy its use can be acceptable for industrial applications only when the overall reactions have very high quantum yields, or when the final products are in any case expensive. This applies in particular to pharmaceutical and cosmetic products, and here photochemical processes can be used within the requirements of commercial considerations.

The industrial synthesis of vitamin D is a perfect replica of the biosynthesis which relies on a key photochemical step of electrocyclic ring closure/ring opening (section 5.6). In this case the photochemical process is essential, since the dark reaction is forbidden by reasons of orbital symmetry considerations.

The second example of industrial photochemical synthesis is that of a compound known as 'rose oxide'. It is used for the manufacture of perfumes on small industrial scales. The photochemical reaction itself is simply the production of singlet (excited) oxygen $^1O_2^*$ through energy transfer from a light-absorbing dye (Figure 6.19). The quantities of products made through processes which involve a photochemical reaction are small by industrial

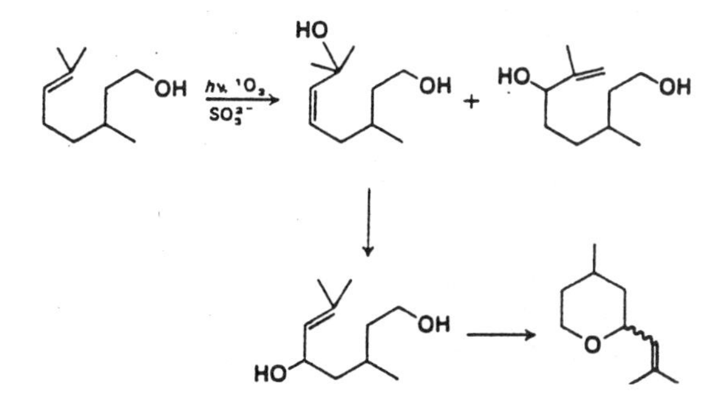

Figure 6.19 *Reaction sequence in the synthesis of 'rose oxide'*

standards. This remark does not apply to laboratory syntheses where the quantities of materials are small and the cost of light energy can be neglected. The future of photochemical synthesis probably lies in this direction; it is well suited for the production of high cost, specialized chemicals which cannot be made easily through dark reactions.

6.3.4 Photochemical Reactors

Various designs of photochemical reactors have been developed to suit synthetic processes in industry. It is essential to use all the light emitted by the lamp(s), so these are placed at the centre of the reactor itself. It should not be forgotten that much of the light is degraded into heat, and this may raise the temperature of the reactants. In general it is therefore necessary to include heat filters, wavelength filters and cooling liquids between the lamps and the sample.

The two broad classes of photochemical reactors are the 'batch' processors and the 'continuous' processors. The batch processor is simple in design, but costly in operation, because it requires the loading of the reactant, the unloading of the product and the cleaning of the reactor vessel; all operations which involve human intervention. Batch processing is used as a rule in laboratory synthesis, but industrial applications prefer continuous systems for reasons of efficiency. Still, it must be accepted that batch processing will be used for many small-scale industrial syntheses.

Figure 6.20 shows some photochemical reactors in common use in indu-

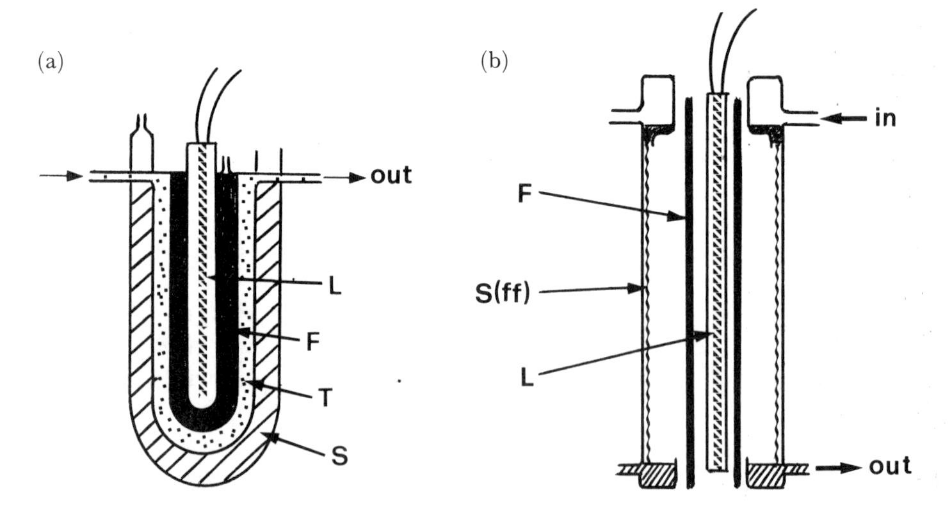

Figure 6.20 *Examples of photochemical reactors: (a) for batch production; the lamp L is placed in the middle of the sample holder S, separated by a filter F and a thermostatted vessel T through which the coolant is circulated. (b) The falling film reactor uses a central lamp L surrounded by a filter F. The sample S(ff) falls slowly as a thin film on the inner wall of the reactor, and the photoproducts are collected at the bottom*

stry. A batch reactor consists in principle of a cylindrical lamp surrounded by a liquid filter, a thermostatic, circulating liquid and finally the reactant itself. Continuous processes use the falling film technique in which the liquid reactant flows slowly around the inner surface of a vessel; the irradiation time must of course be adapted to the photochemical reaction.

6.4 PHOTOCHEMISTRY OF DYES AND PIGMENTS

A 'dye' is a compound which is added to give colour to a normally colourless material, for instance natural or artificial fibres. By their very nature dyes must absorb light in the VIS region of the spectrum, and this can lead to two separate problems.

(1) The dye molecule itself must be photostable, under the conditions of practical use. 'Dye fading' is the name given to the photochemical degradation of the dye molecules, since the colour of the dyed fabric is then slowly lost.

(2) Even photostable dye molecules can in some cases promote the degradation of the fibre on which they are adsorbed. This is called the 'phototendering' of the fabric, and it can be a serious problem quite irrespective of the photostability of the dye itself.

Natural dyes have been known for a very long time. These can be inorganic or organic in origin, the former being coloured minerals such as precious and semi-precious stones which have been used to make mosaics in antiquity and stained glass windows in the Middle Ages. Many of these mineral 'dyes' have the great advantage of being absolutely photostable, since their colour results from atomic transitions. Thus the red colour of ruby is due to the Cr^{3+} ions dispersed in the matrix of the stone in the form of an 'impurity'. Against the photostability of mineral colours, it must be said that their potential applications are extremely limited by their availability (this relies merely on good luck) and the fact that they cannot be incorporated into fabrics and other materials. In addition, there is no way of choosing a particular colour or hue except by selecting a few more or less similar stones among many thousands. For these reasons the more readily available organic dyes of biological origin have been used since antiquity to dye cloth, the chemical structure of the most successful of these being rather similar to those of the many synthetic dyes which have been developed since the advent of industrial chemistry in the last century. One example will be found in Figure 6.21; this was a red dye extracted from some insects, used to make scarlet cloth.

Although there are many hundreds of man-made dyes at the present time, it will be seen presently that most of these fall into a few large groups of structurally related molecules, the general structural features explaining their photostability.

The two major classes of industrial dyes that we shall consider here are the *anthraquinone* dyes and the *azo* dyes. The molecules which form the very extensive range of anthraquinone dyes are all based on the 2-amino- or

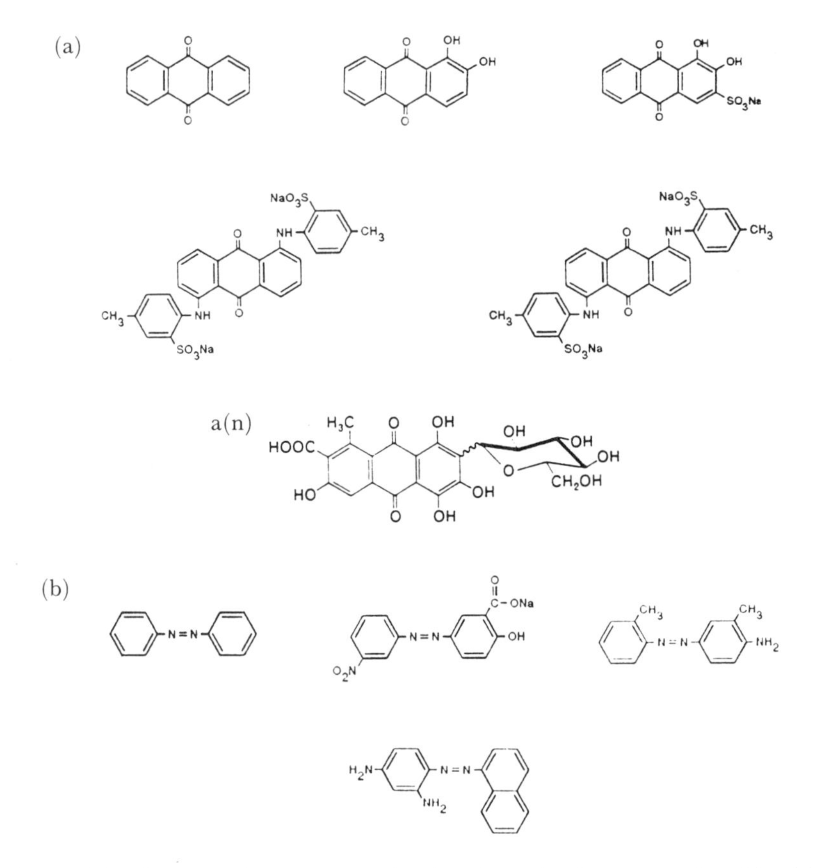

Figure 6.21 *Examples of the structures of anthraquinone dyes (a) and azo dyes (b). a(n) is the natural dye 'carminic acid'*

2-hydroxyanthraquinone structures, with the proximal electron acceptor carbonyl group and the electron donor, NH_2 or OH group held together by an intramolecular hydrogen bond.

In the early theories of colour, before the development of the quantum theory of electronic states and electronic transitions, an empirical rule had been found that a dye molecule had to include a 'chromophore' group and an 'auxochrome' group substituted onto some aromatic system. The carbonyl group, C=O, the nitro group, NO_2, the azo group, N=N, *etc.* were described as chromophores, and the amino group, NH_2 (or more generally NR_2), the hydroxyl group, OH, and the methoyl group, OCH_3, were the auxochromes. With our present knowledge it is readily seen that the so-called chromophores are π electron acceptors and the so-called auxochromes are π electron donors. The colour of these dye molecules is due to a charge-transfer type electronic transition of relatively low energy corresponding to an absorption in the VIS region of the spectrum.

It is noteworthy that the proximal '*ortho*' position of the electron donor and acceptor groups is essential for the photostability of the dye molecule. Thus

2-substituted anthraquinones are also coloured compounds which could be used in principle as dyes, but they are degraded quite rapidly through photochemical reactions. In the great majority of 'fast' dyes (photostable dyes) one can recognize the pattern of internally hydrogen bonded acceptor and donor groups, and this is very similar to the structure of polymer photostabilizers such as 2-hydroxybenzophenones and benzotriazoles (Figure 6.22). These photostable dye molecules are non-fluorescent, the excitation energy being degraded extremely quickly into heat through some non-radiative transition or some reversible photochemical reaction. The latter is very probably operative in these cases, the reaction being an intramolecular proton transfer which would result in electronic deactivation. It is well established that protonation of some aromatic molecules in the excited state leads to quenching, although the presence of the proton-transfer tautomer as a very short-lived product is difficult to demonstrate. In any case, the general requirement of the proximity of the electron donor and acceptor groups suggests a mechanism of this kind.

Processes of photo-oxidation by singlet (excited) oxygen $^1O_2^*$ can also be important for the degradation of dyes, in particular of the azo dyes. Singlet oxygen can be formed by sensitization through a variety of excited molecules, even some dye molecules. It is therefore very important that the excited states of such molecules should have very short lifetimes so that energy transfer to molecular oxygen is very unlikely.

6.5 PHOTOCHROMISM

A 'photochromic' substance is one that changes colour under the influence of light, this change being reversible in the dark. The photochromic substance should therefore adapt its transmission to the intensity of incident light; the stronger the light, the more the photochromic system absorbs it.

Obvious applications of photochromism include glasses which darken according to the intensity of the surrounding light, and such glasses have now been available for some time both in the form of plate glass for windows and in the form of lenses for sunglasses. Their action is very slow, it takes indeed many minutes for the photostationary state to be established. This is due to the fact that the photochromic effect takes place at the molecular level

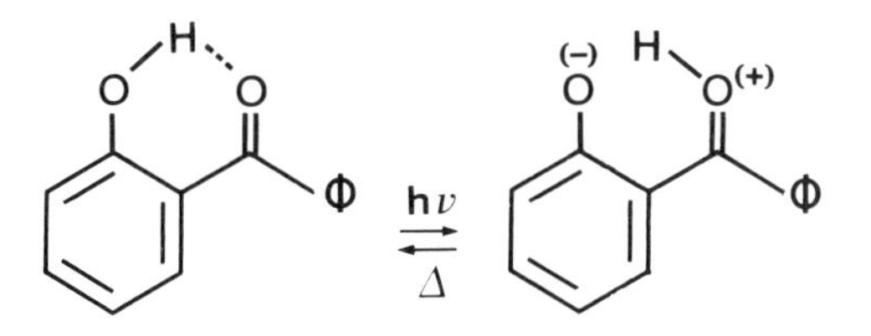

Figure 6.22 *Intramolecular proton transfer in the charge-transfer state of 2-hydroxybenzophenone. The zwitterionic form reverts to the ground state in a time of ps*

and it is therefore relatively insensitive (compared with photography with its built-in amplification during the development stage).

The molecule spiropyran and its derivatives are one of the most widely used photochromic systems. The starting material is colourless, hence near UV light is needed for the photochromic action. The coloured species is a tautomer which reverts to the starting material thermally in the dark (Figure 6.23).

Another photochromic reaction is the photochemical keto–enol tautomerization of 2-methylbenzophenone. The light-induced reaction takes place with high quantum yield (in deoxygenated solutions) but the reverse dark process is very slow, of the order of minutes or hours depending on the temperature. The low rate of recovery is indeed one of the major problems of many photochromic processes of potential practical applications. The development of useful photochromic systems is however very promising for high-resolution imaging systems and for the storage of information in computers. The theoretical spatial resolution of a photochromic imaging or storage system is of the order of molecular dimensions, since the photochemical process itself takes place at the molecular level. The resolution is therefore very much greater than that of the conventional photographic process which depends on the size (and on the size distribution) of the grains of gelatin/silver halide.

The very high spatial resolution of photochromic systems implies a very low sensitivity, so that these systems are most useful in conditions of high light intensity, *e.g.* when used with flash light or laser light. A potential application of photochromic materials is the protection of eyesight against light pulses of extremely high intensity, for example the flash emitted during a nuclear explosion. In such a case the photochromic material must respond 'instantly' and must recover very quickly if it is to be of any use. It has been suggested that some aromatic species, dispersed in a clear polymer (*e.g.* in polymethyl methacrylate, also known as 'perspex' or 'plexiglass'), could act as rapidly reversible photochromic molecules through their triplet–triplet absorptions in the VIS region (Figure 6.24). The darkening would then be effectively 'instantaneous' at high photon fluxes, since the time of population of the triplet state is of the order of ns or less. Recovery takes place by deactivation of the triplet state to the ground state, in times of ms or at the most of a few seconds. This is of course a purely photo*physical* photochromic process, there being no chemical change of the molecules involved.

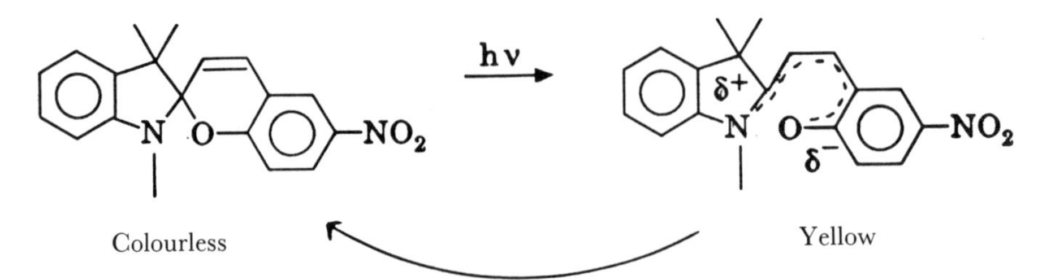

Figure 6.23 *Structures of the two isomers of the photochromic molecule spiropyran*

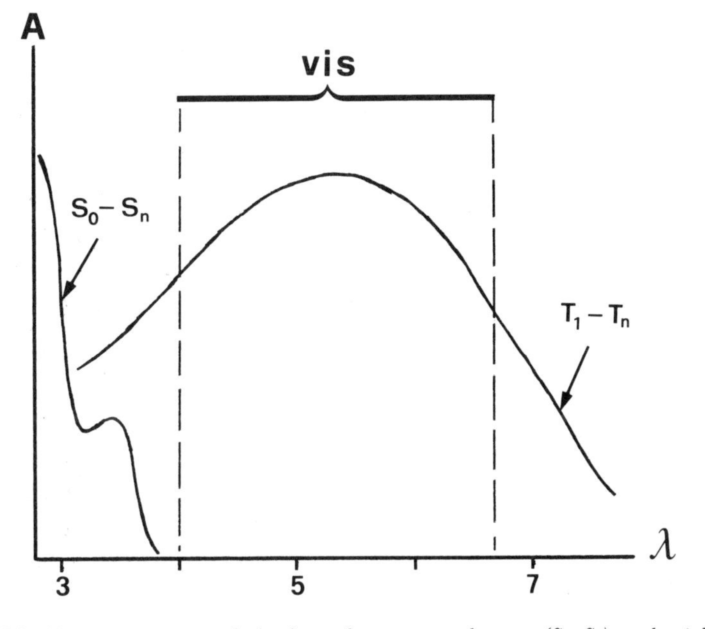

Figure 6.24 *Absorption spectra of the benzophenone ground state (S_0–S_1) and triplet state (T_1–T_n) in the near IR, VIS and near UV regions. Wavelength λ in nm/100*

6.6 ENERGY CONVERSION AND STORAGE

It is well recognized that the energy requirements in modern industrial societies increase in leaps and bounds, and that consequently the problems of the primary sources of this energy supply are of prime concern for the future of mankind. In the wake of the (political) oil crisis of the seventies, and of the (technological) nuclear power crisis of the eighties, there has been renewed interest in the use of solar energy as the primary source. However, unless this energy is simply degraded into heat and used as such, its conversion into more useful forms of energy which can be stored and transported is a major problem.

Electrical energy is the useful form which can in principle be derived from sunlight. This is the readily acceptable form of energy in the home and in industry, but it is not practical for use in most transportation systems (motor cars, aircraft, ships) where a compact chemical energy source is essential (in the form of hydrocarbons to be burned in air). There has been speculation that gaseous hydrogen, stored in high pressure bottles, could provide an alternative; should it be so, the energy of sunlight might be used to produce molecular hydrogen from water.

There are several approaches to the conversion of sunlight into electrical or chemical energy. The first method is the direct transformation into electricity by means of photovoltaic cells. These are based on a purely photophysical process, and they have been used successfully in many small-scale applications; to provide electrical power for artificial satellites, for example. It is unfortunately a very expensive technique, and it cannot be considered to be economically viable on a large scale.

A photochemical process could use the energy of sunlight to split water into H_2 and O_2, the former then being used as a fuel which can be stored indefinitely. The splitting of water requires in principle a photoactivated catalyst dissolved or dispersed in water. The energy requirement for the overall reaction $H_2O \rightarrow H_2 + \frac{1}{2}O_2$ is 1.23 eV (per electron). In conditions of electrolysis it would however proceed extremely slowly at this potential difference which pertains to a thermodynamic equilibrium, and it is well known that a substantial 'overvoltage' of the order of 0.5 V is required in practice to drive this reaction.

In photochemical and photoelectrochemical systems the overall reaction, as shown above, is separated into two processes which are the oxidation of water to form O_2 and its reduction to form H_2.

Much of the research for the photochemical decomposition of water has concentrated on the use of semiconductor surfaces as heterogeneous, photoactivated catalysts. A semiconductor is characterized by its 'bandgap', which is the smallest energy difference between its valence band and its conduction band. The bandgap therefore corresponds to the onset of absorption of light by the semiconductor and it determines the energy available between the positive hole and the excited electron. The photoelectrolysis of water requires a bandgap of at least 2.2 eV because of the overvoltage needed to drive the reaction at a reasonably fast rate, and this limits, already, the choice of suitable semiconductor materials. Not only must the bandgap be sufficiently wide, but also the absolute energies of the valence and conduction bands must correspond to the two redox reactions of water. It must be noted that an extremely wide bandgap (that is, well beyond the minimal 2.2 eV) would have no advantage, since the absorption of sunlight in the VIS region would then become inefficient.

6.6.1 Photoelectrochemical Cells

There are several designs of photoelectrochemical cells that can produce either fuel (H_2) or fuel and electrical power.

6.6.1.1 Photoelectrolysis of Water. Figure 6.25 shows the principle of this type of cell, based on the chemical system

$$\text{n-SrTiO}_2\text{/1M NaOH/Pt} \quad (6.5)$$

The electrodes are connected in short-circuit, so the current is not used to provide power but simply to regenerate the oxidized and reduced species. This is the simplest type of cell for producing fuel, but as it works with aqueous electrolytes it is subject to corrosion problems.

6.6.1.2 Photoelectrochemical Storage Cells. These cells are secondary batteries which are recharged by light. A photochemical reaction produces some photoproduct(s) at the photoelectrode, this product then being stored ready for reaction at the dark electrode. The dark reverse reaction of these products

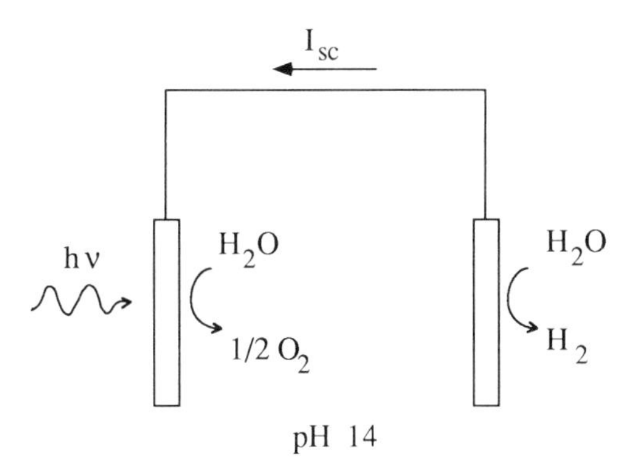

Figure 6.25 *Outline of a cell for the photoelectrolysis of water. The photoexcited semiconductor electrode is made of* $SrTiO_2$

is used to produce electricity in the absence of light. For practical reasons 2-electrode cells are little used and 3- or 4-electrode cells are preferable. Figure 6.26 gives an outline of such a 3-electrode cell as it operates (a) under irradiation and (b) in the dark. The switch is used to disconnect the photoelectrode from the circuit in the dark; the two 'dark' electrodes only are then connected as in a conventional battery. The oxidation and reduction products generated under irradiation revert to the reactants in the dark.

6.6.1.3 Photovoltaic Electrolysis Cells. These cells produce both fuel (*e.g.* H_2) and electrical power by means of a low-energy electrochemical reaction such as the dissociation of HI. The principle of this cell is shown in Figure 6.27.

6.7 ATMOSPHERIC POLLUTION AND THE OZONE LAYER

The composition of our atmosphere is complex and varies with altitude. It also varies with time following both short- and long-term variations, from the daily cycles, to processes which cover many years. We are concerned here

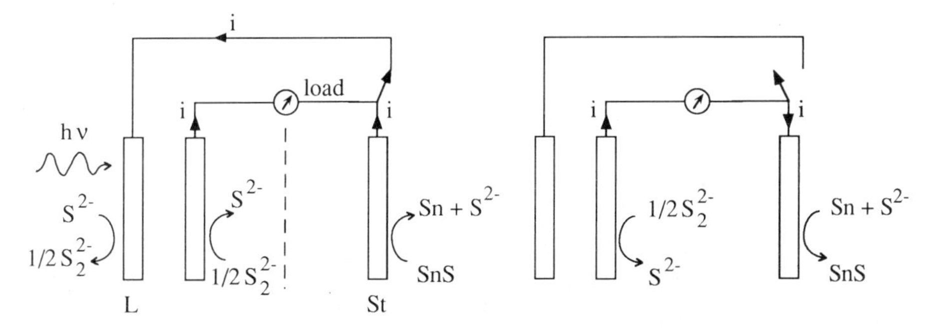

Figure 6.26 *The principles of photoelectrochemical storage cells. L is the photoexcited electrode which, under illumination, provides a current i and builds up photoproducts. These can recombine in the dark, again producing an electric current. St =*

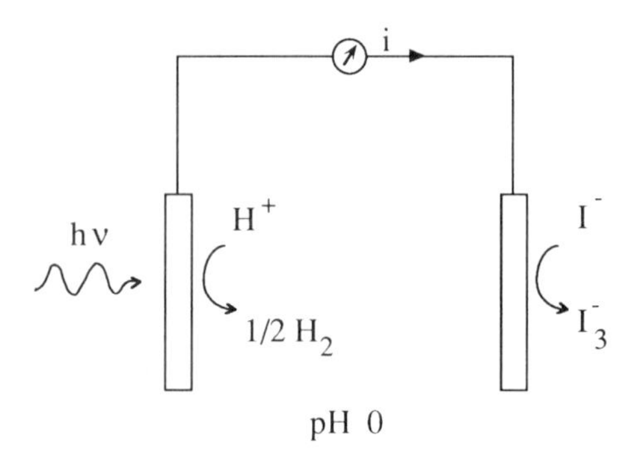

Figure 6.27 *Outline of a photovoltaic electrolysis cell*

only with the photochemical effects such as the notorious photochemical smogs and similar pollutions. In addition, we shall be concerned with the photochemical formation and destruction of the ozone layer which has the important function of absorbing far UV light.

Although air is made up essentially of N_2 (78 vol. %) and O_2 (21 vol. %) there are many minor components which must be taken into account. Of these CO_2 (0.03 vol. %) and the rare gases (*e.g.* Ar, 0.9 vol. %) represent the highest concentrations under normal conditions, but there is also water (of the order of 0.5 p.p.m. depending on the humidity), methane (2 p.p.m.), N_2O (0.5 p.p.m.), and hydrocarbons, as well as oxygen compounds of sulfur and of nitrogen. These result from combustion processes, the hydrocarbons being present because of incomplete combustion.

At the surface of the Earth sunlight contains little UV radiation thanks to the absorption properties of the stratospheric ozone layer. Major sources of pollution are therefore chemical species which absorb in the VIS region, for example NO_2 (Figure 6.28). It is formed in combustion processes and undergoes a photodissociation reaction with VIS light to form a pair of radicals which lead to further reactions.

$$NO_2 \xrightarrow{h\nu} NO^{\bullet} + O^{\bullet} \tag{6.6}$$

$$O_2 + O^{\bullet} \rightarrow O_3 \tag{6.7}$$

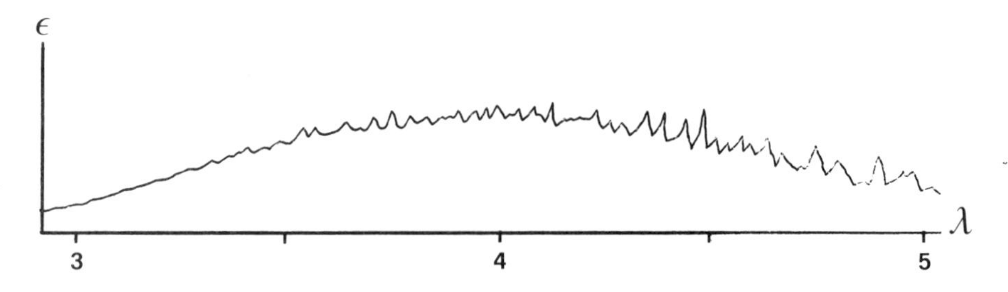

Figure 6.28 *Absorption spectrum of NO_2 (wavelength λ in nm/100)*

Ozone absorbs in the near UV region and can therefore be decomposed by sunlight with the production of singlet (excited) oxygen $^1O_2^*$. The details of the formation of the hydroxyl radicals, $OH^\bullet$, are not yet clear, but these radicals lead to secondary processes with hydrocarbons, RH, and with SO_2 to produce final products like peroxides and sulfuric acid.

$$RH + OH^\bullet \rightarrow R^\bullet + H_2O \tag{6.8}$$

$$OH^\bullet + SO_2 \rightarrow H\dot{S}O_3 \tag{6.9}$$

$$OH^\bullet + H\dot{S}O_3 \rightarrow H_2SO_4 \tag{6.10}$$

The oxygen atom $O^\bullet$ will then react with most organic molecules to form oxidation products; peroxides and carbonyl derivatives in particular. With molecular oxygen it leads to formation of ozone O_3

$$O_2 + O^\bullet \rightarrow O_3 \tag{6.11}$$

6.7.1 The Ozone Layer; Its Photochemical Formation and Degradation Processes

There is around the Earth a layer of ozone in the upper atmospehre which has the important role of absorbing far UV light, in the wavelength region below 320 nm. Sunlight of this wavelength can be quite dangerous for living organisms, since it is absorbed by the essential biological molecules such as nucleic acids and proteins.

Ozone is formed by the photodissociation of molecular oxygen, O_2, with UV light of very short wavelength (around 200 nm)

$$O_2 \xrightarrow{hv} O^\bullet + O^\bullet; \qquad O_2 + O^\bullet \rightarrow O_3 \tag{6.12}$$

Ozone itself absorbs light of longer wavelength, near 300 nm, and undergoes a dissociation reaction

$$O_3 \xrightarrow{hv} O_2 + O^\bullet \tag{6.13}$$

The O atoms then react with ozone to again form molecular oxygen, O_2

$$O_3 + O^\bullet \rightarrow 2O_2 \tag{6.14}$$

In this way an equilibrium is maintained between the three forms of oxygen, O, O_2 and O_3. This equilibrium can be upset by chemical species which are carried to the upper atmosphere, specifically nitrogen oxides and chlorofluorocarbons.

6.7.1.1 Reactions of Nitrogen Oxides. Molecular nitrogen, N_2, is the major component of air, and it is a very stable and chemically inert species. However, it can form various oxides by reaction with oxygen at high temperatures, such reactions being important in combustion processes. In the presence of atomic oxygen $O^\bullet$ an overall destruction of ozone takes place.

$$NO_2 + O^\bullet \rightarrow NO + O_2; \qquad NO + O_3 \rightarrow NO_2 + O_2 \tag{6.15}$$

6.7.1.2 Reactions of Chlorofluorocarbons ('CFCs'). The chemical compounds known as 'CFCs' are widely used as cooling agents in refrigerators because of their remarkable stability under normal conditions (ambient temperature and absence of far UV light). When these compounds are released into the atmosphere, they drift in the winds and air currents to reach, eventually, the ozone layer. At such high altitudes the filtering effect of ozone decreases gradually, so that photodissociation reactions with short-wavelength UV light becomes important.

$$CFCl_3 \xrightarrow{h\nu} CFCl^{\bullet}_2 + Cl^{\bullet} \qquad \text{('Fluorocarbon 11')} \qquad (6.16)$$

$$CF_2Cl_2 \xrightarrow{h\nu} CF_2Cl^{\bullet} + Cl^{\bullet} \qquad \text{('Fluorocarbon 12')} \qquad (6.17)$$

The chlorine atoms, $Cl^{\bullet}$ will then react with ozone:

$$Cl^{\bullet} + O_3 \rightarrow ClO + O_2; \qquad ClO + O \rightarrow Cl^{\bullet} + O_2 \qquad (6.18)$$

These processes, which result in the destruction of the ozone layer, are at the centre of the concern about the continued use of CFCs.

6.7.2 Time Dependence of Pollutant Concentrations

It can be reasonably expected that the concentrations of pollutants of general household and industrial origins should follow the scale of human activities, broadly speaking population density and industrial activity. Climatic factors should also be taken into account, as shown by the notorious cases of

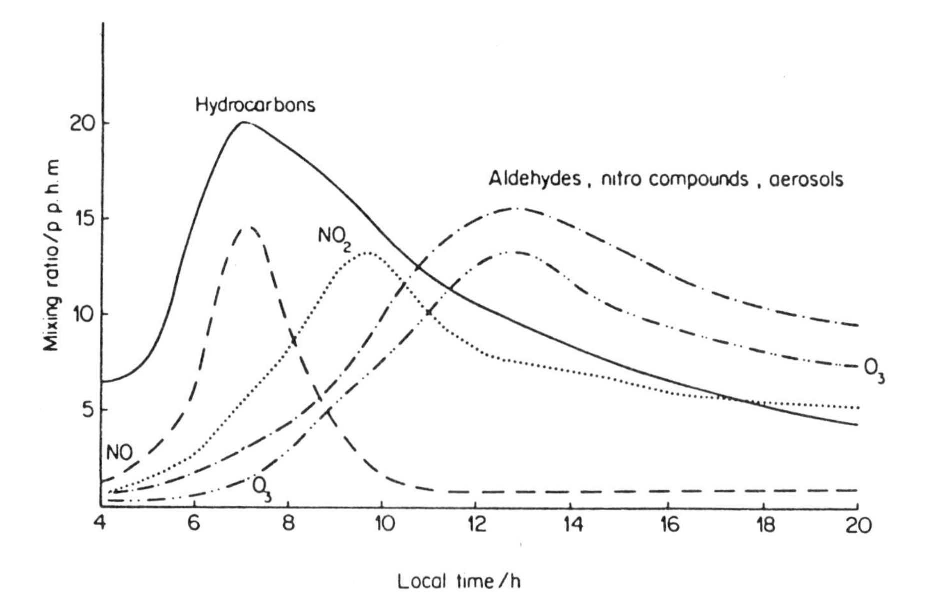

Figure 6.29 *Relative concentrations of hydrocarbons, nitrogen oxides (NO_2) and hydrocarbon oxidation products between 4 am and 8 pm in a city*

photochemical smogs in Los Angeles and Athens; to take but two well known examples.

It is hazardous to attempt any general long-term forecast, for instance over a time-scale of many years. However, there are reliable data concerning the short-term variations of pollutant concentrations related to human activities, and Figure 6.29 shows a typical concentration *versus* time function of three classes of pollutants over a 1 day period in a major city somewhere in the industrial society. Things are on the whole quiet during the small hours of the morning, although even then the concentration of unburnt hydrocarbons (H) remains rather high. This results mostly from domestic heating during the night. In early morning the rush-hour traffic brings a sharp increase in hydrocarbons and especially nitrogen oxides (NO) which can also be traced to incomplete combustion. Photochemical reactions then build up oxidation products which reach a peak of concentration at around midday.

The sequence of dark reactions which follow the photochemical processes during the night are not shown here. The light-induced reactions are then stopped, and much human activity also declines sharply. The graph in Figure 6.29 shows that great care must be taken when comparisons of pollutant concentrations are made without proper reference to the time of day and time of year.

CHAPTER 7

Experimental Techniques

7.1 LIGHT SOURCES, FILTERS AND MONOCHROMATORS

7.1.1 Sunlight, the First Light Source

Sunlight plays a most important role of course in photobiology as well as in many photochemical processes of industrial significance. For quantitative experimental photochemistry, it is however far from being an ideal light source because of its variability. It is a trivial remark that the intensity of sunlight varies with time, as well as with the weather. It may be less obvious that its spectrum also shows variations, as the concentrations of absorbing species in the atmosphere vary with the time of day and the season of the year. An example of the spectrum of sunlight was shown in Figure 5.1, p.163.

In the early days of modern photochemical investigations, sunlight was often used as the source of light. There are indeed some photographs dating back to the 1920s that show the Italian pioneers Ciamiciam and Silber with rows of flasks on the roof of their laboratory in Bologna. Sunlight can still be useful for comparative experiments, inasmuch as all the samples will experience the same variation in light intensity, but the real problem resides in the absorption spectra of the samples. If these spectra are different, then the overlap between the spectrum of sunlight and the sample's absorption spectrum must be taken into account according to the integral

$$W = \int_{-\infty}^{+\infty} I(\nu)A(\nu)\,\mathrm{d}\nu \tag{7.1}$$

where $I(\nu)$ is the intensity of sunlight and $A(\nu)$ is the relative absorption of the sample for all frequencies of light ν. The quantity W is a flux of photons absorbed if I and A are given in the proper units.

7.1.1.1 The Importance of Sunlight in Industry. Sunlight is hardly ever used nowadays for quantitative laboratory photochemical research, but it remains the most important light source for many indirect industrial applications which involve the light-induced degradation of chemicals.

There are some standard tests for the assessment of the photostability of dyes and pigments, for example, which must be realistic with respect to the actual use of such chemicals. The complex conditions of exposure to sunlight, to air and to moisture are difficult to reproduce in the laboratory, and for this reason field tests are widely used.

There is yet another very important use of sunlight, *i.e.* its conversion into electricity or hydrogen. In this case also it is obvious that for practical applications sunlight is the only acceptable light source.

7.1.2 Incandescent Lamps and Arc Lamps

The most simple and cheapest lamps are those of the incandescent type in which a metal wire is heated by an electric current. These give continuous spectra which correspond to the temperature of the wire, as shown in the example of Figure 7.1. The disadvantages of these lamps are of two kinds. In the first place, their output in the UV region is relatively low, and this restricts their use in photochemical research. In the second place, the optical quality of the beam is poor, because of the relatively large size of the incandescent wire. Such beams cannot be focussed to sharp points, so that in many optical devices there are important losses.

The most useful light sources in photochemistry are the arc lamps. These are made of two (sometimes three) metal electrodes encased in a glass or silica bulb (Figure 7.2) filled with a gas or a gas mixture. When an electric current of sufficient intensity passes through the electrodes, the gas is ionized to a plasma which acts as an electrical conductor. The spectrum of the light depends on the gas, as shown by a few examples in Figure 7.3. The low-pressure mercury arc gives practically a single line at 254 nm and is used in many applications, for instance as bactericidal lamps. The medium-pressure mercury arc produces many lines in the UV and the VIS regions and is probably the most useful light source for photochemistry. As the pressure of mercury is increased, a continuum appears below the atomic emission lines, and this offers of course a wider choice of wavelengths. The last example in Figure 7.3 is the spectrum of the xenon arc. This is a nearly

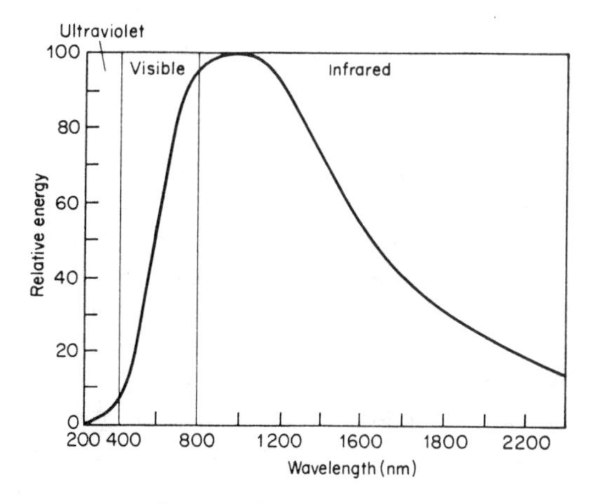

Figure 7.1 *Emission spectrum of an incandescent lamp*

Figure 7.2 *Outline of an arc lamp*

continuous spectrum which covers a very wide range of wavelengths, between 200 nm and the far IR.

The advantage of the arc lamp is that the actual light source is very small. It is concentrated between the electrodes in a volume of a few mm^3, and the outgoing beam can be focussed accurately by optical devices such as lenses and curved mirrors. However, the power supplies needed for arc lamps are much more complex and expensive than those for incandescent lamps. Arc lamps are not electrically conducting when cold, so that a high voltage pulse is required to start them. The current must be controlled very accurately to ensure a steady light intensity.

7.1.3 Some Other Light Sources

In photochemical research and its applications, lasers and arc lamps are very important. Lasers will be considered in section 7.2. A few other light sources of interest will be mentioned here.

The deuterium arc is an important light source for the production of far UV light, its spectrum being shown in Figure 7.4. The intensity of these lamps is however rather low, and they are seldom used for irradiation or excitation of luminescence.

There are many atomic emission lamps which give very precise line spectra. These are little used in photochemical applications, but are useful as wavelength calibration standards. A small selection of available wavelengths is listed in Table 7.1.

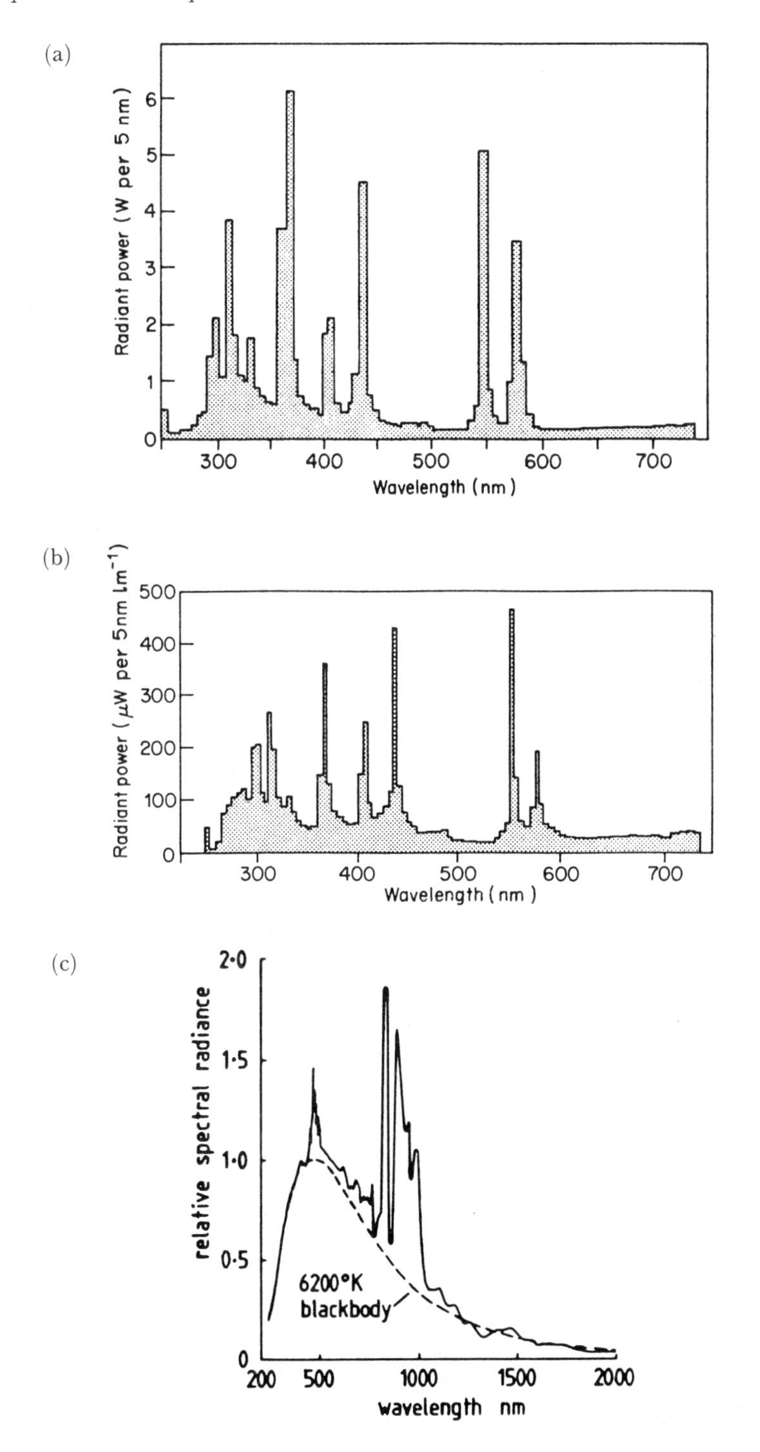

Figure 7.3 *Examples of emission spectra of some arc lamps: (a) medium-pressure mercury arc; (b) high-pressure mercury arc; (c) xenon arc*

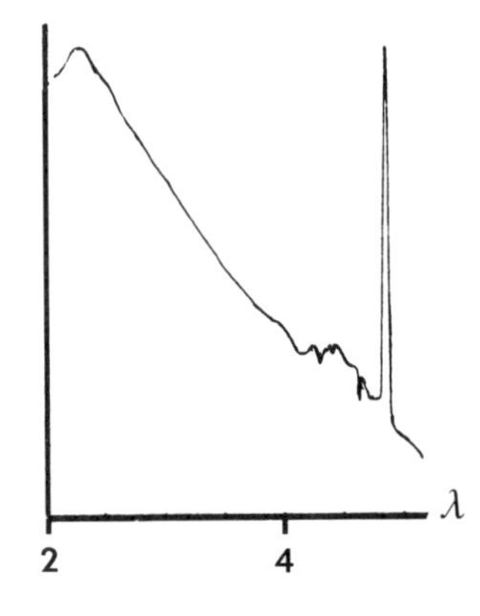

Figure 7.4 *Emission spectrum of the deuterium lamp (λ in nm/100)*

Table 7.1 *Wavelength calibration standards (nm)*

Cd	509, 644
Zn	468, 472, 481, 636
Tl	535
Pb	364, 368, 406

7.1.4 Optical Filters

Most light sources emit light of several wavelengths. Many give continuous spectra over wide wavelength ranges. For most applications in photochemistry it is then necessary to isolate a narrow wavelength region which is defined by its 'nominal' wavelength, λ, which corresponds to the maximum intensity, and the bandwidth, $\Delta\lambda$, which is a measure of the 'purity' of the light (Figure 7.5). There are two types of filters which work on quite different principles. The first type is the 'colour filter' which simply absorbs the unwanted wavelengths and lets through the required wavelengths of light. Colour filters can be gases contained in glass or silica cells, liquid solutions, or solid films. There is a wide selection of dyes incorporated into thin polymer films that can be used as filters. A small selection of spectra is shown in Figure 7.6.

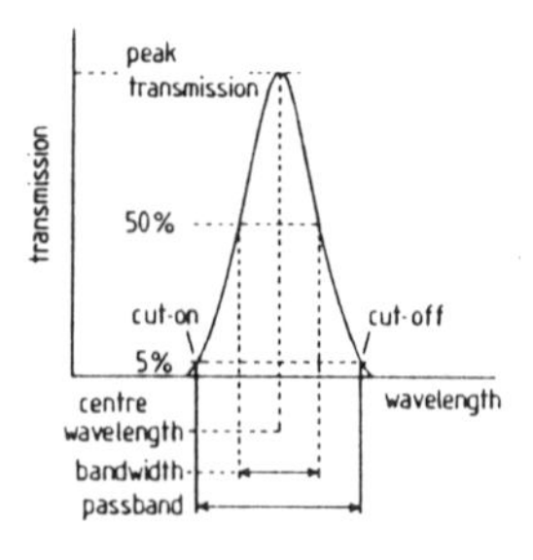

Figure 7.5 *Definitions of the resolution of a monochromator*

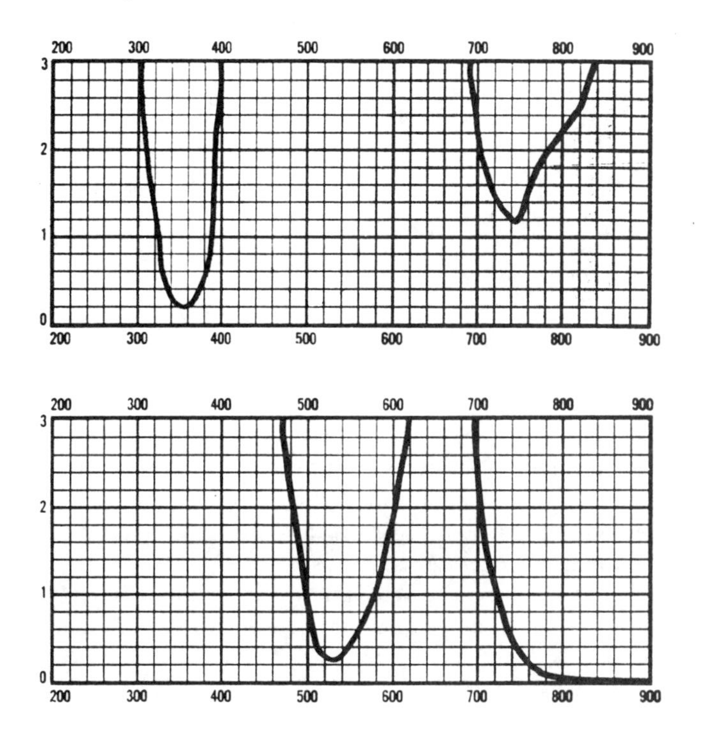

Figure 7.6 *Examples of transmission spectra of colour filters. Horizontal axis, wavelength in nm; vertical axis, absorbance*

These filters are quite cheap and very easy to use. Their spectral characteristics are however not very precise, and their photochemical stability is often limited. Several filters can be superimposed to obtain light of higher purity.

The second type is that of 'interference filters'. These work on the principle of interference of light waves, and consist of layers of reflecting coatings deposited on a substrate of glass or silica. Light is transmitted when the difference in the optical path between two beams leads to constructive interference. This principle applies also to gratings and plays an important role in optical instruments (Figure 7.7). Depending on the spacing of the reflecting layers, interference filters can be classed as 'broadband' or 'narrow-band'. The narrowest bandwidths are still of the order of 15 nm, and this may be very large for some spectroscopic applications.

Unlike colour filters, interference filters and gratings can lead to some optical artefacts of which all workers in this field should be aware. All processes of interference of light waves depend on the difference in optical path between the beams, as illustrated in Figure 7.7. The general law for constructive interference is then

$$\lambda = 2nt/m \tag{7.2}$$

and this shows that all wavelengths $m/2$ will be transmitted by the filter,

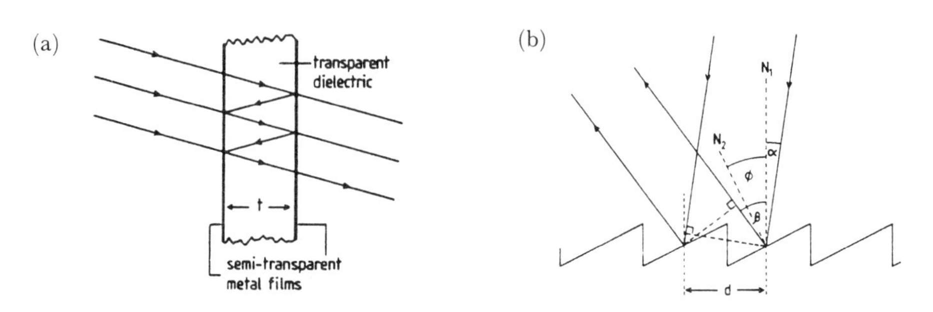

Figure 7.7 *Principle of interference with an interference filter (a) and a diffraction grating (b)*

where m is an integer. For $m = 1$, the light corresponds to the fundamental wavelength; other values of m correspond to the various harmonics. When an interference filter is made to transmit light at 300 nm, it will also transmit light at 600 nm, at 900 nm, and so on. It may then be important to eliminate this stray light by means of coloured filters.

Finally, it can be mentioned that there are 'graded' interference filters which can be used as very simple monochromators. The spacing of the reflecting layers varies with the length of the filter, so that different wavelengths are transmitted as the filter is moved in front of a slit. The wavelength resolution is not very high (some 15 nm) but such simple monochromators can still be useful, in particular for some teaching instruments.

In the case of a grating, a slightly different form of this eqn. 7.2 is used, the path difference between two lines or grooves of distance d being expressed as a function of the angles α and β.

$$m\lambda = d(\sin\alpha \pm \sin\beta) \tag{7.2a}$$

The angular dispersion of the grating is then

$$\mathrm{d}\beta/\mathrm{d}\lambda = m/d\cos\beta \tag{7.2b}$$

7.1.5 Spectrographs and Monochromators

Most light sources give complex line spectra or continuous spectra. Spectrographs are used to disperse the light in space, so as to obtain a spectrum of the intensity of light against wavelength or frequency. Monochromators are used to isolate a narrow wavelength range from the whole spectrum.

Figure 7.8 shows the outline of a spectrograph. The polychromatic light is focussed on the entrance slit S and falls on a spherical mirror M_1 which makes it roughly parallel. This beam then reaches the dispersing element D which can be a prism or a diffraction grating. Different wavelengths then follow different directions and are refocussed by the spherical mirror M_2. The diffracted spectrum is formed on the wall W of the instrument. In a spectrograph a photographic plate or a diode array is placed here, so that an entire spectrum can be recorded. In a monochromator there is only a narrow

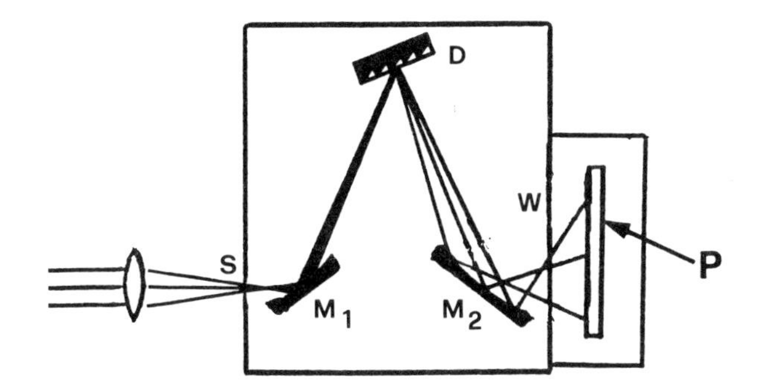

Figure 7.8 *Outline of a grating monochromator. S, slit; M_1, M_2, spherical mirrors; D, diffraction grating; W, wall; P, photographic plate*

slit in the wall so that a light detector placed behind the slit will measure the light intensity over a narrow wavelength range.

There are many different designs of spectrographs and monochromators which follow the general outline of Figure 7.8. The main consideration is the choice of the dispersing element, prism or grating. Most modern instruments use diffraction gratings which have the advantage of linear dispersion, that is, the angular dispersion is independent of the wavelength of light. In contrast, the refractive index of glass or silica used to make a prism is wavelength dependent, so that rather elaborate corrections are needed to obtain spectra which are linear in wavelength or frequency (Figure 7.9). In the case of monochromators, the prism has the added disadvantage of not having a constant bandwidth. The slit widths of the monochromator should be adjusted automatically to correct for this effect.

For all these reasons diffraction gratings are used in most dispersive optical instruments. It is however essential to bear in mind the artefacts that can arise from harmonic transmissions.

Figure 7.10 shows a simplified emission spectrum of anthracene as it would be observed on a grating fluorimeter over a wide wavelength range. The real emission at around 400–500 nm is reproduced with lower intensity in the 800–1000 nm region which corresponds to the first harmonic. Higher order harmonics are not observed because of the limitations of the sensitivity

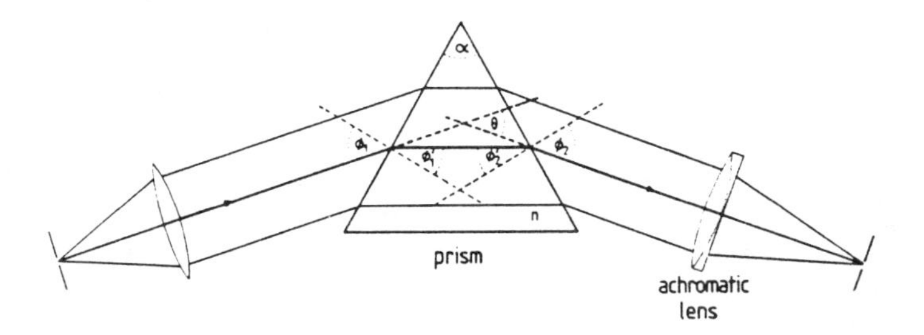

Figure 7.9 *Outline of a prism monochromator*

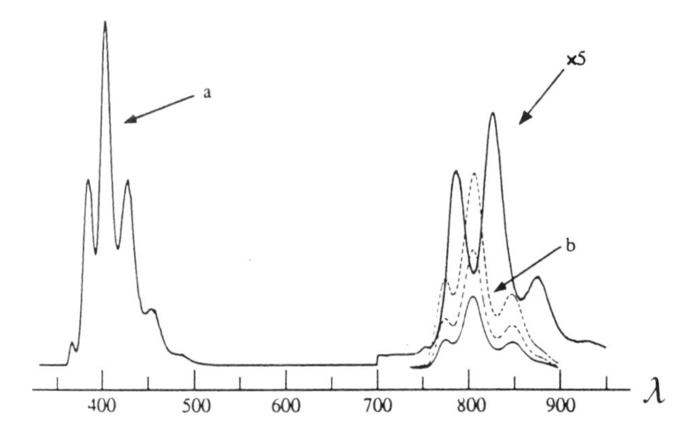

Figure 7.10 *First- and second-order fluorescence spectra of anthracene observed through a grating monochromator. Horizontal axis, wavelength in nm; vertical axis, fluorescence intensity in arbitrary units*

of light detectors. Nevertheless the first harmonic transmission could easily be mistaken for a long wavelength emission.

From the point of view of the photochemist there are three important features in the choice or design of a monochromator.

(1) The spectral range. Diffraction gratings can be used only over a limited wavelength range around the 'blaze' wavelength which corresponds to the maximum efficiency. The overall wavelength range is usually of the order of 500 nm, so it may be necessary to have several interchangeable gratings to cover the vacuum UV to the near IR.

(2) The linear dispersion. This is the spectral width in nm transmitted through the exit slit of the monochromator. It is given in units of nm/mm. For example 5 nm/mm would mean that the nominal bandwidth of the 'monochromatic' light is 5 nm if the slit width is 1 mm. This defines the purity of the monochromatic light.

(3) The numerical aperture. This defines the angle of the beam of light which will be accepted by the monochromator. It is defined by the ratio of the diameter of the focussing element (*e.g.* the spherical mirror, M_1, in Figure 7.8) to its focal distance. The intensity of the light obtained at the exit slit of the monochromator depends on this value, the smaller numerical apertures corresponding to higher illumination. The spectral purity of the 'monochromatic' light is better with higher numerical apertures. For photochemical irradiations or luminescence excitation, high intensity, low numerical aperture instruments are preferable.

7.1.5.1 Double and Multiple Monochromators. If monochromatic light of very high purity is required, it is often necessary to use two or several monochromators in series, such that the exit light of the first one becomes the entrance light of the second. The reasons for using such complex optical arrangements are two-fold.

(1) All real optical elements (mirrors, lenses, gratings, prisms) scatter some

of the light. This scattered light forms a spectral background which approximates a weak spectrum of the light source. This can be most troublesome with line sources such as mercury arcs.
(2) The dispersing elements (gratings or prisms) always have minor physical defects which produce spectral aberrations known as 'ghosts'. The real spectrum is followed by a slightly displaced ghost spectrum of much lower intensity and these can cause problems in the measurement of low-intensity light (weak luminescence or Raman scattering), or when very high spectral resolution is needed. The use of double or multiple monochromators reduces these spectral defects, at the expense of the final, exit light intensity.

7.2 LASERS

The word 'laser' stands for light amplification by stimulated emission of radiation. The process of stimulated emission is the release of a photon by an excited atom or molecule under the impact of an incoming photon. This can be written as

$$M^* + h\nu \rightarrow M + 2h\nu \tag{7.3}$$

Under normal conditions this process is negligible because the population of ground state molecules, M, far exceeds that of excited molecules, M*. The incoming photons are then absorbed by the ground state species

$$M + h\nu \rightarrow M^* \tag{7.4}$$

Stimulated emission therefore requires the condition of 'population inversion', in which the concentration $[M^*] > [M]$. This can be achieved in many different ways. A few lasers of special interest in photochemistry will be considered here although it cannot be an exhaustive review of laser technology.

7.2.1 General Principles: Two-, Three- and Four-level Lasers

The heart of a laser is the 'active material' which will emit the light. This will be produced through some form of excitation, it can be a chemical reaction, the absorption of UV light or the passage of an electric current. In all cases excited states of atoms or molecules in the active material are formed, and the population of these excited states will increase until population inversion is reached. The excitation of the ground state species is called 'pumping'. In optically pumped lasers the energy levels of the active material fall into the classes of three- or four-level systems. A simple energy-level diagram provides an illustration of these two processes (Figure 7.11). In the three-level laser the emission light $h\nu_L$ is lower in frequency than the excitation light $h\nu_P$. The emission light cannot be absorbed by the ground state species, M. The situation is rather similar in the four-level laser, except that emission does not go back directly to the ground state molecule, M, but to some intermediate state, which can be the unrelaxed ground state.

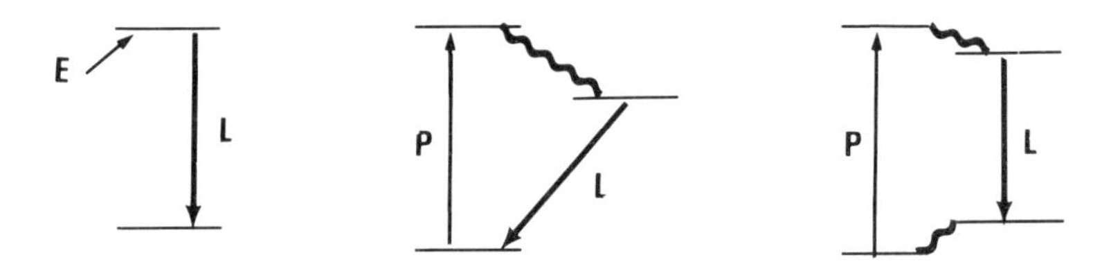

Figure 7.11 *Principles of lasers and energy levels. E is the input of electrical or chemical energy, P the 'pump' light and L the laser emission*

7.2.2 Solid State Lasers: Ruby and Nd/YAG

These are some of the most important lasers used in photochemical research. As they are rather similar; only the ruby laser is described here in detail. The active material of the ruby laser is a dispersion of Cr^{3+} ions in alumina, Al_2O_3, in the form of a glass rod. This is in fact a synthetic 'ruby', not the natural half-precious stone which would not have the required degree of purity; the details of the synthetic process are outside the scope of this book.

As shown in Figure 7.12, the synthetic ruby rod is placed along a flash lamp, L, inside the laser cavity. The walls of this cavity are reflecting, curved mirrors so that most of the flash light will reach the active material. In the axis of the ruby rod there are two mirrors which define the length of the 'resonator', a most important concept in laser physics. One of these mirrors is totally reflecting, the 'output' mirror being only partially reflecting. These mirrors are the boundaries for the light waves produced inside the laser cavity. They keep the light within the cavity while the excited state population is being built up inside the active material. The pump light is a flash of white light similar to that of a photocamera. The Cr^{3+} ions are promoted to an excited state which decays very rapidly to the metastable excited state which will emit the laser light. In such a laser the light emission would simply follow the time-scale of the pump flash over several milliseconds, and this would not be useful for the study of fast photochemical reactions. There are two additional devices which allow the generation of very short pulses of laser light in the ns and ps time-scales.

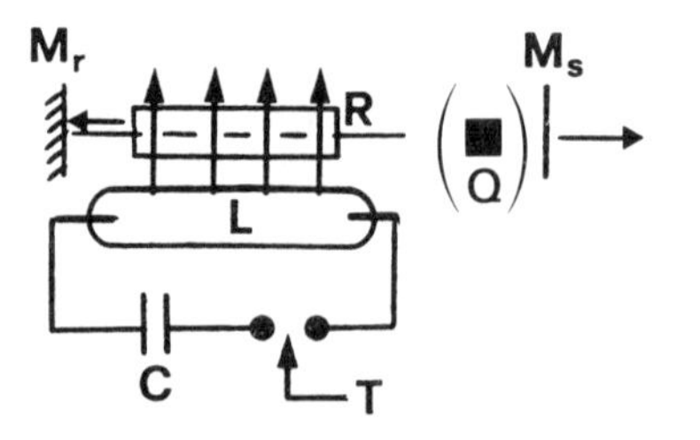

Figure 7.12 *Outline of a pulsed solid-state laser. L, flash lamp for pump light; C, high-voltage capacitor; T, trigger spark gap; M_r, reflecting mirror; M_s, partially reflecting mirror; R, rod of lasing material; Q, Q-switch cell or crystal*

7.2.2.1 Q-switching. The simplest device consists of a dye cell placed inside the laser resonator cavity (Q in Figure 7.12). The dye absorbs the laser light, so long as this light is weak; but when it reaches a very high intensity the dye is bleached and the laser light is reflected between the mirrors, so that stimulated emission becomes the major process of radiative deactivation of the emitting species in the active material. If the dye concentration is correct, the entire output of the laser can be concentrated into a single pulse. This is called the 'giant pulse'. All excited molecules emit within a very short time through a kind of chain reaction. The giant pulse has a time span of a few ns (Figure 7.13).

This technique is known as the 'passive' Q-switch. The dye acts as an absorber for weak light, so that the population of excited atoms or molecules in the active material can increase until its maximum level is reached. The dye cell is in fact a high-speed shutter.

There is another way to obtain giant laser pulses of a few ns duration, known as 'active' Q-switching. The shutter is an electro-optical cell which is triggered at some preset time after the pump flash. These electro-optical shutters are Kerr cells or Pockels cells.

7.2.2.2 Mode Locking. The two mirrors which are placed in the axis of the active material define the 'resonator' of the laser. The light waves which are built up in the active material must be stationary waves in the cavity, so that the allowed wavelengths are given by eqn. (7.5) (see also Figure 7.14).

$$\nu = (c/2Ln)q; \qquad \lambda = c/\nu \tag{7.5}$$

n is the refractive index of the cavity, and q is an integer. These wavelengths are the 'longitudinal' modes of the laser. Although they give in principle all the allowed wavelengths, only a small number actually contribute to the laser light emission, according to the spectrum of the active material. This light of different wavelengths travels at different times through the laser cavity and produces the ns pulse which is in fact made up of a burst of much shorter pulses. The 'passive' mode locking is somewhat similar to the passive Q-switch. A saturable absorber dye is placed inside the cavity. The weaker pulses are absorbed, and finally only the most intense pulse is transmitted,

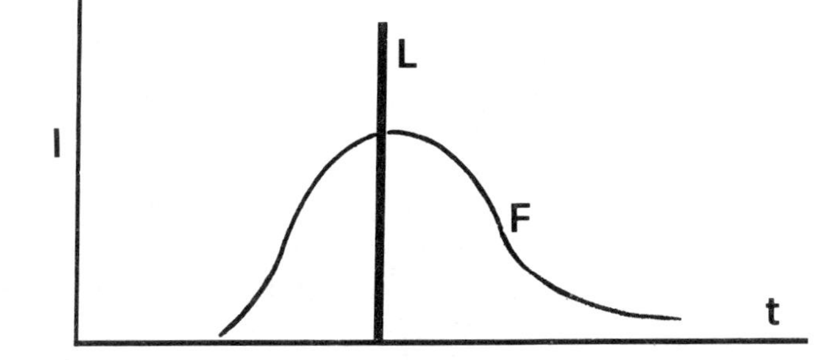

Figure 7.13 *Shape of the pump flash light spectrum and of the Q-switched laser light. t, time; I, light intensity*

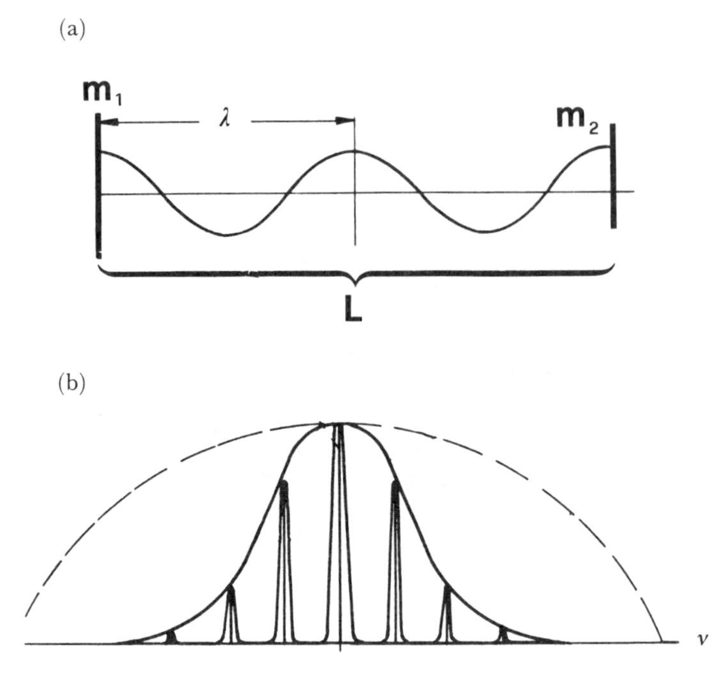

Figure 7.14 *Principles of mode-locking. (a) Standing waves in a cavity of length L defined by two mirrors m_1, m_2. (b) Selection of frequencies within the emission spectrum of the lasing material; ν is the frequency*

corresponding to a single longitudinal mode. The output pulse is highly monochromatic and has a duration of a few ps.

7.2.3 Frequency Conversion of Laser Light

Most solid-state lasers emit light of long wavelengths which is not very useful in photochemistry. The ruby laser has an emission at 694 nm and the Nd laser emits at 1065 nm, both in the far VIS or near IR regions. Light of such wavelengths in not useful for most photochemical reactions, since the molecules of interest absorb only in the UV. For most applications, this 'fundamental' laser light is therefore transformed into light of higher frequencies which are the second, third or fourth harmonics of frequencies 2ν, 3ν and 4ν respectively, ν being the fundamental frequency.

The process of frequency 'multiplication' is in fact a frequency addition which takes place in 'non-linear' materials. Figure 7.15 gives a simple picture of this technique. If the photon energy is not sufficient to promote the atom or molecule to an excited state, it will still reach a 'virtual' state S_ν. This state is not stable, and it has a very short lifetime before it falls back to the ground state with the re-emission of the photon. If a second photon happens to hit the molecule promoted to its virtual state, then the molecule may

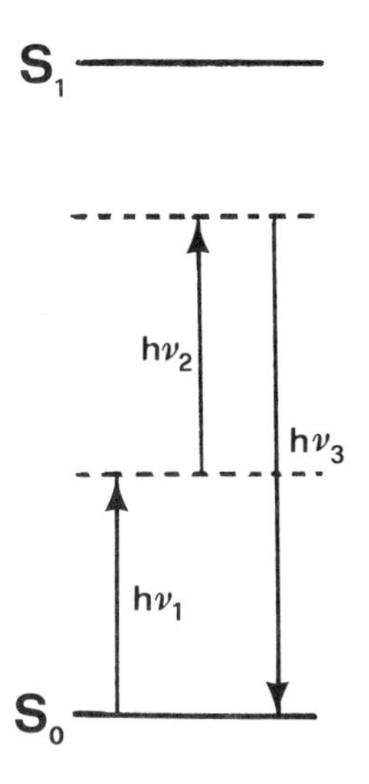

Figure 7.15 *Principle of frequency addition through a virtual state (shown by the dotted line)*

reach a higher virtual state which can emit a photon of frequency $h\nu_3 = h\nu_1 + h\nu_2$.

In the wave picture of electromagnetic radiation, the intensity I of the light transmitted by any material depends on the value of the electric field ε, according to the equation

$$I = c_1\varepsilon + c_2\varepsilon^2 + c_3\varepsilon^3 + \ldots \tag{7.6}$$

In principle this series goes to infinity, but in fact the coefficients c_i decrease rapidly and at low light intensities all terms higher than c_1 can be neglected. This is the 'linear' region. At high light intensities the higher terms become important, so that the transmitted beam will contain radiation of frequencies 2ν, 3ν, *etc.* This is the 'non-linear' region, since the light intensity is no longer proportional to ε. The incoming light of a single frequency ν comes out partly as light of frequency 2ν, the conversion efficiency from ν to 2ν being of the order of 10% in the best cases. In theory there could be some light of frequencies 3ν, 4ν, *etc.*; in practice these are negligible as the conversion factors become very small.

The conversion of laser light to the third or fourth harmonics requires its passage through two non-linear optical elements (Figure 7.16). In the first element frequency doubling takes place; in the second element the fundamental can be added to the frequency-doubled beam to produce light of 3ν, or the frequency-doubled light can be added to itself to produce light of frequency 4ν.

7.2.3.1 Up-conversion. It is also possible to add the laser light of frequency ν_L to some other light, for instance luminescence light of frequency ν_F to obtain a whole spectrum of frequencies $\nu_L + \nu_F$. The two beams of light are mixed

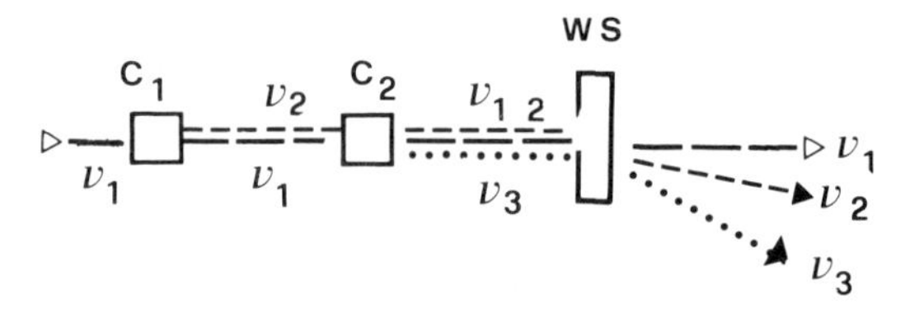

Figure 7.16 *Outline of a frequency multiplier for laser light. C_1, C_2, second harmonic generators; WS, wavelength separator. The frequencies of light are shown as ν_1 (fundamental), ν_2 first harmonic and ν_3 second harmonic*

in the non-linear optical element which in many diagrams is noted simply as SHG, for 'Second Harmonic Generator'.

7.2.3.2 Non-linear Optical Materials. The most widely used non-linear optical elements are inorganic crystals such as KDP. There is a growing interest in the use of organic molecules for these applications, for example derivatives of 4-nitroaniline (Figure 7.17). These molecules have exceptionally large excited state dipole moments and it is probable that their virtual states have rather long lifetimes. In the wave picture of light, it would be said that such molecules have exceptionally large hyper-polarizabilities; a concept discussed in Appendix 7A.

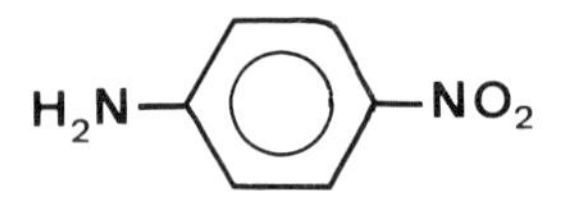

Figure 7.17 *Structure of 4-nitroaniline*

7.2.4 Gas Lasers

These lasers are also called—incorrectly—'excimer' lasers. It will be clear that they could be called 'exciplex' lasers. The active material is a gas mixture which contains a halogen (F_2 or Cl_2 in most cases) and a 'rare' gas such as Kr, Ar or Xe. These cannot form any stable compounds in their ground states, but excited state species do exist and can fluoresce. These excited state species (*e.g.* KrF) are formed through the recombination of ions, for instance

$$Kr^+ + F^- \rightarrow (KrF)^* \xrightarrow{h\nu} KrF \rightarrow Kr + F \tag{7.7}$$

The molecule $(KrF)^*$ is the exciplex (all too often incorrectly named 'excimer') which decays to its dissociative ground state with emission of light. The gas laser consists of a cavity which contains two metal electrodes (Figure 7.18). A high-voltage electrical pulse is passed through the electrodes to ionize the gas mixture. The cavity boundaries are two mirrors, one of them being totally reflecting and the output mirror being partially reflecting; a common feature of most lasers.

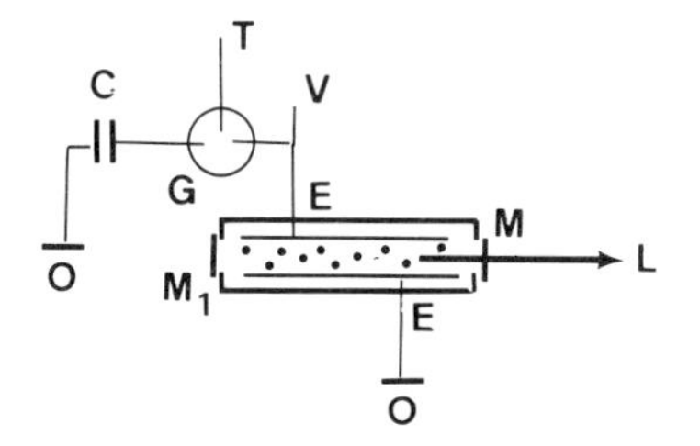

Figure 7.18 *Outline of a gas laser. C, high voltage capacitor; T, trigger spark gap; E, electrode; M_1, M, mirrors; L, laser beam; V, high voltage; O, ground*

The optical quality of the light of these lasers is poor. The beam shows considerable divergence, while the beams of solid-state lasers are nearly parallel. The exciplex lasers are however very useful in photochemistry, as they give the highest light intensities in the UV region. By changing the gas mixture, a number of useful wavelengths of light can be obtained, of which a selection is given in Table 7.2.

Table 7.2 *Laser light wavelengths (nm) and energies (mJ)*

	Wavelength	*Energy*
ArF	193	400
KrCl	222	60
KrF	248	800
XeCl	308	500
XeF	351	300

7.2.5 Dye Lasers

The active material in a dye laser is a liquid solution of a dye in a solvent. The fluorescence of the dye is excited by a 'pump' light, which can be a flash of white light (seldom used nowadays) or a UV laser pulse from an Nd/YAG or exciplex laser. Figure 7.19 shows the principle of such a dye laser. The cavity consists of a diffraction grating, G, which plays the role of a totally reflecting mirror, and of a partially reflecting mirror which will let the laser

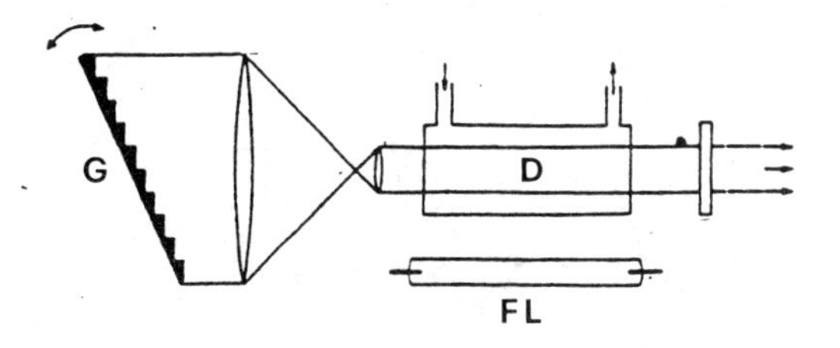

Figure 7.19 *Principle of a flash-lamp pumped-dye laser. FL, flash lamp; D, dye cell; G, diffraction grating*

light through. The grating can be tilted to various angles which will define the wavelength of the light that can be transmitted through the cavity. The laser emission wavelength can therefore be tuned within the fluorescence spectrum of the dye. There is a large selection of laser dyes which cover a wide spectral range (Figure 7.20).

7.2.6 The Properties of Laser Light

For the photochemist the laser is just a source of light, though of a special kind. There are two quite different types of lasers known as 'continuous wave' (cw) and 'pulsed' lasers. In the former the light keeps a steady intensity, like in an ordinary lamp; in the latter the light comes out as single pulses of high intensity and short duration, or in trains of pulses. Laser light presents three particular properties which distinguishes it from other light sources. These are *parallelism*, *monochromaticity* and *coherence*.

7.2.6.1 Parallelism. In laser light, the light is emitted only along the resonator axis, whereas in ordinary lamps and arc lamps light is emitted in all directions. Even so, laser light must have some angular divergence, but this can be very small. The divergence is expressed in radians or milliradians.

7.2.6.2 Monochromaticity. The light emitted by a laser corresponds to an atomic or molecular luminescence. Atomic emissions such as those of Cr^{3+} (ruby laser) or Nd^{2+} (Nd/YAG laser) are naturally very sharp so that the light is highly monochromatic. The fluorescence of molecules used in dye lasers is much broader, but in practice the emission is still only a narrow wavelength range selected by the grating (Figure 7.19). Dye lasers are 'tunable', while the ruby or Nd lasers are not.

The monochromaticity of laser light has many advantages in photochemistry. The absorbance of a sample at the laser wavelength is defined very accurately, so that Beer's law applies quite strictly even when the absorption spectrum of the sample is highly structured.

Some lasers emit light of different wavelengths. The 'argon ion' laser is an important example. Nevertheless the different lines are still highly monochromatic, even if the complete emission may appear to be white to the human eye.

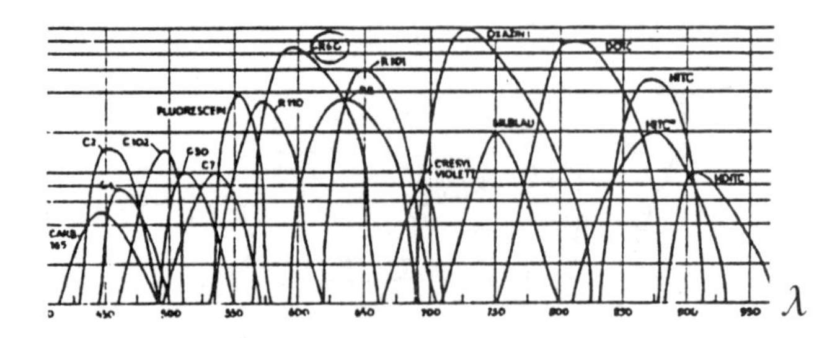

Figure 7.20 *Selection of emission spectra of laser dyes. Horizontal axis, wavelength in nm; vertical axis, relative emission intensity on a logarithmic scale*

7.2.6.3 Coherence. In the wave picture of light the laser resonator cavity must be filled with stationary light waves. These light waves therefore preserve their coherence when they leave the laser. Figure 7.21 gives an illustration of coherent and incoherent waves. Ordinary light sources (lamps, arc lamps) emit incoherent light, as the many excited atoms or molecules emit quite independently in a random manner. The laser emits through the process of stimulated emission, so that the time when atoms or molecules emit is synchronized. The property of coherence of light waves is of little interest in photochemistry, but has interesting applications in optics.

7.2.7 'Self-phase' Modulation: White Light from Monochromatic Lasers

When an intense pulse of monochromatic laser light is focussed on a transparent liquid or solid, there is an emission of 'white' light over a wide continuous spectral range. This process is known as 'self-phase modulation'. We will not consider its physics. For our purpose it is important to note its photochemical implications. On the one hand, this pulse of white light can be used to provide a probe light in ps and fs flash photolysis (sections 8.1 and 8.2). On the other hand, it can be a source of stray light in some luminescence measurements. This comes as a surprise to many users of lasers for luminescence kinetics measurements, but it is an unavoidable problem.

7.3 LUMINESCENCE MEASUREMENTS

'Luminescence' is the radiative deactivation of atoms and molecules. In practice it is the emission of light by a sample which can be a gas, a liquid, a solid, or biological matter, *etc.* The main measurements concern *emission* and *excitation* spectra, luminescence *kinetics* (or lifetimes) and *time-resolved emission spectra*. The basic principle of a spectrofluorimeter is the same for all these experiments (Figure 7.22). A light source L provides the excitation light, through a filter or monochromator [Mn (ex)]. This light is focussed on the sample, S, and the luminescence light is collected by a spherical mirror or by

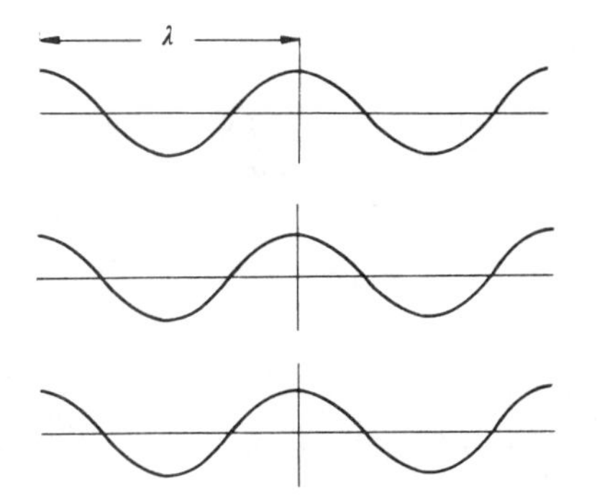

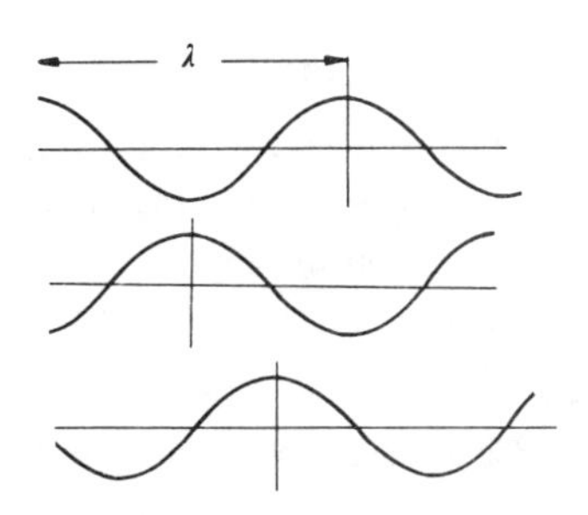

Figure 7.21 *Incoherent and coherent waves*

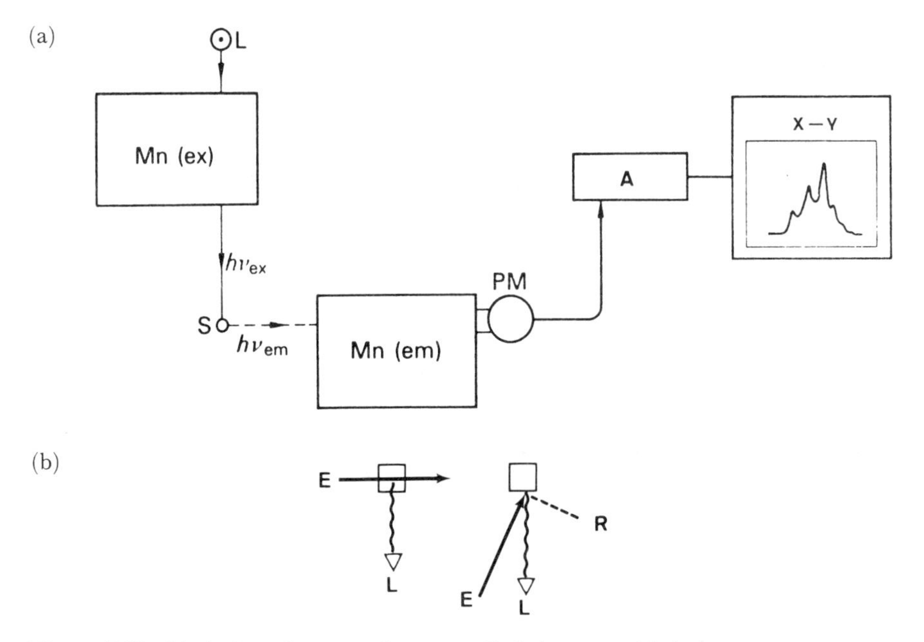

Figure 7.22 *(a) Outline of a spectrofluorimeter. L, light source; Mn (ex), excitation monochromator; S, sample; Mn (em), emission monochromator; PM, photomultiplier tube; A, amplifier; X–Y, recorder. (b) Right-angle (left) and front-face (right) excitation. E, excitation beam; L, luminescence; R, reflection*

a lens usually at right angle from the excitation beam. This luminescence light goes through a second monochromator, Mn (em), (or filter) and its intensity is measured by a detector which is usually a photomultiplier (PM) tube but sometimes a photodiode or a phototransistor.

'Luminescence' covers all emissions of light in the near IR, VIS and near UV spectral regions. The origin of the luminescence can be specified as 'photoluminescence', 'electroluminescence', 'chemiluminescence', or 'bioluminescence' for example. These definitions depend on the mode of formation of the excited molecule which eventually emits the luminescence.

7.3.1 Photoluminescence

The excited states are produced through the absorption of light. This is a major process of luminescence in photochemistry. As mentioned in section 3.4, a distinction is made between 'fluorescence' and 'phosphorescence'. The conditions required for their observation are discussed below.

7.3.2 Electroluminescence

In some cases excited states are formed when two ions recombine to form neutral product molecules, according to

$$M^+ + N^- \rightarrow M^* + N$$

followed by the radiative deactivation

$$M^* \rightarrow M + h\nu$$

Such reactions take place in electrolytic cells, at the boundary of the fluids which carry the ions M and N.

7.3.3 Chemiluminescence and Bioluminescence

These have been described in sections 4.8 and 5.8. As in the case of electroluminescence, there is no need for an excitation light source. The luminescence is observed when two reacting samples are mixed together. The excitation and emission beams are generally at right angles, so it is necessary to use samples of reasonably low absorbance at the excitation wavelength (less than 0.3). If a very concentrated sample must be used, then the arrangement known as 'front face illumination' is preferred (Figure 7.22b). The angle of 45° should of course be avoided since it corresponds to the reflection of the excitation light in the direction of the emission monochromator. Very high purity of the excitation light is required for this technique because of the high intensity of scattered light.

There is another artefact that can arise with concentrated samples, and this is the reabsorption of emission light by the sample itself. Light which is emitted at the centre of the cell must travel through a pathlength of a few mm of the sample, and absorption will take place in the wavelength region of overlap of the emission and absorption spectra. This problem can be serious when the Stoke's shift of these spectra is small. Reabsorption then results in an apparent red shift of the emission maximum with increasing concentration.

7.3.4 Correction of Emission and Excitation Spectra

When a luminescence spectrum is obtained on an instrument such as that used to produce the spectra in Figure 7.23, it will depend on the characteristics of the emission monochromator and the detector. The transmission of the monochromator and the quantum efficiency of the detector are both wavelength dependent and these would yield only an 'instrumental' spectrum. Correction is made by reference to some absolute spectra. Comparison of the absolute and instrumental spectra then yields the correction function which is stored in a computer memory and can be used to multiply automatically new instrumental spectra to obtain the corrected spectra. The calibration must of course be repeated if the monochromator or the detector is changed.

Similarly, the instrumental excitation spectrum depends on the wavelength characteristics of the excitation monochromator and of the light source. Mercury arcs which emit line spectra are not suitable and xenon arcs are normally used. Correction is made against a reference such as a solution of Rhodamine which has a wavelength-independent fluorescence quantum yield.

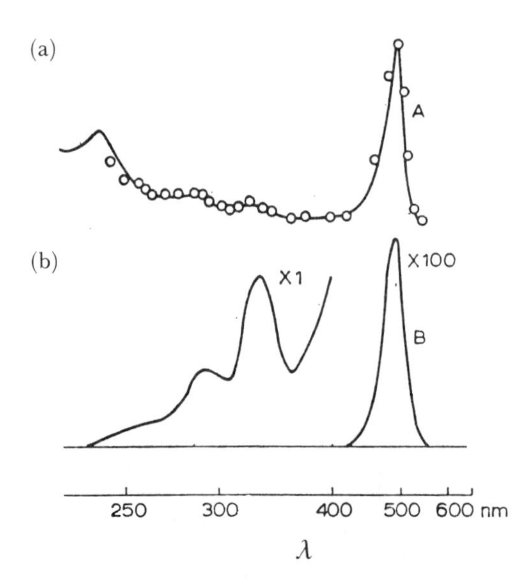

Figure 7.23 *Absorption and excitation spectra of Rhodamine G in ethanol. (a) Corrected excitation (open circles) compared with absorption (full line). (b) Uncorrected excitation spectrum. Vertical axes, fluorescence intensity in arbitrary units*

The absolute excitation spectrum is then identical to the absorption spectrum. The correction factor can again be stored in a computer memory and used to multiply other instrumental excitation spectra (so long as the lamp is not changed).

Instrumental excitation spectra are so badly distorted that they are practically useless. This results from the very sharp drop in light intensity of xenon arcs in the UV region (see Figure 7.23). The Figure shows the apparent (instrumental) excitation spectrum of Rhodamine compared with the absorption and corrected excitation spectra.

7.3.5 Light Detectors

The most important light detector in photochemistry is the 'photomultiplier' (PM) tube. It is based on the photoelectric effect (section 2.1), but the 'primary' electrons released by light are accelerated over a number of 'dynodes' to produce an avalanche of secondary electrons (Figure 7.24). A single photon can produce a pulse of some 10^6 electrons at the anode. Each of these pulses lasts about 5 ns, so that when the light intensity is rather high these single pulses combine to form a steady electric current. This current is amplified and displayed on a chart recorder or computer.

The limitation in the sensitivity of a PM tube used in this way is the 'dark current', the current which passes at the anode in the absence of light. This current results from the thermoelectric emission of electrons and can be reduced by cooling the PM; in the best cases the dark currents are of a few nA.

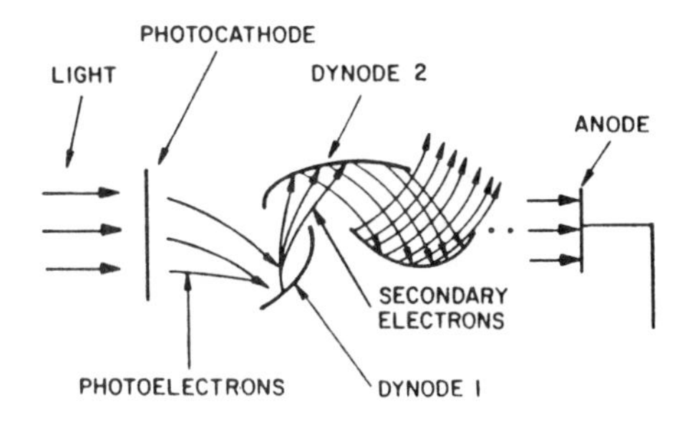

Figure 7.24 *Components of a photomultiplier tube*

The real limitation of detection in spectrofluorimetry is not the sensitivity of the detector, but rather the stray light which result from imperfections of the monochromators and emissions by impurities in the solvents. The limiting quantum yields of luminescence detection are of about 10^{-4} in optimal conditions.

There are other light detectors of interest in luminescence measurements such as photodiodes and other solid-state devices. Their sensitivity is very low compared with PM tubes, and there is a requirement for considerable amplification of the photocurrents. The solid-state photodetectors have the advantage of broad wavelength response in the IR region, beyond the range of PM tubes.

7.3.5.1 Wavelength Ranges of Light Detectors. Figure 7.25 shows the relative sensitivities of some common light detectors as a function of wavelength. PM tubes have excellent sensitivity in the UV and VIS regions, between 200 and 600 nm. Some PM tubes can still detect near IR light up to 1200 nm, but here the sensitivity is lower. The wavelength range of a PM tube depends on the nature of the photocathode material, and these spectral ranges are

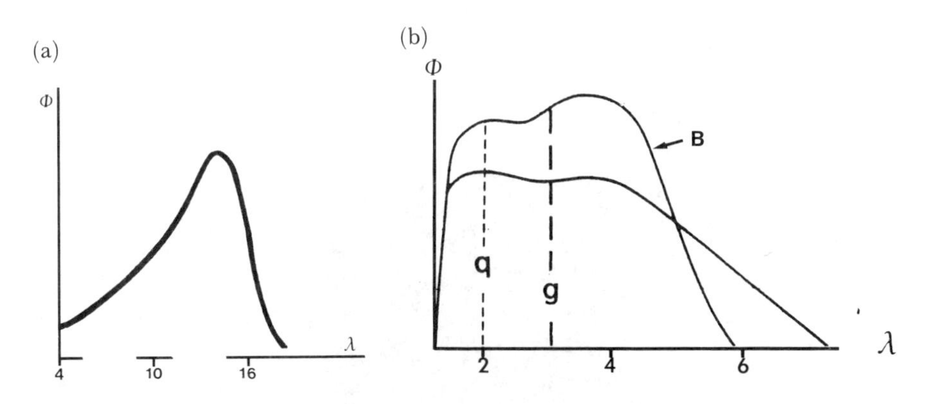

Figure 7.25 *Examples of spectral sensitivities of a photodiode (a) and of two types of photomultipliers (b). The window transmissions are shown as q (silica) and g (optical glass). Horizontal axes, λ in nm/100; vertical axes, φ relative quantum yield*

commonly referred to as 'S1, S4, S20', *etc*. These labels have no photophysical meaning but are useful for selecting a tube for a specific application.

Solid-state light detectors are best in the near IR and IR regions. Here also the spectral response depends on the nature of the material, *i.e.* the semiconductor. In the very far IR photons cannot be detected directly and the only answer is to convert light into heat. The pyroelectric detectors work on this principle.

A full response in the UV and vacuum UV is accessible only with detectors fitted with special fluoride windows. The cut-off wavelengths of optical glass (g) and of silica (q) are shown in Figure 7.25. For some special applications windowless detectors are used and the sensitivity is then greatly extended in the vacuum UV region and beyond.

7.3.5.2 Time Response of Light Detectors. Any light detector has a 'risetime' and a 'falltime' which define its time response. This can be very important in direct kinetic measurements, because the rise- and falltimes define the ultimate limits that are accessible to luminescence lifetime determinations. In an ideal experiment, an instantaneous square pulse of light would be observed, and the rise- and falltimes would be zero, but the detector cannot follow these. The actual response of the detector is defined as the 'half-life' of an exponential rise or fall, and this varies greatly with the detector. PM tubes have response times of a few ns, but there are some special tubes (microchannel plate PM tubes) which are faster. The fastest light detectors are the high-speed photodiodes which reach the ps region. An example of a laser pulse seen by such a detector is shown in Figure 7.26.

7.3.5.3 Sampling and Accumulation of Repetitive Signals. When a luminescence signal is very fast (typically under 1 ns), the acquisition system may not be able to record many points in a single experiment. If the experiment can be repeated many times, the time delay can be changed each time so that the signal is recorded over many experiments, each single measurement giving only the signal (*e.g.* luminescence intensity) at one time after the excitation

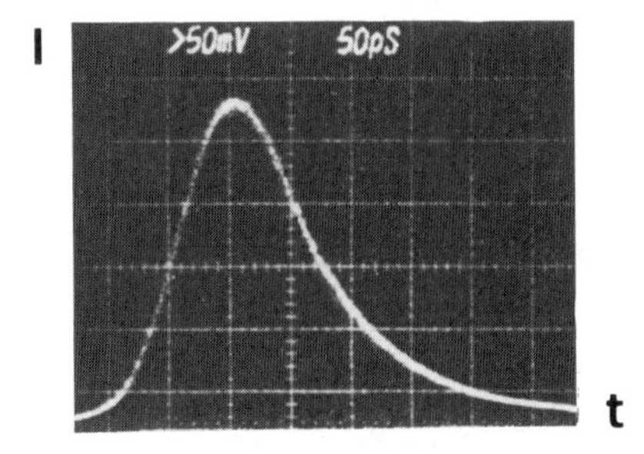

Figure 7.26 *Oscillogram of a short laser pulse seen by a fast photodiode. Horizontal axis, time in 50 ps/division*

light pulse. There are two requirements for the use of this sampling technique:
(1) the experiments must follow at a high repetition rate, of the order of kHz at least;
(2) the signals must be well reproducible from one experiment to another.
The problem of reproducibility can be solved through the method of signal accumulation, so long as the variations of the signal (the 'noise') follow a random pattern. Signal accumulation applies to single shot experiments as well as to sampling methods. In both cases the same experiment is repeated *n* times, and the final signal is the point-by-point addition of all the separate signals. The concept of 'signal-to-noise ratio' is discussed in Appendix 7B.

7.3.6 Single Photon Counting

The 5 ns pulses of about 10^6 electrons released at the anode by a photon absorbed by the photocathode of a PM tube can be used to count photons. In such instruments the intensity of light is displayed as a 'count per second' which varies between about 15 (dark count) and 10^5. A photon-counting detector system is of course much more complex than the simple PM/amplifier used in conventional spectrofluorimeters. Figure 7.27(a) is a block diagram of such a photon counter; (b) gives a simple illustration of the important process of pulse selection through a discriminator. The output of

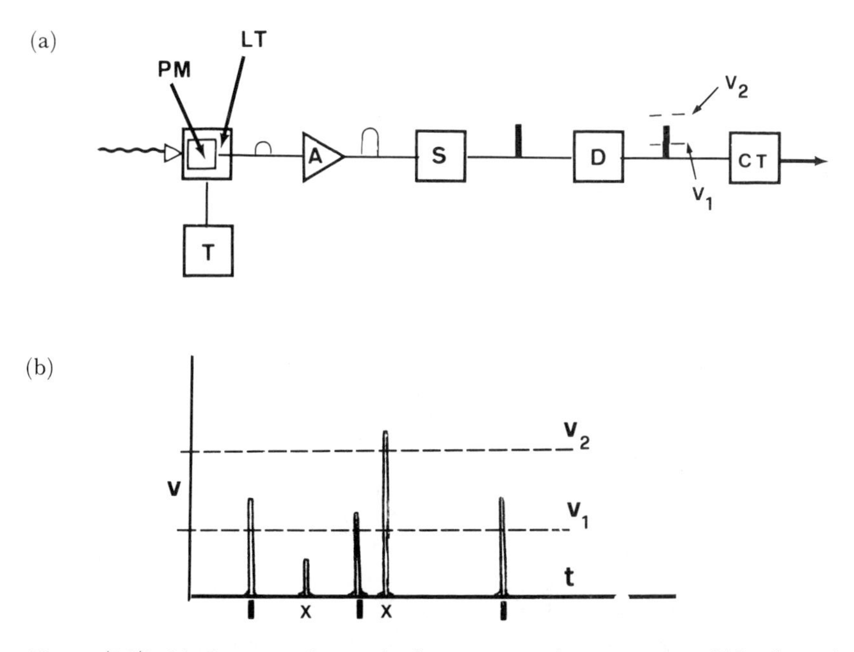

Figure 7.27 *(a) Diagram of a single-photon-counting detection system. PM, photomultiplier; LT, low-temperature housing; A, amplifier; S, pulse shaper; D, discriminator; CT, counter; T, timer. (b) Principle of the discriminator;* V_1 *and* V_2 *are the low and high signal limits*

the PM tube goes through a preamplifier (A) which gives an electrical signal of about 1 V. These signals pass through the discriminator which will select 'acceptable' pulses. The discriminator is a two-level comparator which passes only pulses of voltages between V_1 and V_2. The former is known as the 'thermoionic' level and the latter is the 'cosmic ray' level; this is perhaps not a very good term since the high pulses also include radioactivity which has nothing to do with cosmic rays.

The low level comparator will reject pulses of low intensity which are generated by thermoionic emission in the dynodes. It is obvious that an electron emitted by dynode 5 for instance will produce a much smaller avalanche at the anode than an electron emitted by the photocathode; but of course thermoionic emission from the photocathode itself would still appear as a genuine pulse, and for this reason PM tubes used in photon counting applications must be cooled down.

The high level of the discriminator rejects pulses that are much higher than those produced by primary electron emission from the photocathode. Such pulses occur when ionizing radiation passes through the PM tube, because electrons can then be released from several dynodes at (nearly) the same time.

Photon counting detection reaches the ultimate limits of sensitivity in light detection at the present time. It is useful for the detection of very weak luminescence of quantum yields below 10^{-4}; some phosphorescence emissions in liquids at ordinary temperatures can be measured in this way (Figure 7.28).

7.3.7 Experimental Conditions for Luminescence Measurements

The distinction between 'short-lived' and 'long-lived' luminescences is not totally arbitrary from the point of view of the experimental conditions required for their observation. It can be set at a lifetime of around 1 μs which corresponds to the quenching time of excited states by molecular

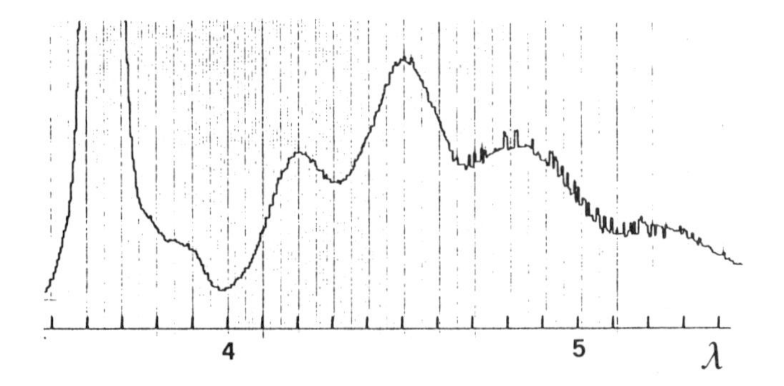

Figure 7.28 *Phosphorescence spectrum of benzophenone in acetonitrile at room temperature. Horizontal axis, wavelength in nm/100; vertical axis, light intensity in arbitrary units*

oxygen. The quantum yield of luminescence is

$$\Phi_e = k_e/(k_e + k_q[O_2]) \tag{7.8}$$

if the observable emission rate constant is k_e, the oxygen concentration is $[O_2]$, and k_q is the quenching rate constant. It is difficult to measure luminescence quantum yields Φ_e lower than 10^{-3}; not because of sensitivity problems of detectors, but because of stray light known as the 'luminescence blank' which results from luminescence of impurities in the solvents and of the walls of the containers. This generally presents no problems for fluorescence detection since $k_e = 10^8\ s^{-1}$ and the concentration $[O_2] = 10^{-4}$ in liquids under normal conditions. When the luminescence lifetimes are long—over the μs time-scale—it is necessary to remove the dissolved oxygen (Appendix 7C), or to reduce the quenching rate constant by increasing the viscosity of the solvent. For these reasons the long-lived phosphorescence emissions are observed only in solid matrices (frozen solvents at low temperatures or polymers at room temperature), or in liquid solvents which have been thoroughly deoxygenated.

7.3.7.1 The Phosphoroscope. This is a simple mechanical device which allows the separation of long-lived emissions (phosphorescence) from short-lived emissions which consist of scattered light and fluorescence. It is a disc or drum in which there are holes or slots placed in such a way that the excitation and emission beams reach the sample and the detector respectively at different times. With the fastest practicable rotation velocities of the phosphoroscope, the cut-off time is of the order of 1 ms.

Long-lived emissions of very low intensities can be observed against very high intensities of short-lived emissions. It is also possible to observe the decay kinetics of long-lived emissions directly on an oscilloscope, by varying the speed of rotation of the phosphoroscope.

7.3.8 Luminescence Quantum Yield Measurements

Absolute luminescence quantum yield measurements are not made in photophysical practice and are left to specialized laboratories such as the National Physical Laboratory (UK) or the National Bureau of Standards (USA). These provide the quantum yields of a variety of 'primary standards' that are used in practice to determine an unknown quantum yield Φ_e. First the luminescence spectrum of the primary standard is measured, and then that of the unknown sample is compared with it as the ratio of the integrated spectra.

$$\int_{\lambda_1}^{\lambda_2} I_e(\lambda)\, d\lambda \Big/ \int_{\lambda_1}^{\lambda_2} I_s(\lambda)\, d\lambda = \Phi_e/\Phi_s \tag{7.9}$$

This ratio gives the relative quantum yield, or the absolute quantum yield so long as that of the standard is accurate. There are some essential conditions to be met.

(1) The excitation light should be highly monochromatic and of high purity.

(2) The absorbance of the standard and of the sample should be very similar and should not exceed 0.1. As they cannot be identical a correction can then be made according to the relative intensities of light absorbed, assuming of course a linear relationship between the intensities of excitation and emission lights.
(3) The emission spectra of the sample and of the standard must be wavelength-corrected spectra, or (if uncorrected) must cover the same wavelength range.
This method of luminescence quantum yield measurement against a standard emitter is simple and easy to implement with computer-linked fluorimeters, in particular for the integration of the spectra. Its accuracy should however not be over-rated. It is, in the best cases, of the order of 5% (and often far worse). It remains at this time the most widely used technique in photophysics.

7.4 FLASH PHOTOLYSIS

The different techniques of flash photolysis are used to detect 'transient' species, that is atoms, molecules and fragments of molecules which have very short lifetimes. These cannot be observed by usual experimental techniques which require rather long observation times. For example, the measurement of an absorption or fluorescence spectrum takes several seconds, and this is of course far too long in the case of transient species which exist only for fractions of a second. In some cases these transient species can be stabilized through inclusion in low-temperature rigid matrices, a process known as 'matrix isolation'.

7.4.1 'Conventional' (Microsecond) Flash Photolysis

There are three major detection techniques in flash photolysis: transient absorption, transient luminescence and transient photocurrent. Figure 7.29 shows the principle of a conventional flash photolysis apparatus. The sample C is contained in a cylindrical tube set next to the flash lamp. This flash lamp is similar to a photographic flash and will emit a short, very intense light pulse when a high-voltage capacitor C is discharged through it. The flash light must be intense enough to produce a detectable concentration of transient species in the sample cell, typically of around 10^{-6} M, but this depends of course on the absorbance. There are two distinct methods used for the observation of transient absorption.

7.4.1.1 The 'Spectrographic' Technique. A short flash of white light is passed through the sample after a delay time following the 'photolytic' flash. This white light is dispersed in the spectrograph to produce an absorption spectrum on a photographic film (or on a diode array in some modern instruments). In this way the entire absorption spectrum of the sample at the set delay time is obtained over a wide wavelength range, usually from 300 to

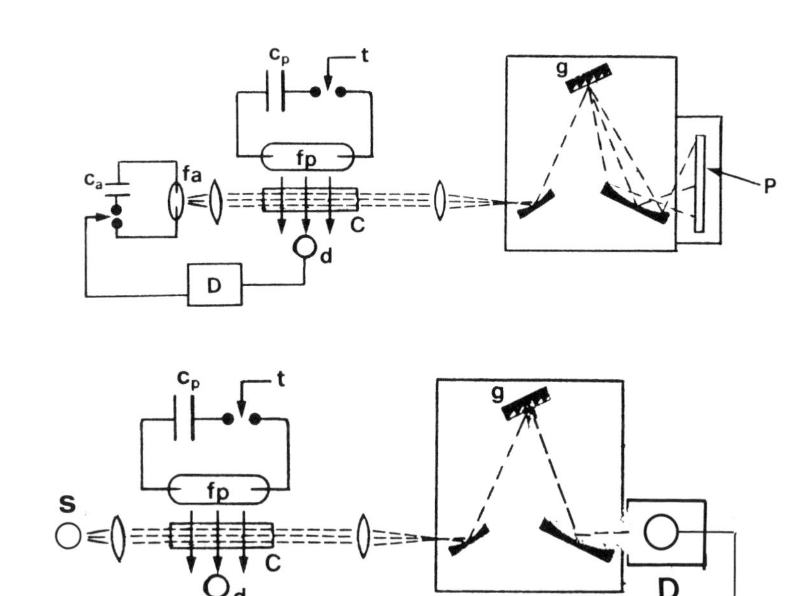

Figure 7.29 *(a) Diagram of a spectroscopic flash photolysis apparatus. C, sample cell; fp, photolytic flash lamp; fa, analytical flash lamp; Cp, Ca, high voltage capacitors; t, trigger spark gap; d, light detector; D, delay unit; g, diffraction grating; P, photographic plate. (b) Diagram of a kinetic flash photolysis apparatus. S, light source; D, light detector; X, trigger of oscilloscope time-base; Y, oscilloscope vertical signal*

1000 nm. This technique is still important for the measurement of high-resolution absorption spectra of free radicals and excited molecules in the gas phase. In the 'conventional' technique which is limited to the μs time-scale, the delay between the photolytic flash and the monitoring flash is set by an electronic device (called the delay unit).

7.4.1.2 The 'Kinetic' Technique. In this case a steady beam of light passes permanently through the sample and falls on a detector D. A single wavelength of this light is selected by a monochromator or a filter, so that the detector will display the intensity of transmitted light as a function of time following the photolytic flash. In the kinetic technique a transient absorption is observed at one single wavelength over all time. In the spectrographic technique a complete transient spectrum is observed at one single time. Figure 7.30 shows two examples of experimental observations made with these two techniques. In (a) the absorption spectra of an anthracene derivative are displayed at various delay times between the photolytic and monitoring flashes. These are due to the triplet excited state which has in this case a lifetime of some 70 μs. In (b) the kinetics of a longer-lived transient is given as an example of second-order kinetics. This is the recombination of the free radicals formed in the photoreduction of benzophenone.

7.4.2 Nanosecond Laser Flash Photolysis

With the photographic flash lamp the light pulse has a duration of several microseconds at best. The Q-switched pulsed laser provides pulses some thousand times faster, and the kinetic detection technique remains similar since photomultiplier tubes and oscilloscopes operate adequately on this time-scale. The situation is different with the spectrographic technique; electronic delay units must be replaced by optical delay lines, a technique used mostly in picosecond spectroscopy. This is discussed in Chapter 8.

One important difference in the design of a μs conventional flash photolysis apparatus and the ns laser flash photolysis system is the size of the sample. The energy of laser pulses is usually very much lower than that of photographic flashes, typically 0.1 J as against 10^3 J. For this reason the laser light must be focussed on very small samples (0.1 ml for example).

A diagram of a kinetic, ns, laser flash photolysis apparatus is shown in Figure 7.31. Transient absorption changes are similar to those obtained on conventional μs instruments but the time-scales are of course much shorter.

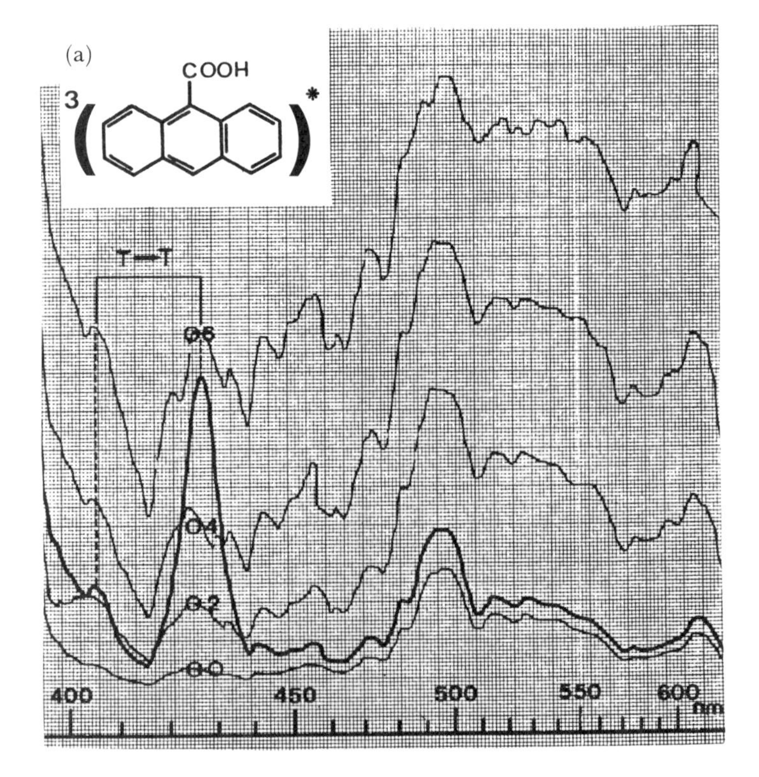

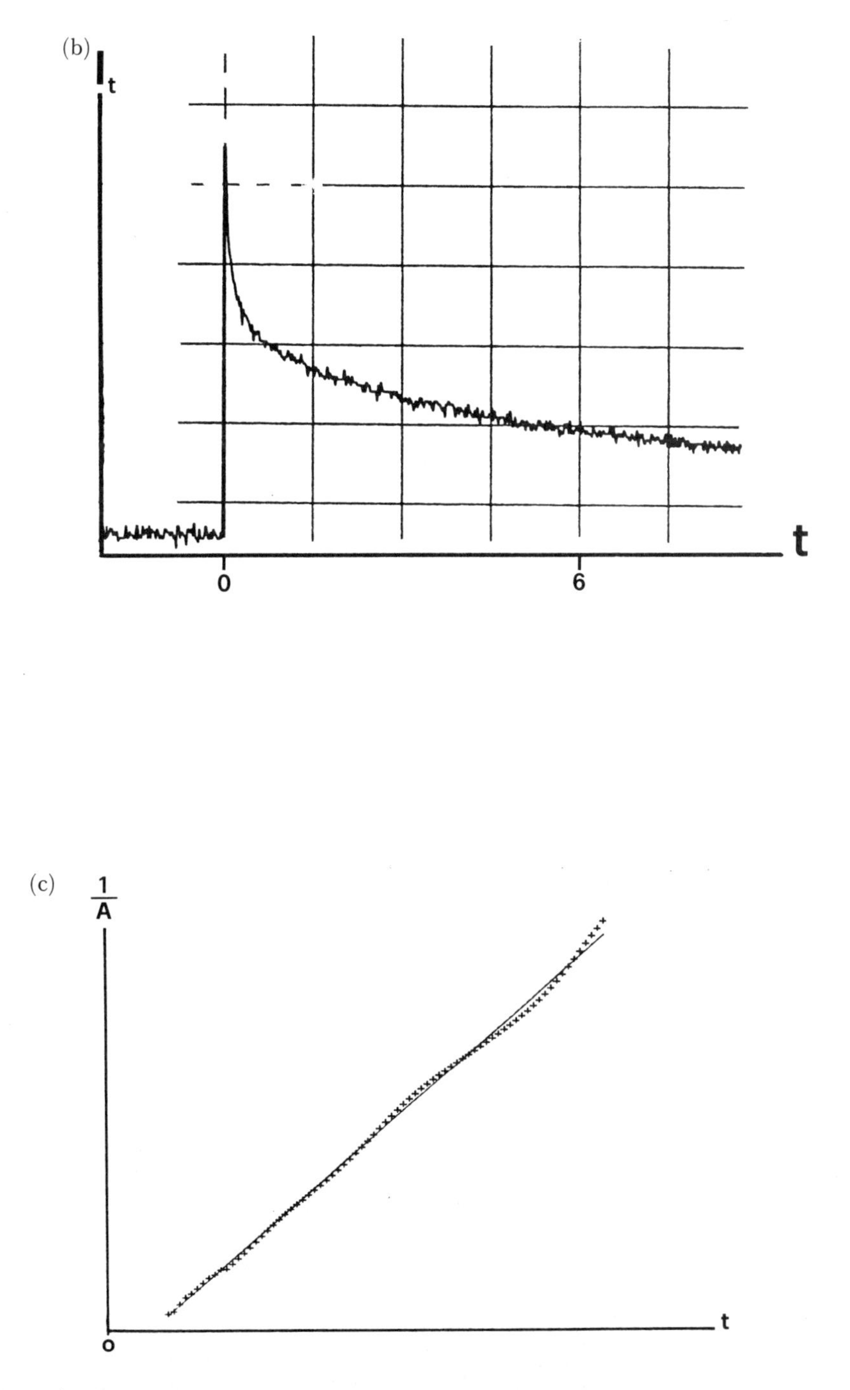

Figure 7.30 *(a) Microdensitometer records of a photographic plate showing the absorption of the triplet state of anthracene-9-carboxylic acid. Horizontal axis, wavelength in nm; vertical axis, absorbance. (b) Oscilloscope trace of the absorption of the ketyl radical of benzophenone in ethanol. (c) Second-order plot of the decay kinetics against time t*

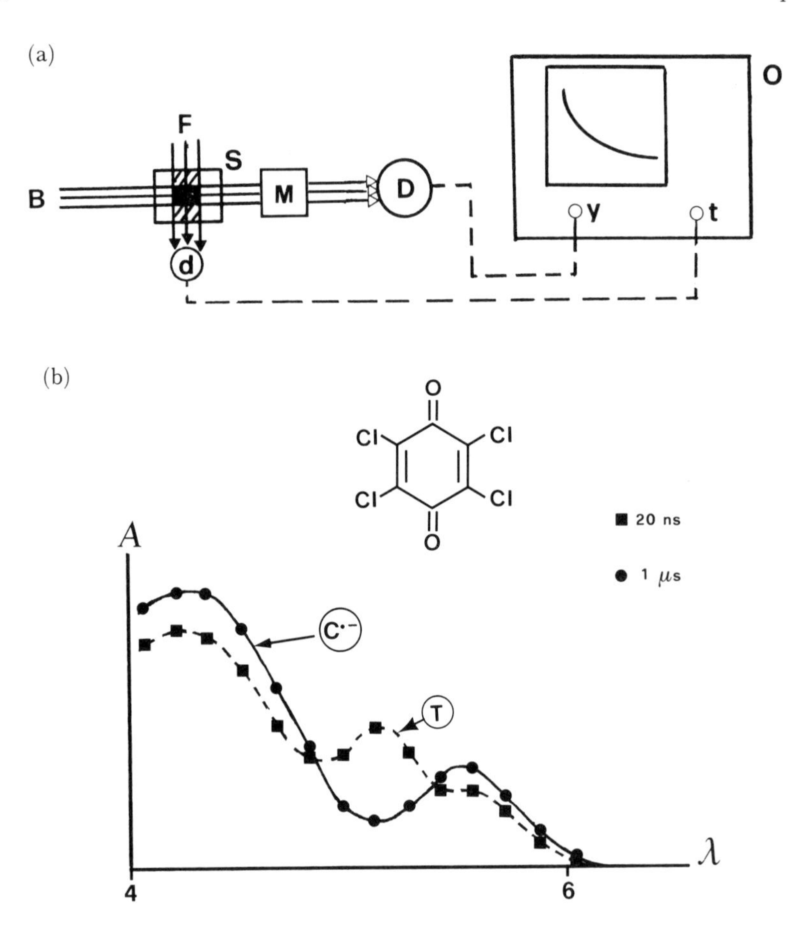

Figure 7.31 *Diagram of a ns, kinetic, laser flash photolysis apparatus. F, photolytic laser beam; B, continuous analytical beam; S, sample cell; d, light detector; M, monochromator; D, photomultiplier; O, oscilloscope with t (time-base trigger) and y (vertical signal) inputs. (b) Point-by-point absorption spectra of chloranil in acetonitrile at 20 ns, 1 μs after excitation. T corresponds to the absorption by the triplet state, $C^{\bullet-}$ by the radical anion*

Transient spectra are usually taken 'point-by-point', the monochromator being set at a different wavelength for each flash. Figure 7.31(b) gives an illustration of such point-by-point spectra.

7.4.3 Luminescence Kinetics; Deconvolution

The decay of phosphorescence emissions can be observed easily with conventional flash photolysis instruments, since they last between ms and seconds. However, fluorescence lifetimes are of the order of ns and such kinetics can be measured only by laser flash photolysis or by time-resolved single photon counting.

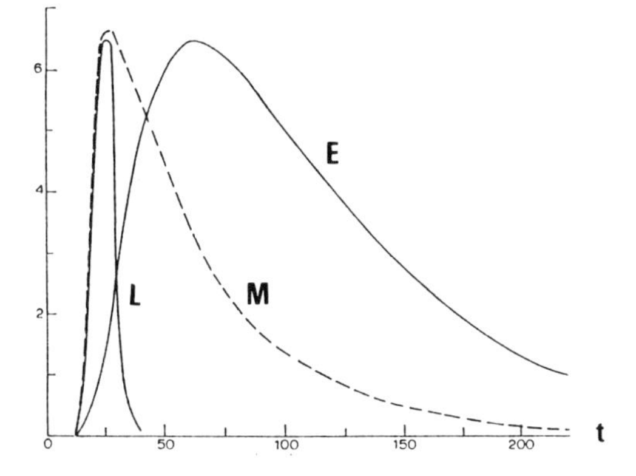

Figure 7.32 *Kinetics of luminescence of pyrene following laser flash excitation. L, laser pulse profile; M, monomer emission, E, excimer emission rise and decay. Horizontal axis, time in ns; vertical axis, light intensity in arbitrary units. The three kinetic curves are normalized to a common maximum*

7.4.3.1 Laser Flash Photolysis. The instrumental set-up is similar to that used for transient absorption, except that there is no need for a monitoring beam. Figure 7.32 shows the rise and decay of the pyrene excimer in solution.

7.4.3.2 Deconvolution. When the luminescence lifetime is close to the laser pulse duration, the kinetics can still be obtained by a process known as deconvolution, so long as the luminescence decay is exponential. The deconvolution program is a computer simulation of the shape of the laser pulse modified for various luminescence lifetimes. An example is given in Figure 7.33. The kinetics of fluorescence decay could not be resolved directly, but it is clear that the emission pulse shape differs from the laser pulse shape

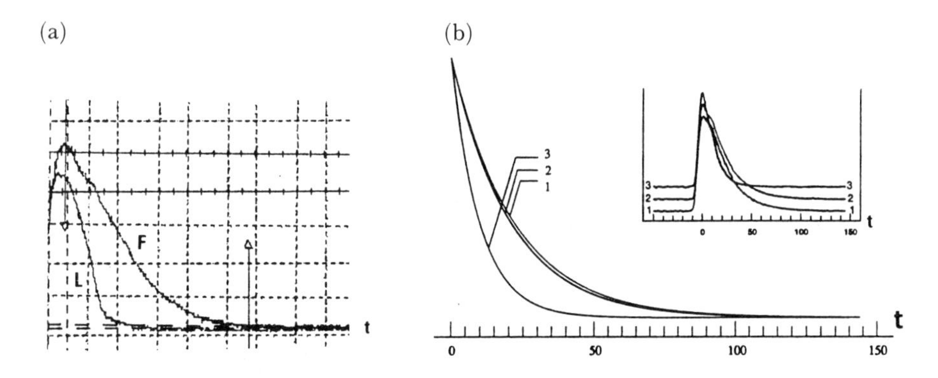

Figure 7.33 *(a) Records of an excitation laser pulse (L) and of a short fluorescence (F) without deconvolution. Horizontal axis, time in 10 ns/division. (b) Deconvoluted exponential fluorescence decays obtained from three experimental traces (inset). These are the emissions of 4-aminophthalimide in three different solvents. Horizontal axes, time in ns*

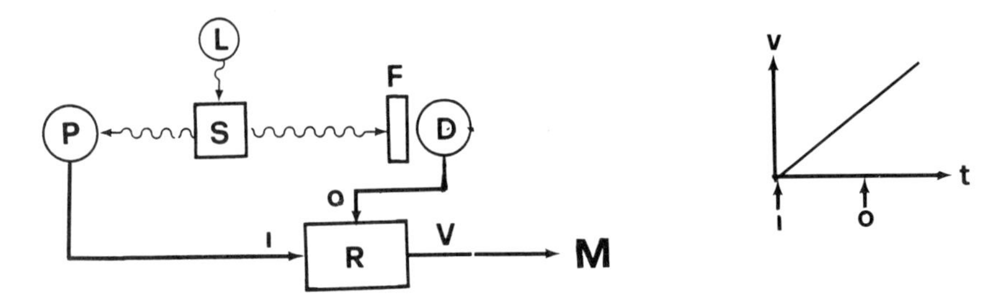

Figure 7.34 *Principle of kinetic single-photon counting. L, pulsed light source; S, sample; P, photodiode; F, interference filter or monochromator; D, photomultiplier; R, voltage ramp generator (1 to start the ramp, 0 to stop it). The voltage V is fed into a multichannel analyser M. Inset: voltage ramp V(t)*

by being a little longer. The deconvolution program allows the resolution of luminescence kinetics which are shorter than the laser pulse by a factor of 10 in the best cases.

7.4.3.3 Luminescence Kinetics from Single Photon Counting. The kinetics of very weak luminescence emissions can be obtained by this technique which measures the time between the excitation pulse and the emission of the first photon by the luminescence sample. Following the simplified instrumental set-up of Figure 7.34, a photomultiplier P detects the excitation light pulse (laser or pulsed flash lamp) to give the '0' time. This trigger pulse starts a linear voltage ramp known as the time-to-voltage or time-to-amplitude converter (inset of Figure 7.34). When the first photon is detected by the emission photomultiplier D, the voltage ramp is stopped and the corresponding channel of a multichannel analyser is incremented by one unit.

A multichannel analyser is a type of digital x–y recorder in which x is the channel number (in this case the time after the excitation pulse) and y is the number of events which occur for the relevant channel (in this case the detection of a photon). For each excitation pulse the first emitted photon only is counted, but as the experiment is repeated many times with excitation pulse trains of thousands per second the whole kinetics of luminescence is reconstructed.

7.4.4 Time-resolved Spectroscopy

In conventional luminescence measurements the observation pertains to a large number of molecules which emit at different times following excitation. The emission spectrum obtained in this way is in fact an integration of many instantaneous spectra, because in general these instantaneous spectra change in time. To take one example, consider molecules which have a much larger dipole moment in the excited state than in the ground state. The solvent will then reorganize in the course of time, and the better solvated molecules will emit at longer wavelengths (Figure 7.35). This important information about

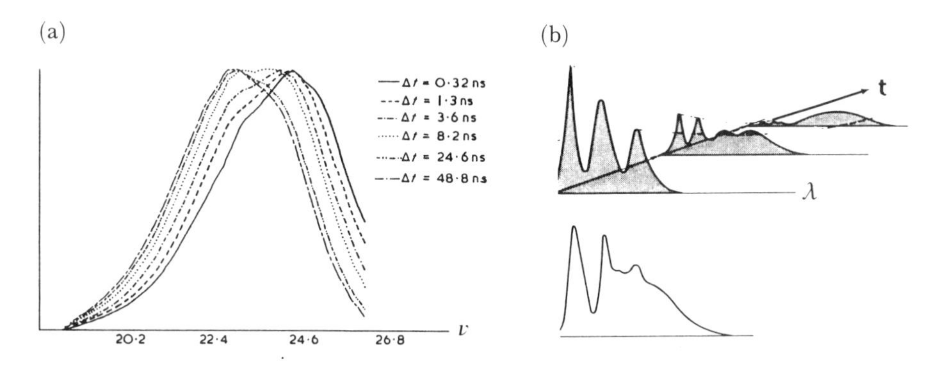

Figure 7.35 *(a) Time-resolved spectra of a molecular emission in the course of solvent relaxation. Horizontal axis, frequency in 1000 cm^{-1}; vertical axis, emission intensity in arbitrary units, with normalized maxima. (b) Outline of the spectral changes of an excimer fluorescence; λ wavelength, t time*

solvent relaxation kinetics is lost in conventional luminescence spectroscopy which observes all molecules at the same time.

Time-resolved spectra can be obtained by laser flash photolysis or by single photon counting. Both these techniques will yield 'point-by-point' spectra, so the wavelength resolution must be defined to fit the experiment.

One of the most spectacular observations in time-resolved emission spectroscopy is the rise and decay of molecular and excimer (or exciplex) spectra, illustrated in Figure 7.35(b). The structured molecular emission decreases immediately while the excimer emission increases up to a time of tens of ns, depending on the concentration. At longer times only the broad red-shifted excimer spectrum is observed. In Figure 7.35(b) the steady-state spectrum is shown in white; this represents of course the integration of all the instantaneous spectra which can be obtained only through time-resolved spectroscopy.

7.4.4.1 Time-correlated Single Photon Counting. In the steady-state technique of single photon counting, photons are detected continuously in the course of time. Time-resolved spectra can be obtained by counting photons within a defined time window after the excitation flash. The layout of such an apparatus is shown in Figure 7.36. The excitation light is a laser pulse or a short flash of UV light (*e.g.* from a pulsed deuterium arc lamp) of sub-ns duration. A part of this light is detected by a photomultiplier P, which provides the trigger for the time-base. This time-base is a voltage ramp (the time-to-voltage converter) which passes through two comparators that set the time window between the lower level t and the higher level $t + \Delta t$. The output of these two comparators provides a 'gate' for the acquisition system (usually a multichannel analyser) so that only photons emitted by the sample, S, within the set time window are counted.

The advantage of this technique is its high sensitivity, since it is based on single photon detection. However, the observation time is defined only within the width of the window, Δt, which is of several ns.

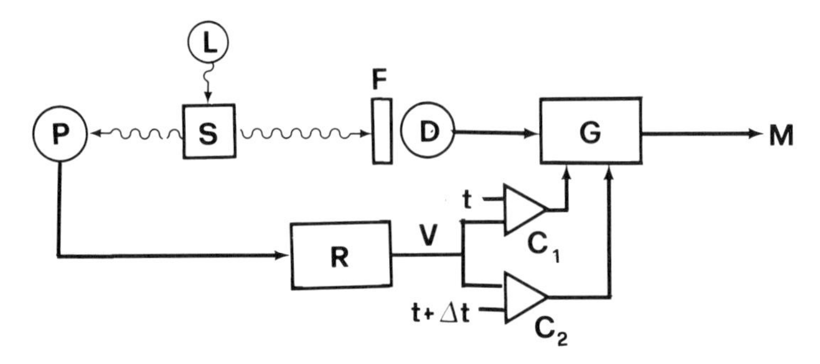

Figure 7.36 *Diagram of a single photon counting apparatus for time-resolved spectroscopy. L, pulsed light source; S, sample; P, photodiode; F, filter or monochromator; D, photomultiplier; R, voltage ramp driver; G, gate;* C_1, C_2, *comparators; M, multichannel analyser*

7.4.5 Special Detection Techniques in Flash Photolysis: Photoconductivity, Thermal Lensing, Photoacoustics, *etc.*

The photolytic flash must have enough energy to prepare, in a very short time, a detectable concentration of transient species. The lowest 'detectable' concentration depends on the probe technique, and here the methods of UV/VIS/near IR absorption and emission spectroscopy are the best. Their drawback is that they provide very little structural information about the nature of the transient species. IR and Raman spectra are much more informative, but they present many problems in fast reaction kinetics because of the weakness of the signals.

There are other spectroscopic techniques that can be used as probes in flash photolysis experiments, in particular electron spin resonance (ESR), which detects free radicals and radical ions, and microwave transmission, which detects dipolar transients. We shall not consider these special techniques in this text, references being made at the end of this chapter.

Three special detection methods have gained increasing importance in flash photolysis experiments, and these are discussed below.

7.4.5.1 Transient Photoconductivity. A solution of neutral molecules in a polar solvent shows only ohmic conductivity, but if ions are formed by the action of the photolytic flash these charge carriers generate an additional current which is proportional to the ion concentration. The observation of such transient photocurrents is the most direct experimental evidence for the formation of free, solvated ions in electron transfer reactions. The quantum yield of ion formation can be obtained through proper calibration procedures and the kinetics of ion recombination can be determined. Figure 7.37 gives an example of such transient photocurrent rise and decay.

7.4.5.2 Thermal Lensing. This technique is used for the measurement of the heat released (or taken up) by the sample after the photolytic flash. It is in

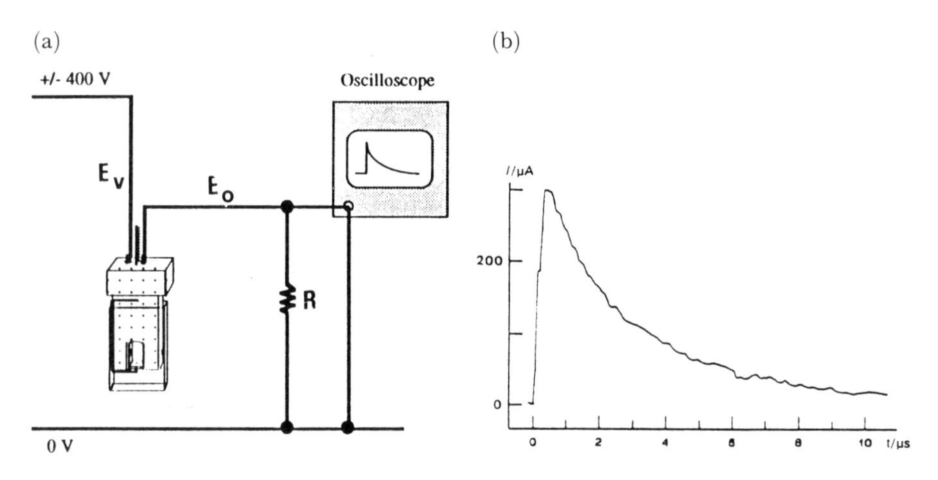

Figure 7.37 *(a) Diagram of a transient photocurrent detection system. E_v, E_0 high voltage and virtual ground electrodes; R, load resistance. (b) Oscillogram of the rise and decay of the photocurrent in an electron transfer reaction*

this respect a fast calorimetry method, its useful time-scale being restricted to about 1 μs to 1 ms.

Figure 7.38 shows the principle of a thermal lensing experiment which uses a liquid sample contained in a square cell, SC. The probe light is a narrow cw laser beam which falls on a pinhole after passage through the sample.

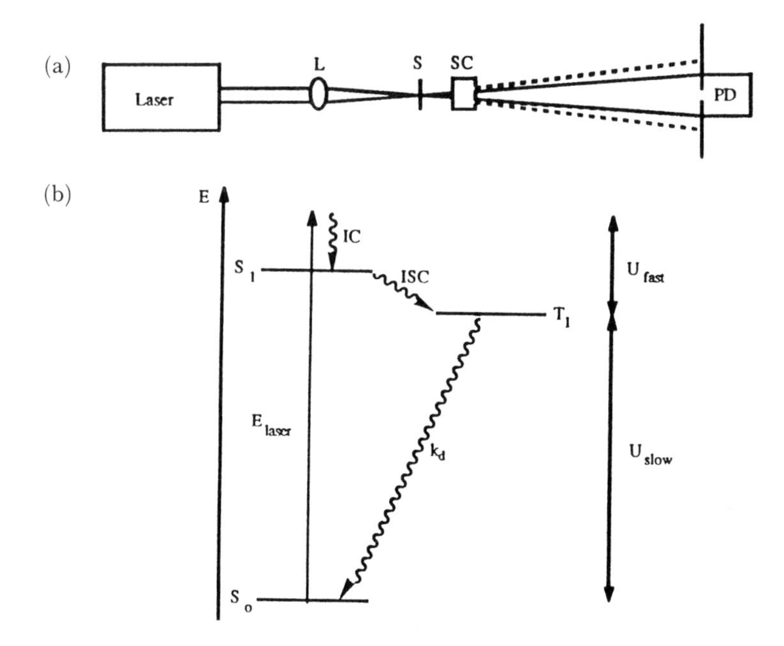

Figure 7.38 *(a) Principle of thermal lensing. L, lens; S, slit; SC, sample cell; PD, photodetector. (b) Energy diagram of fast and slow processes in thermal lensing*

When heat is produced in the sample after the photolytic flash, the refractive index of the liquid changes and the probe beam is deflected. The intensity of this probe beam measured by a photomultiplier tube placed behind the pinhole decreases as the temperature of the irradiated volume increases (then its density and its refractive index decrease). The total optical signal change is a measurement of all the heat produced in the sample, *i.e.* the sum of non-radiative transitions, chemical reactions and solvation energies. Luminescence does not contribute to this signal (nor does scattered light) and for this reason thermal lensing can be used to determine luminescence quantum yields.

Thermal lensing can also provide kinetic information within its restricted time-scale. In many cases the signal shows a fast change followed by a slow change, these representing different times of heat evolution in the sample. The fast change is then limited by the time resolution of the technique and includes the fast non-radiative transitions such as internal conversion (IC in Figure 7.38) and intersystem crossing (ISC), as well as other sub-microsecond processes. The slow signal corresponds to the decay of the triplet states, to recombinations of radicals and ions, and to other bimolecular reactions.

Figure 7.39 shows a thermal lensing oscillogram in which the fast and slow components can be well separated to give the relative heats evolved in these processes.

7.4.5.3 Time-resolved Photoacoustic Spectroscopy. In photoacoustic spectroscopy (PAS) the heat evolved by the absorption of light in the sample is transformed into sound waves which are detected by a microphone. In steady-state spectroscopy the light is continuous, but it is also possible to use a pulsed laser and to observe the change in the intensity of the sound signal with time. In this respect time-resolved PAS is somewhat similar to thermal lensing, but both techniques have different limitations and advantages.

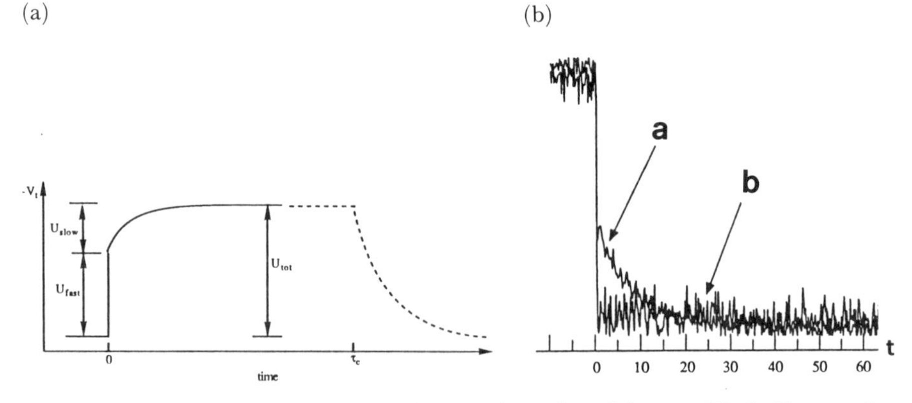

Figure 7.39 *(a) Separation of fast and slow signals in thermal lensing. (b) Oscillogram of a thermal lensing experiment (electron transfer between a ketone and an amine in degassed, a, and oxygenated, b, solutions. Horizontal axis, time in μs; vertical axis, detector signal in arbitrary units*

Time-resolved PAS can be faster in optimal cases, but the shapes of the signals are more complex and require elaborate computer processing.

7.5 QUANTUM YIELDS OF PHOTOCHEMICAL REACTIONS: ACTINOMETRY

The definition of the quantum yield of a photochemical reaction has been considered in section 4.1 as the ratio of the number of molecules transformed to the number of photons absorbed

$$\Phi_r = \Delta n \Big/ \int_0^t x(t) I_0(t)\, dt \tag{7.10}$$

In practice a sample is irradiated with monochromatic radiation of wavelength λ_r during a time t. The concentrations of reactants and products are determined by usual analytical methods. The second problem concerns the measurement of the number of photons absorbed. This is calculated from the number of photons incident on the sample, multiplied by the fraction of light absorbed x,

$$I_a(t) = x(t) I_0(t) \tag{7.11}$$

In practice both the incident light intensity I_0 and the fraction of light absorbed I_a vary with time, the former because of the instability of the output of the light source, and the latter because of the change in the absorbance of the sample at λ_r in the course of irradiation. Although it is possible in principle to measure and integrate the light intensity by means of electronic detectors, the technique of chemical actinometry is still widely used in practice.

The (chemical) actinometer A (Figure 7.40), is a photoreactive sample of known quantum yield Φ_a. For the determination of the quantum yield of the unknown sample S, this sample and the actinometer are irradiated simultaneously through the beam splitter B, which divides the incoming beam in a ratio r_λ, which is independent of time and of light intensity. The change of concentration, $\Delta c(\mathrm{A})$, of the actinometer gives the number of photons absorbed, so that the quantum yield of the unknown reaction is obtained as the ratio of the concentration changes

$$\Phi_r = \Phi_a r_\lambda \Delta c(\mathrm{S}) / \Delta c(\mathrm{A}) \tag{7.12}$$

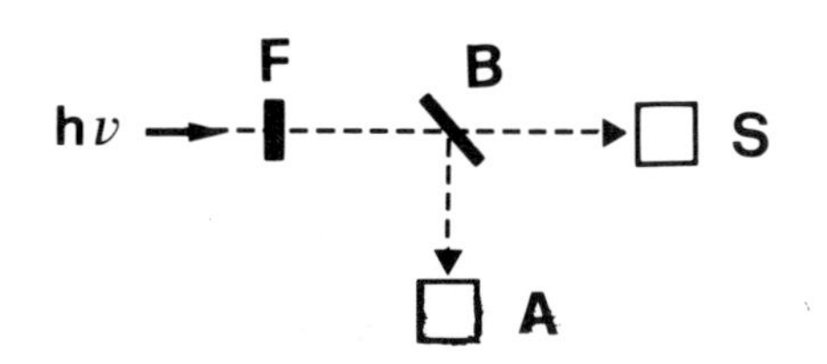

Figure 7.40 *Diagram of an irradiation apparatus for quantum yield measurements. F, filter; B, beam splitter; A, actinometer; S, sample*

7.5.1 Types of Actinometers

In principle any photochemical reaction of known quantum yield can be used as an actinometer, the choice depending on the irradiation wavelength(s). It is convenient to measure the concentration of the actinometer by absorption spectrophotometry, and in many cases the same technique can be used to determine the concentration of the sample itself.

7.5.1.1 The Anthraquinone Actinometer. The first example is the photoreduction of anthraquinone in ethanol (Figure 7.41).

Figure 7.42 shows the change in the absorption spectrum of this actinometer during the course of irradiation. One minor complication in its use is the necessity to deoxygenate the solution. In the presence of oxygen the dihydroxyanthracene reverts to anthraquinone so that the solution can be reused after deoxygenation.

The quantum yield of the primary hydrogen abstraction process is unity, and is independent of irradiation wavelength, light intensity and temperature. Its wavelength range follows of course the absorption spectrum of anthraquinone so that this actinometer is particularly well suited for the UV region.

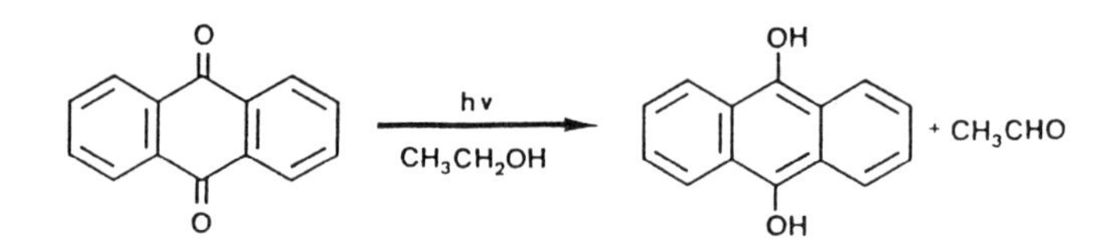

Figure 7.41 *Reaction of the anthraquinone/ethanol actinometer*

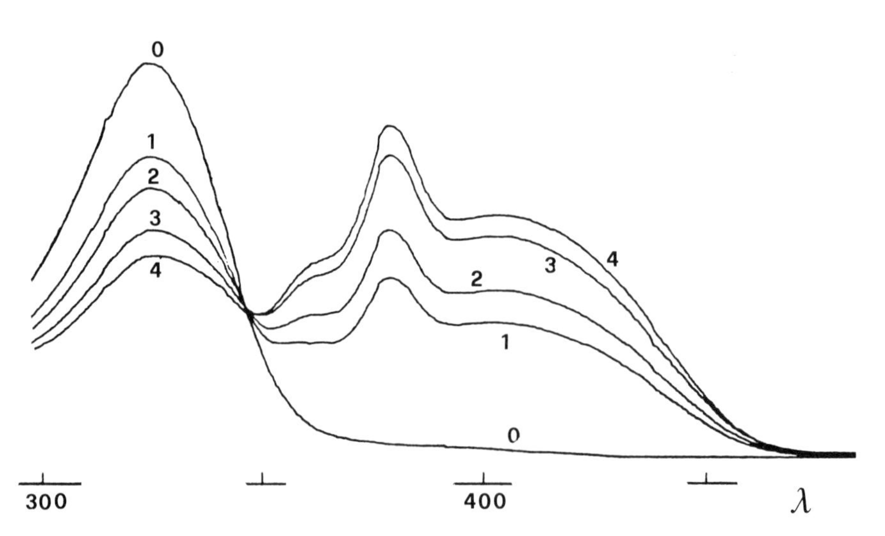

Figure 7.42 *Absorption spectra of anthraquinone in ethanol during the course of irradiation. 0–4 =*

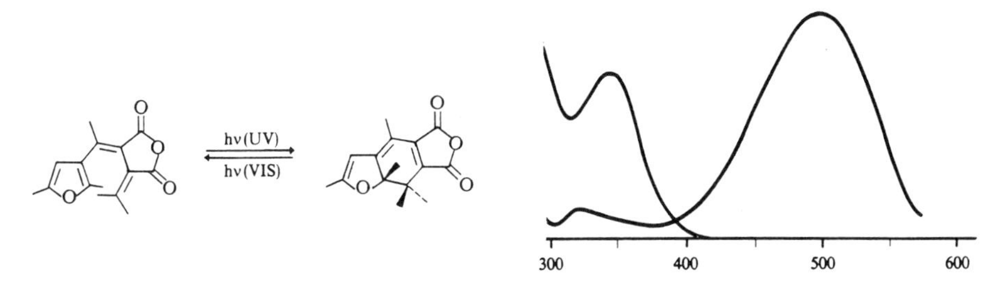

Figure 7.43 *Photochromic reaction (left) and the spectra of a fulgide (right)*

7.5.1.2 The Fulgide ('Aberchrome') Photochromic Actinometer. This is based on the reversible isomerization reaction of fulgides, which are photochromic materials. These make convenient actinometers for a wide wavelength region which covers the UV and much of the VIS (Figure 7.43). The quantum yields are generally wavelength dependent, since there are two opposite photochemical reactions. These actinometer solutions do not require deoxygenation, and they can be used many times since both reactions are photoreversible.

7.5.1.3 The Ferrioxalate Actinometer. The photochemical reaction of this organometallic complex is often used for actinometry, although it presents the drawback of requiring a 'development' stage after irradiation. It cannot therefore be used for the continuous monitoring of integrated light intensity in the course of time. The chemical formula of the crystalline compound is $K_3Fe(C_2O_4)_3.3H_2O$, the photochemical reaction involving the reduction of the metal centre followed by dissociation.

$$2Fe^{3+} + C_2O_4^{2-} \xrightarrow{h\nu} 2Fe^{2+} + 2CO_2 \tag{7.13}$$

In the presence of 1,10-phenanthroline, Fe^{2+} forms a complex which has a broad absorption band centred at 510 nm. The quantum yield of the reaction is wavelength dependent, but the useful wavelength range is quite wide. The quantum yields are 1.25 at 254 nm, 1.21 at 365 nm and 1.01 at 436 nm (these being important emission lines of mercury arcs).

CHAPTER 8

The Frontiers of Photochemistry

Photophysics and photochemistry are relatively young sciences, a real understanding of light-induced processes going back some 50 or 60 years. The development of quantum mechanics was an essential step, as classical physics cannot account for the properties of excited states of atoms and molecules. In the past 30 years the advent of new experimental techniques has given a major impetus to research in new areas of photochemistry, and these are the subject of this final chapter. It must of course be realized that these developments advance all the time, and that we talk here of a moving frontier, as it is in 1992.

In the first place, we shall take a look at the recent advances in fast reaction photochemical kinetics and spectroscopy, in particular at picosecond laser flash photolysis and femtosecond observations. Next, photophysics and photochemistry in molecular beams will be considered. Here observations are made under single molecule–single photon conditions, and these experiments provide insight into the most fundamental unimolecular gas phase reactions.

On the opposite scale of molecular size there is growing interest in the field of 'supramolecular' photochemistry which takes place in groups of loosely associated molecules. The largest artificial molecular assemblies are more or less organized arrays of molecules in micelles and in multimolecular layer films. With these systems we are getting a little closer to biological properties, but it is only fair to say that there remains a very long way to go before anything like model photosynthetic centres may be made. Finally, we shall consider a fringe area of photochemistry, namely the chemical effects of intense, laser IR light. This has been described as 'vibrational photochemistry' and it held at one time the promise of bond-selective dissociation of polyatomic molecules. These promises have not been fulfilled, but isotopic separation is a reality.

It would be presumptuous to attempt to forecast the future developments in this very wide field of light-induced processes which, as we have seen, covers photophysics, photochemistry and photobiology, with many practical applications. There is no set limit to knowledge, for we have learned since the last century that nature is infinitely more complex than we like to imagine. The advance of knowledge is then like an inflating balloon. As the content of knowledge increases with its volume, so the frontier with the unknown also increases as its surface.

8.1 PICOSECOND FLASH PHOTOLYSIS

The picosecond (1 ps = 10^{-12} s) time-scale is important in photochemistry because it includes processes such as vibrational relaxation, solvent reorganization, electron and proton transfer as well as some of the fastest chemical reactions such as unimolecular dissociations. The light source for ps flash photolysis is the mode-locked laser (section 7.2.2.2), but the real problem resides in the techniques of detection. Light travels only 0.3 mm in 1 ps, so it is not surprising that electronic detection devices and amplifiers cannot cope with such fast signals (with the exception of the 'streak camera' which will be described further on). In most cases the experimental techniques rely on optical delays between two pulses of light, known as the 'pump' and 'probe' pulses. This is somewhat similar to the conventional spectrographic flash photolysis technique, the optical delay replacing the electronic delay unit between the photolytic (pump) flash and the analytical (probe) flash.

8.1.1 Spectrographic Picosecond Laser Flash Photolysis

Figure 8.1 shows the major components of such an instrument. The ps pump pulse from a mode-locked laser is divided by a beam-splitter into a pump and a probe beam or, as shown here, it falls directly onto the sample cell C; some of this beam (lf) is absorbed and forms excited states in the sample, the transmitted beam being focussed on a solvent S, after an optical delay path. The overall length of this optical path can be adjusted by changing the distance L of the mirrors. A flash of white light from S is obtained by the process of self-phase modulation, this probe light passing through the sample C some time after the laser pump pulse. It is then dispersed in the spectrograph (g is the diffraction grating) on a photographic plate P. In many modern instruments this plate is replaced by a diode array which gives the absorption spectrum directly as an electrical signal. The response time of

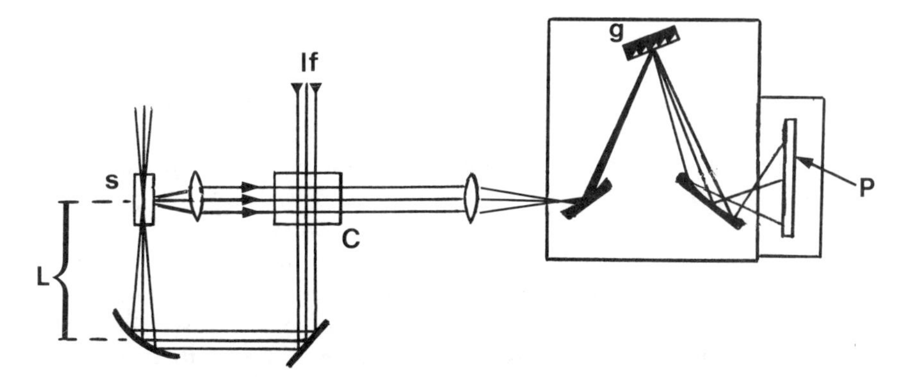

Figure 8.1 *Schematic diagram of a spectrographic ps flash photolysis apparatus. lf, laser flash pulse; C, sample cell; S, continuum pulse generator; g, diffraction grating; P, photographic plate or diode array*

the diode array is of no consequence, it can be quite slow. The electrical charges produced by the ps light pulse are stored almost indefinitely and can be read at leisure and transferred to a computer.

In this technique a whole absorption spectrum is obtained for one set delay time between the pump and probe pulses. The delay time can be varied to obtain spectra at different times.

8.1.2 Kinetic Picosecond Laser Flash Photolysis

In the nanosecond (ns) time-scale the use of kinetic detection (one absorption or emission wavelength at all times) is much more convenient than spectrographic detection, but the opposite is true for ps flash photolysis because of the response time of electronic detectors. Luminescence kinetics can however be measured by means of a special device known as the streak camera (Figure 8.2). This is somewhat similar to the cathode ray tube of an oscilloscope, but the electron gun is replaced by a transparent photocathode. The electron beam emitted by this photocathode depends on the incident light intensity $I(h\nu)$. It is accelerated and deflected by the plates d which provide the time-base. The electron beam falls on the phosphor screen where the trace appears like an oscillogram in one dimension, since there is no 'y' deflection. The thickness of the trace is the measurement of light intensity.

In practice the trace obtained in this way is too faint to be seen when the electron beam scans the screen in a time of 1 ns. An image intensifier camera is needed to make it visible, and in a complete system this would be followed by a CCD camera to store the trace and transmit it to a computer.

The streak camera gives a time resolution of about 5 ps. It requires a rather complex calibration procedure since the incident light intensity appears as the thickness of trace on the screen. It is used mostly for luminescence kinetics measurements, one of its advantages being that it can record single events.

8.1.3 Sampling Techniques for Repetitive Events

In many cases luminescence kinetics can be made repetitive when the excitation laser light intensity is very stable. If the pulse repetition rate is high enough (at least 1000 pulses per second) it is possible to use a sampling technique to obtain the variation of light intensity with time using fast photodiode detectors and special sampling oscilloscopes. In the sampling technique the light intensity is measured at one delay time only for each laser pulse, so it is necessary to have many pulses to obtain a good kinetic trace. It is also necessary to use a laser with a high pulse repetition rate.

Sampling techniques are not as fast as the streak camera because the response time of the detectors is a limiting factor. The interpretation of the data is however much simpler and does not require complex computer programs.

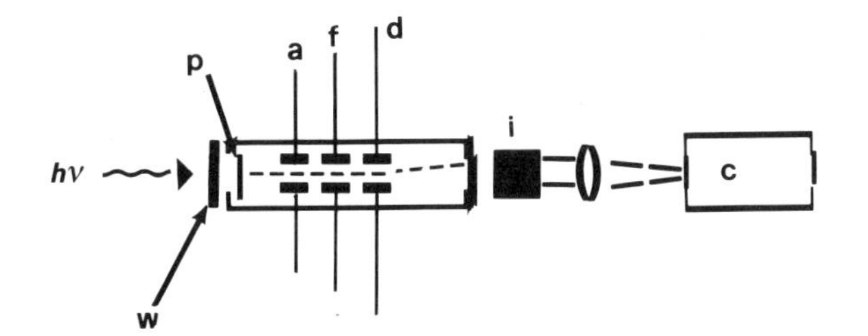

Figure 8.2 *Principle of the streak camera. w, window; p, photocathode; a, accelerating plates; f, focussing plates; d, deflection plates; i, image intensifier; c, camera*

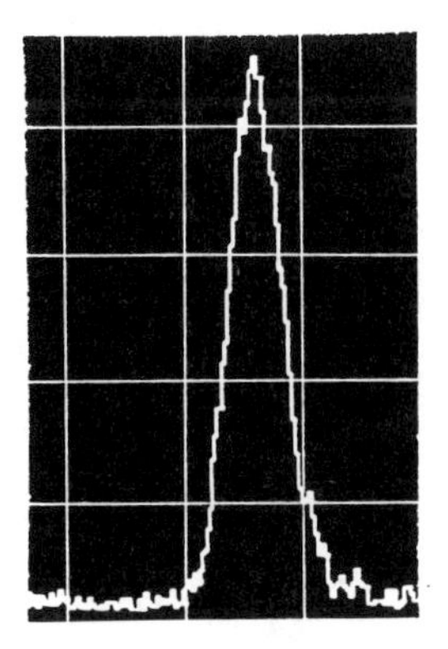

Figure 8.3 *Streak camera record of a ps laser pulse. Horizontal axis, time in 50 ps/division; vertical axis, light intensity in arbitrary units*

8.1.4 Indirect Optical Methods: Autocorrelation and Up-conversion

We have seen that the limitations of the time characteristics of electronic devices requires the use of optical delays between the pump and probe pulses in ps flash photolysis. There are also indirect ways of using optical properties to measure the kinetics of laser pulses and of fluorescence, known as 'autocorrelation' and 'up-conversion'. These rely on the non-linear properties of certain materials or chemical systems, *i.e.* they are based on fast biphotonic processes.

8.1.4.1 Autocorrelation. This experimental method is used to determine the shapes of ps laser pulses. In its simplest form the experimental set-up consists of a mirror M, placed behind a non-linear crystal SHG (second harmonic generator), a filter F, and a photomultiplier tube (PM; Figures 8.4 and 8.5). In places where the incoming pulse meets the reflected pulse frequency addition takes place and light of frequency 2ν is detected by the PM tube. The reflecting mirror is moved with respect to the crystal and the intensity of the 2ν light, plotted against distance, gives the shape of the laser pulse.

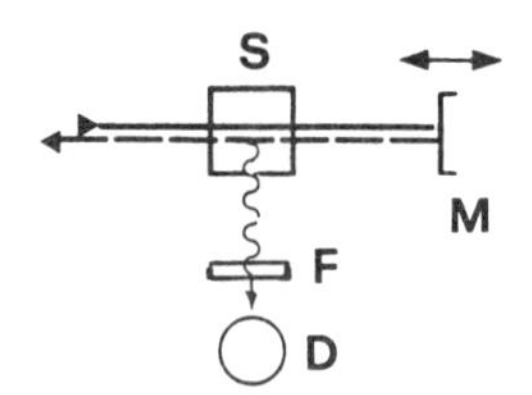

Figure 8.4 *Principle of autocorrelation. S, second harmonic generator crystal or sample cell; M, movable mirror; F, filter or monochromator; D, photomultiplier*

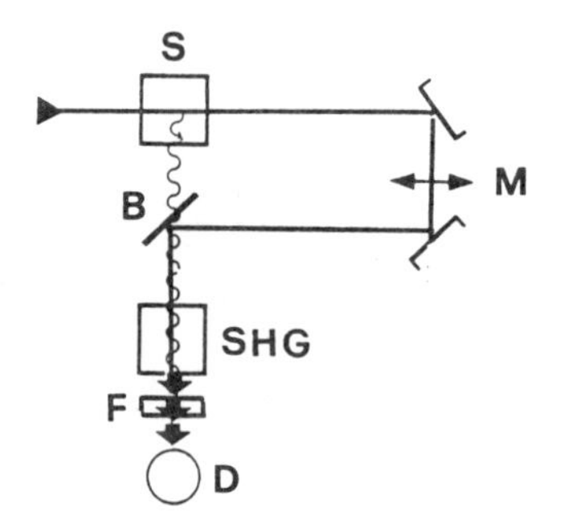

Figure 8.5 *Principle of up-conversion. S, sample cell; M, mirrors; B, beam-splitter; SHG, non-linear crystal; F, filter or monochromator; D, photomultiplier*

8.1.4.2 Up-conversion. This is similar to autocorrelation, but the laser pulse is mixed with the luminescence light of a sample in an SHG crystal to generate light of frequency $\nu_L + \nu_E$. The time between the laser pulse (ν_L) and the luminescence emission (ν_E) is varied by means of an optical delay path, and the light of added frequency is detected by a usual PM tube via optical filters.

There are many variations on these basic systems of autocorrelation and up-conversion. In the former the SHG crystal can be replaced by a solution of molecules which can be excited only by light of frequency 2ν; anthracene for example cannot be excited directly by a laser light of 532 nm, but will fluoresce when excited by two photons.

8.1.5 Picosecond Light-induced Processes

The time-scale of molecular vibrations is of the order of 10^{-13} s, just outside the ps range. Internal conversions and in particular vibrational cascades therefore fall into the femtosecond (10^{-15} s) time-scale. However, the spin-forbidden processes of intersystem crossing take place in times of a few ps to several ns. The case of benzophenone is a good example of the compensation between spin and orbital angular momentum. The rise of the triplet state absorption shows that intersystem crossing is completed within some 20 ps.

Many fluorescence kinetics fall into the ps time regime when non-radiative deactivation competes efficiently with the radiative transition. The natural

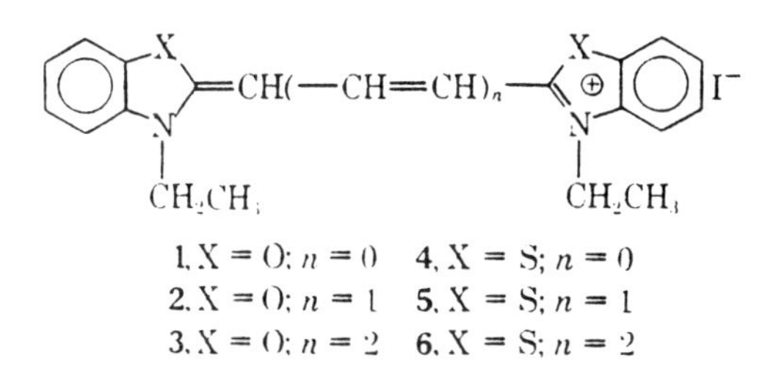

Figure 8.6 *Structures of cyanine dyes*

Table 8.1

Dye	ϕ	$\tau_{intr}(10^9)$ (s)	$\tau_{pred}(10^{10})$ (s)
1	0.0037	1.72	0.064
2	0.053	2.02	1.1
3	0.43	2.37	10
4	0.00059	2.94	0.017
5	0.048	3.09	1.5
6	0.33	2.87	9.5

radiative lifetimes of polyatomic molecules are of the order of ns, but the observable lifetimes are often much shorter when the emission quantum yield is low, and this requires measurements in the ps time-scale. The case of the cyanine dyes has been chosen here as an example. These are important molecules for the photographic industry as they are used as sensitizers which absorb light in the far visible (VIS) and in the near IR regions.

Some of the fastest photochemical processes occur on the ps time-scale, for instance electron transfer reactions. In the case of intermolecular electron transfers the actual reaction rate constants cannot be obtained when the diffusion of the reactants is the limiting factor. High concentrations must be used to ensure that encounters are faster than the reactions. Figure 8.7 shows the ps transient absorption spectra of the electron transfer between benzophenone and DABCO in acetonitrile. The triplet excited state of benzophenone is seen to decay at 525 nm while the radical anion grows at about 700 nm to reach a maximum concentration after 1 ns. The decay and growth kinetics are shown in (b) of the same Figure.

Intermolecular interactions between two molecules can be observed in diffusionless conditions when they form some kind of complex, of which the so-called charge transfer complexes are an example. In the ground state these complexes are held together mostly by van der Waals forces, but they show a new absorption band at long wavelengths which corresponds to an electron transfer. In the example of Figure 8.8 the complex is formed between 1-ethylnaphthalene and molecular oxygen dissolved in some organic solvent, and the triplet state is formed in ps times. Many of the faster inter-system crossings occur on this time-scale. For benzophenone for

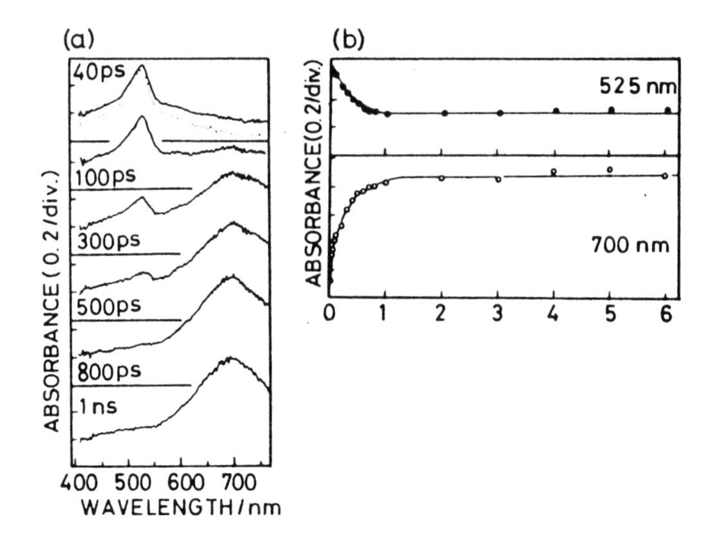

Figure 8.7 *Transient absorption spectra of the system benzophenone/DABCO/acetonitrile, and kinetics of absorbance changes at 525 and 700 nm (horizontal axis, time in ns)*

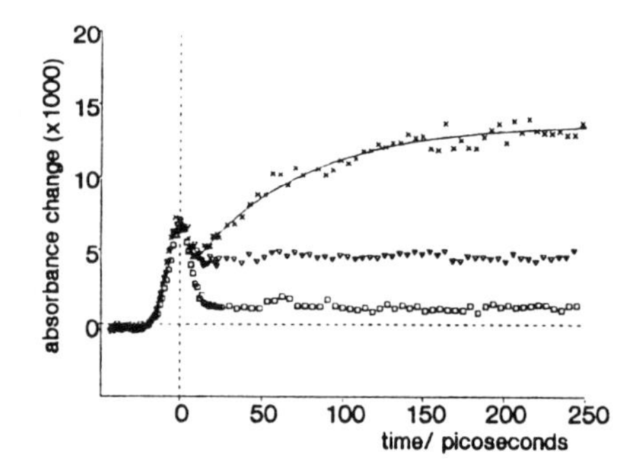

Figure 8.8 *Kinetics of formation of the triplet state of 1-ethylnaphthalene after excitation of its charge-transfer complex with O_2*

instance the triplet rise time is observed over 20 ps following excitation to the singlet state S_1.

The rotational relaxation of low-viscosity solvents takes place in times of ps. This can be observed through time-resolved spectroscopy when the dipole moment of the excited molecule differs substantially from that of the ground state species. Table 8.2 gives a few values of these relaxation times; the alcohols show rather slow relaxation because of the hydrogen bonds which associate the solvent molecules.

Photoinduced unimolecular reactions often have kinetics of the order of ps. One example of isomerization in ps times is shown in Figure 8.9. This is the photochromic reaction of a spiropyran. The photoinduced process takes place

Table 8.2 *Debye relaxation times of some liquids (in ps)*

Dimethylformamide	10
Dimethylsulfoxide	20
Ethanol	140
1-Butanol	480

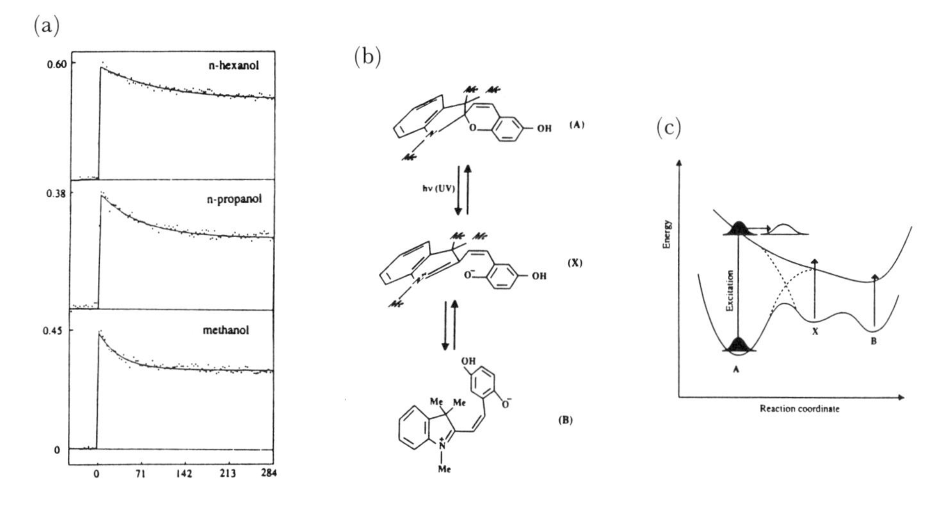

Figure 8.9 *(a) Kinetics of absorbance changes of a spiropyran in various alcohols. Horizontal axis, time in ps; vertical axis, absorbance. (b) Structures of a spiropyran and its photochemical intermediate. (c) Potential energy diagram of the isomerization reaction*

in two stages, an intermediate X being formed before the stable isomer B, which absorbs in the VIS region. The ps decay curves would correspond to the vibrational relaxation of X to yield B, the kinetics of this relaxation being solvent-dependent.

Some fluorescence lifetimes are observed in ps times, although these are unusual cases. In organic molecules the S_1–S_0 fluorescence has natural lifetimes of the order of ns but the observed lifetimes can be much shorter if there is some competitive non-radiative deactivation (as seen above for the case of cyanine dyes). A few organic molecules show fluorescence from an upper singlet state (*e.g.* azulene) and here the emission lifetimes come within the ps time-scale because internal conversion to S and intersystem crossing compete with the radiative process. To take one example, the S_2–S_0 fluorescence lifetime of xanthione is 18 ps in benzene, 43 ps in iso-octane.

8.2 FEMTOSECOND FLASH PHOTOLYSIS

The femtosecond (10^{-15} s; fs) is arguably the final limit in the time-scale of 'chemical' processes. The duration of a molecular vibration is of the order of 100 fs, and even an electronic transition takes some 1 fs (this is of course the duration of the electron jump from one orbital to another, and has nothing to do with the lifetime of excited states.) It is therefore clear that at this time-scale the final frontier of photochemistry is in sight. That is not to say that there can be no faster processes still, but such processes fall within nuclear physics rather than chemistry.

8.2.1 Femtosecond Detection Methods

We shall not consider here the techniques of generation of fs laser pulses, but only the main detection methods used to analyse transient species formed in these very short time-scales. The general principle common to these detection methods is illustrated in Figure 8.10, which shows that the incoming fs laser pulse is divided in two parts by a beam-splitter, these forming eventually the 'pulse' and the 'probe'. Since it is desirable to vary both the pulse and the probe wavelengths, the laser pulses can be converted to a white light continuum through the process of self-phase modulation in the cells CG (continuum generators). The required wavelength is then selected by an interference filter (F) to produce a fs pulse of VIS light. The photolytic pulse must usually be in the UV, so it must pass through a non-linear crystal (SHG) where it is doubled in frequency. The optical path of one of the pulses can be varied by moving a reflector R, which can be a totally reflecting prism or a pair of mirrors fitted to a high-precision linear actuator. The pulses are then recombined by another beam-splitter to produce a pair

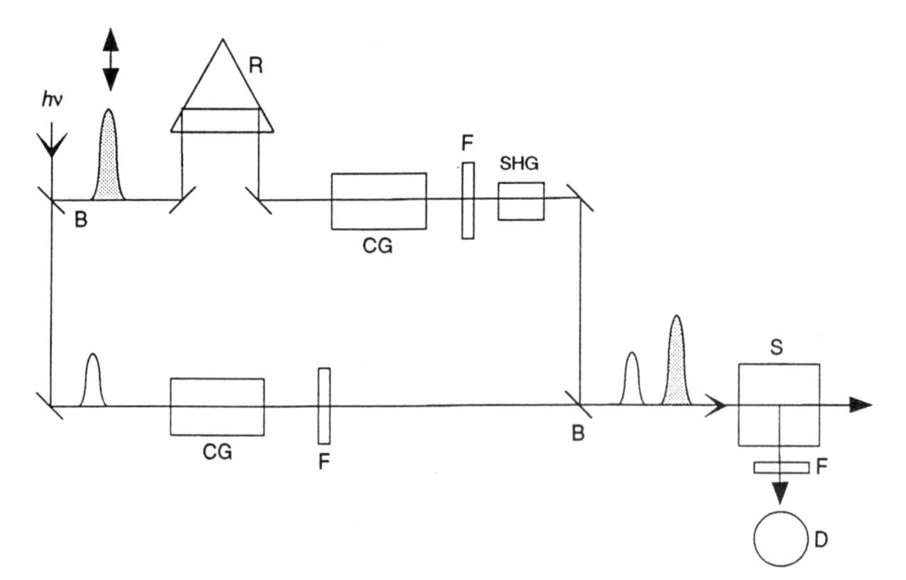

Figure 8.10 *Schematic diagram of a fs spectrometer. hν, incoming fs laser pulse; B, beam-splitter; R, reflector (prism or mirrors); CG, continuum generator; F, filter; SHG, second harmonic generator; S, sample cell; D, photomultiplier*

of photolytic and probe pulses with a delay set by the reflector. The delay between these pulses can be positive or negative or nominally zero when they overlap exactly.

The photolytic and probe pulses are colinear when they reach the sample. The photolytic pulse produces excited states and photofragments, and the probe pulse which follows closely behind must be used to analyse the concentration and/or the chemical nature of the transients. The major detection processes are known as 'laser-induced fluorescence' (LIF) and 'multiphoton ionization' (MPI). Transient absorptions can also be used in some cases, and this is similar to ps spectroscopy.

8.2.1.1 Laser-induced Fluorescence. The probe wavelength λ_p can be adjusted to excite one of the photofragments or the excited complex in the process of dissociation. Consider for instance the dissociation of a molecule AB according to

$$AB \xrightarrow{h\nu} (AB)^* \rightarrow (A \cdots B) \rightarrow A + B \quad (8.1)$$

The potential energy diagram of Figure 8.11 illustrates the variation of the relative energies of these species with increasing internuclear separation (r). A probe wavelength λ_1 will select the dissociating complex, while a wavelength λ_2 monitors one of the free photofragments. The concentration of the reactant AB can also be followed at yet another probe wavelength, and this concentration would decrease in time.

8.2.1.2 Multiphoton Ionization. In some cases the probe pulse provides sufficient energy to ionize the fragments to form a positive ion and an electron. The positive ions can be detected in a mass spectrometer which replaces the

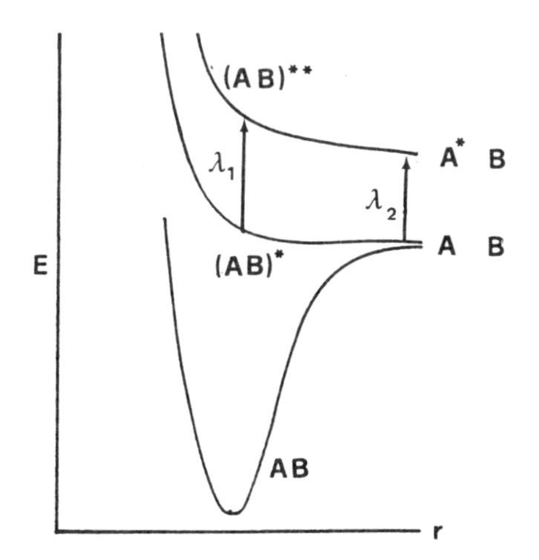

Figure 8.11 *Potential energy diagram of the dissociation of a diatomic molecule AB. Horizontal axis, interatomic distance r. λ_1, λ_2, absorption wavelengths of the excited molecule and its dissociation products*

light detector system in the diagram of Figure 8.10. The advantage of MPI is that it provides structural information about the nature of the fragments through their molecular weights.

8.2.2 Femosecond Photochemical Processes

These are essentially unimolecular reactions of dissociation and of isomerization, studied mostly in the gas phase. We shall consider here a few examples of such reactions. The dissociation of ICN can be written as

$$\mathrm{ICN} \xrightarrow{h\nu} (\mathrm{ICN})^* \rightarrow (\mathrm{I} \cdots \mathrm{CN})^* \rightarrow \mathrm{I}^\bullet + \mathrm{CN}^\bullet \tag{8.2}$$

The free radical $CN^\bullet$ has an absorption at 388.5 nm, corresponding to the wavelength λ_2 in Figure 8.11. The rise of this absorption is continuous with a lifetime of around 400 fs, and this shows that the excited state reached by the photolytic pulse is a dissociative state. The same applies to a molecule like NaBr, but something quite remarkable is observed in the case of NaI. In the reaction scheme

$$\mathrm{NaI} \xrightarrow{h\nu} (\mathrm{NaI})^* \rightarrow (\mathrm{NA} \cdots \mathrm{I})^* \rightarrow \mathrm{Na}^\bullet + \mathrm{I}^\bullet \tag{8.3}$$

it is possible to follow the concentration of free Na atoms at 589 nm, and this rises in a sequence of plateaux (Figure 8.12). When the intermediate is monitored at a different wavelength, an oscillatory behaviour is observed. This results from the fact that there is an avoided crossing between the ionic and covalent states (Na^+ I^- and $Na^\bullet$ $I^\bullet$) on the potential energy surface. The fragments then come back together several times before separating.

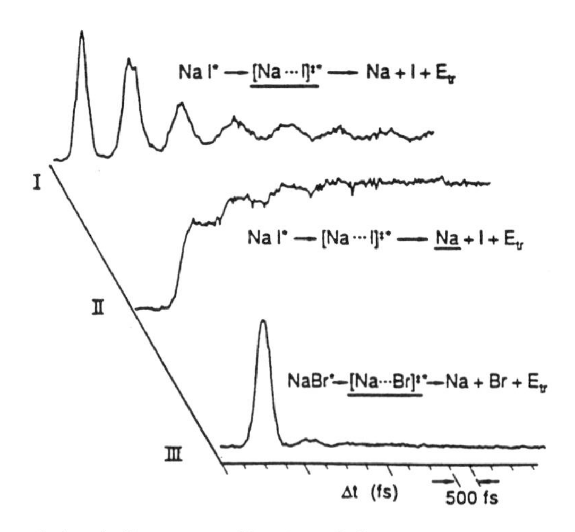

Figure 8.12 *Laser-induced fluorescence kinetics of dissociation reactions. I, reaction of NaI followed at the absorption wavelength of the excited molecule; II, same reaction monitored at the absorption wavelength of the free Na atom; III, reaction of NaBr, followed at the absorption wavelength of the excited molecule*

An important example of femtosecond isomerization is the intramolecular proton transfer in some internally hydrogen bonded hydroxybenzotriazoles which are used as light stabilizers of polymers (section 6.2). Figure 8.13 shows the reaction scheme and the energy levels (left) of the two forms of 'tinuvin P', and the time dependent transmission changes (right) at four different wavelengths in the near UV and VIS. The whole sequence of reversible proton transfer is complete within some 2 ps, and here femtosecond spectroscopy provides the direct evidence for the zwitterionic intermediate ('keto' form). At some wavelengths there is a transient increase in transmission, at others there is a transient decrease. These correspond to the differences in the absorption spectra of the two forms of the molecule. It can be seen also from curves (b) and (c) that there is a slow component in the kinetics which is attributed to vibrational relaxation of the enol ground state formed in a vibrationally excited state after the reverse proton transfer.

There are other examples of such fast bond dissociations in the liquid phase, not necessarily hydrogen bonds. Thus the first step in the photochromic isomerization of spiropyrans (the ps events are shown in section 8.1.) is complete within 100 fs, that is within a single vibration of the bond.

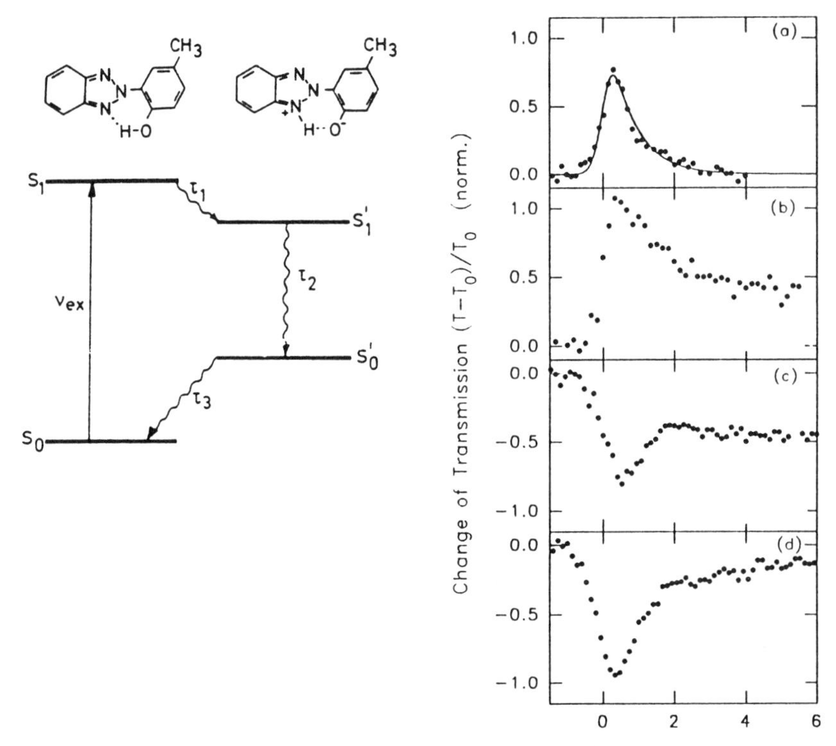

Figure 8.13 *Left: Structures and energies of a hydroxybenzotriazole in the course of intramolecular proton transfer. Right: Kinetics of transmission changes at 325 nm (a), 355 nm (b), 385 nm (c) and 400 nm (d); horizontal axis, time in ns*

8.3 SUPRAMOLECULAR PHOTOCHEMISTRY

A 'supramolecular' assembly is a system of two (or more) molecules which are held together in a restricted geometry. These molecules must conform to a set stoichiometry—generally 1:1—but the bonding pattern can be unspecific (*e.g.* electrostatic interactions) or specific (*e.g.* hydrogen bonding). It is helpful to distinguish between the 'guest' molecule (which can be an ion) and the 'host' molecule which must be much larger in order to accept the guest into its internal cavity. In Figure 8.14 a simplified picture of a cyclodextrin host shows that the guest molecule must fit inside the host molecule in order to form a true supramolecular assembly. From the point of view of light-induced reactions of the guest molecule, the shielding provided by the host is of prime importance. Quenching processes are slowed down, and the photochemical stability of molecules that would normally dissociate in the excited state can be greatly enhanced.

Cyclodextrin cavities form the early models of host molecules involved in supramolecular assemblies. There are many other molecules known as 'cryptands' which can be designed to offer a cavity of fairly precise dimensions to accommodate various ions or metal complexes. It may be possible to locate not just one, but two, guest molecules inside a cryptand cavity, and this may lead to new electron transfer reactions in restricted environments; another step towards synthetic photoinduced biochemical reactions.

The case of benzoin alkyl ethers illustrated in Figure 8.15 is a remarkable example of the effect of complexation with cyclodextrins. Such molecules normally undergo homolytic dissociation in solution (the 'Norrish type 1' process described in section 4.4) and there is practically no intramolecular hydrogen atom abstraction (Figure 8.15). When the benzoin alkyl ethers are complexed with a cyclodextrin to form a 1:1 association, it can be shown that one of the phenyl rings fits inside the cyclodextrin cavity in aqueous solution. When the solid complex is irradiated only the photoproducts resulting from hydrogen atom transfer are detected; the opposite behaviour from irradiation of the crystal of benzoin alkyl ether as well as of solutions in benzene.

There are two possible conformations of the complex, shown as A and B in Figure 8.16. B only can react through intramolecular hydrogen atom

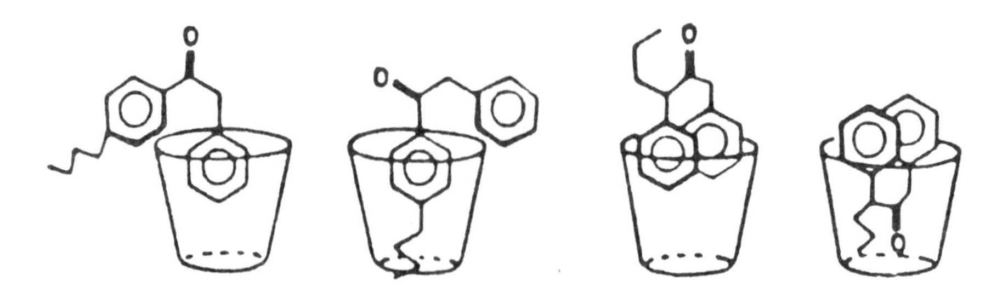

Figure 8.14 *Schematic structures of complexes of aromatic molecules with cyclodextrins*

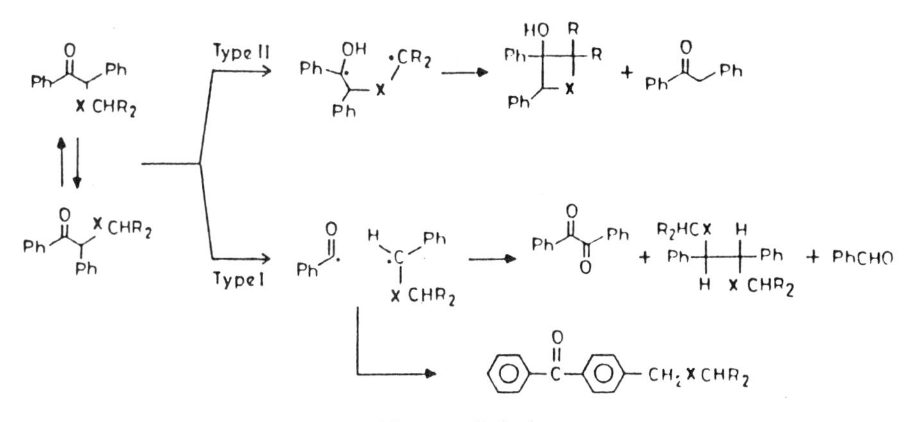

Figure 8.15 *Photochemical reactions of benzoin alkyl ethers*

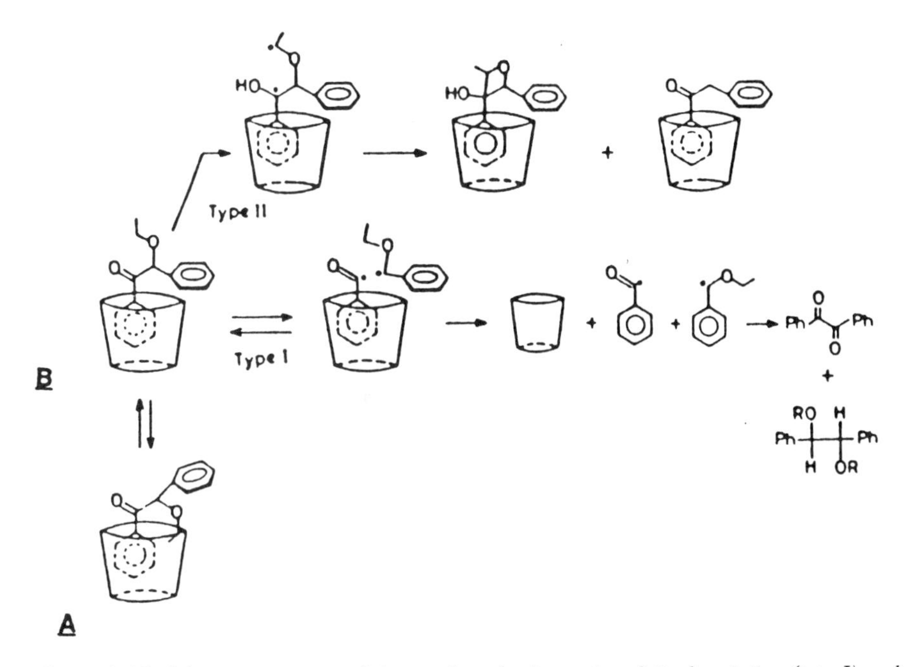

Figure 8.16 *Schematic structures of the complexes leading to homolytic dissociation (type I) and intramolecular hydrogen atom abstraction (type II)*

abstraction. This is not to say that homolytic dissociation cannot take place, but the radicals are trapped inside the cavity and therefore recombine unless some radical scavenger is also present; thus irradiation of the solid complex in the presence of oxygen yields the oxidation products expected from the dissociation reaction.

There are many other types of inclusion compounds of undefined stoichiometry in which the guest molecules are confined in tunnels and cages within the host. The host is then not a single molecule like a cyclodextrin but is for instance a zeolite or an organic crystal. Figure 8.17 is a computer drawing of the tunnels of a special type of urea crystal and of the inclusion compound with a molecule of *n*-hexadecane.

The restricted motion of molecules and of fragments such as free radicals formed by photodissociation results in interesting differences in the photochemistry of some molecules in solution or as guests in inclusion compounds. To take one example, the aliphatic ketone 5-nonanone can yield fragmentation or cyclization products via the biradical formed through intramolecular hydrogen atom abstraction (Figure 8.18). In the photolysis of the inclusion compound the cyclization is the preferred reaction, and there is a marked selectivity in favour of the *cis*-isomer of the cyclobutanol.

Another aspect of supramolecular photochemistry concerns the use of covalently linked bichromophoric and multichromophoric molecules. In the example of Figure 8.19 a two-step sequential, photoinduced electron transfer is observed in the trichromophoric species D–D–A. Local excitation of the cyanonaphthalene acceptor results in charge separation first to the amine and then to the methoxyaniline donor in times of a few tens of ns

$$\mathrm{DDA} \rightarrow \mathrm{DDA}^* \xrightarrow{h\nu} \mathrm{DD}^+\mathrm{A}^- \rightarrow \mathrm{D}^+\mathrm{DA}^- \tag{8.4}$$

This type of sequential electron transfer can be a model for photosynthesis.

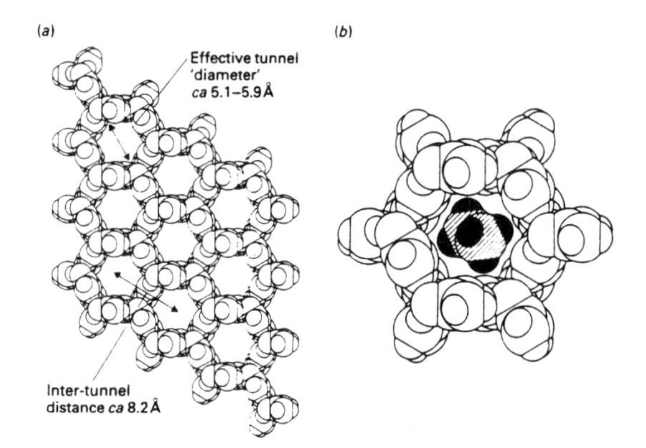

Figure 8.17 *Drawing of the tunnel structure of a urea crystal (a), and of the inclusion compound with hexadecane (b)*

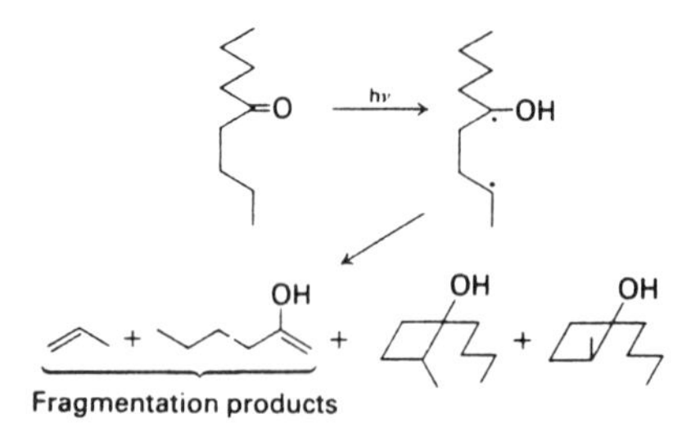

Figure 8.18 *Photochemical reaction of 5-nonanone and secondary reactions of its biradical*

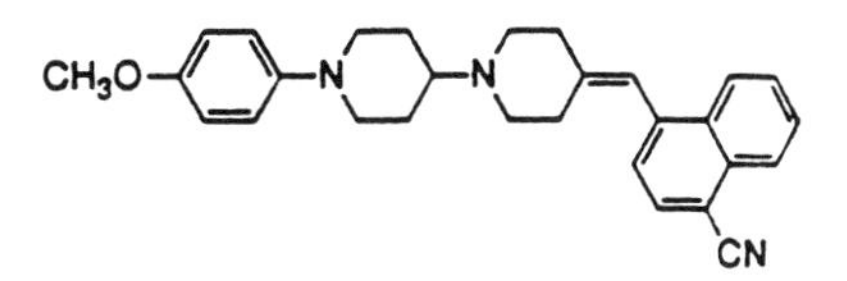

Figure 8.19 *Structure of a trichromophoric DDA molecule*

8.4 PHOTOCHEMISTRY IN MOLECULAR FILMS

Most experiments in photophysics and photochemistry involve statistical assemblies of molecules in the gas phase or the liquid phase. Even in rigid environments the distribution of molecules remains essentially random, and very small, well defined molecular assemblies can be obtained only in molecular beams or in supramolecular systems. It is however clear that the properties of biological systems, such as the photosynthetic centre, rely on the precise spatial distibution and orientation of molecules, and for this reason there is nowadays a growing interest in the photophysical and photochemical properties of organized molecular assemblies. There is a very long way to go before anything like an actual photosynthetic centre will be made in the laboratory, but significant progress has been made in the fields of monomolecular and multimolecular films. These so-called 'Langmuir–Bloggett' (L–B) films come closest (at the present time) to biological assemblies.

An L–B film is formed by the dispersion of amphoteric molecules at an air–water surface (Figure 8.20). These molecules have a polar group at one end, something like a carboxy substituent (in this respect they resemble the surfactant molecules which make micelles), and a long non-polar aliphatic chain. The polar group stays in the polar water phase, and the aliphatic chain stays in the non-polar air environment. The L–B film at the water surface is then made by the controlled compression of these molecules by means of a floating barrier. The molecules then line up to form a monomolecular layer on the water surface.

Under optimal conditions this layer can be transferred to a solid substrate (glass or metal) and several monomolecular layers can be deposited in this way. These L–B films represent highly organized molecular assemblies on a macroscopic scale, since not only the distance between neighbouring molecules but also the relative orientations of their chromophores can be determined. The distance dependence of photoinduced energy and electron transfers have been investigated in L–B films. The R^6 dependence of the Förster dipole–dipole mechanism has been confirmed, but it must be realized that some questions remain concerning the possible role of defects in the film structures.

Figure 8.21 gives a schematic representation of the type of photophysical experiments which can be performed in L–B films. The distance dependence of a quenching action through energy transfer or electron transfer can be

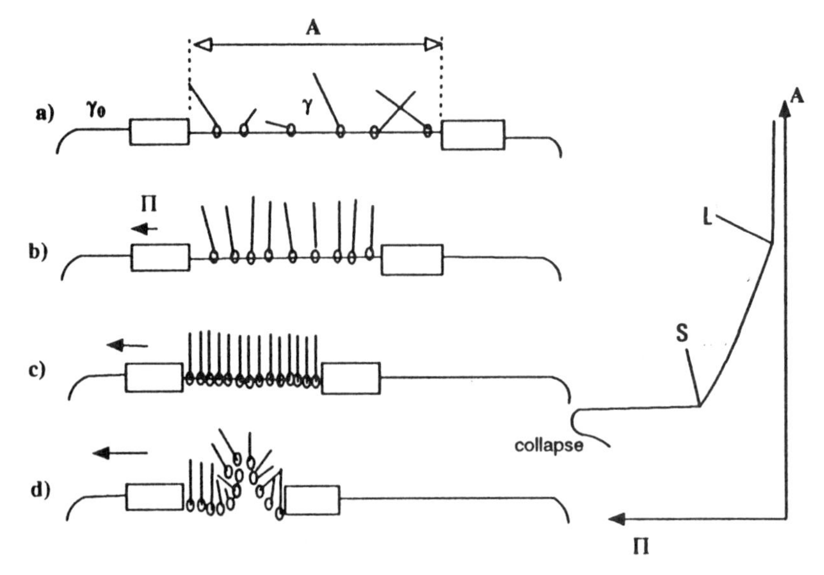

Figure 8.20 *Sequence of the preparation of a Langmuir–Bloggett film on water: (a) before compression, (b) during compression, (c) at limiting pressure, (d) film collapse at higher pressure. Right: Pressure–surface diagram showing the phase changes L (gas–liquid) and S (liquid–solid). π, pressure and A surface area*

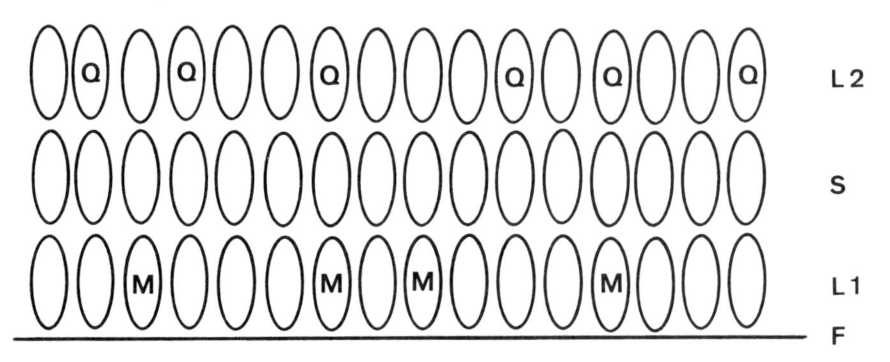

Figure 8.21 *Diagram of a multilayer Langmuir–Bloggett film. F, metal or glass support; L1, layer of chromophoric molecules M; S, spacer layer; L2, layer of quencher molecules Q*

studied by the deposition of layers of chromophoric molecules M (layer L1), and quencher molecules Q (layer L2), separated by one or several spacer layers of varying thickness. In general the molecules M and Q are diluted with non-active molecules in their respective layers L1 and L2 in order to avoid effects of concentration quenching or exciton delocalization. These non-active molecules which are also used in the spacer layers are usually derivatives of arachidic acid.

An L–B film deposited on a solid surface consists in fact of many microdomains within which the molecules are ordered. For example, all chromophores have the same orientation, but this orientation can be different in another microdomain. The ordering of chromophoric molecules on a macroscopic scale results in polarization effects for the absorption and emission of light. Figure 8.22 gives an example of the absorption spectra of a uracil derivative in an L–B film. Note that the first band at around 250 nm shows little change with the angle of polarization of linearly polarized light, but the short-wavelength band at around 210 nm displays a marked dependence. This depends on the orientation of the film with respect to the electric vector of the light, and allows the determination of the direction of the transition dipole moment for each absorption band.

The fact that molecules in L–B films keep fairly rigid relative orientations (at least within a microdomain) is potentially useful for regio- and stereo-selective addition reactions (Figure 8.23). Here a derivative of cinnamic acid forms the head-to-head dimer in the photochemical cycloaddition reaction.

Of the many studies of applications of L–B films, we shall mention at last the development of non-linear optical materials. A molecule such as 4-nitroaniline has a very large hyperpolarizability (section 7.2) and can be used in principle for frequency conversion in solution if a strong electric field is used to provide an orientation of the dipoles. There is no need for such an external field if similar molecules are used in an L–B film, and hyperpolarizability values far in excess of those of commonly used inorganic crystals have been obtained. Figure 8.24 shows two examples of the molecules used in

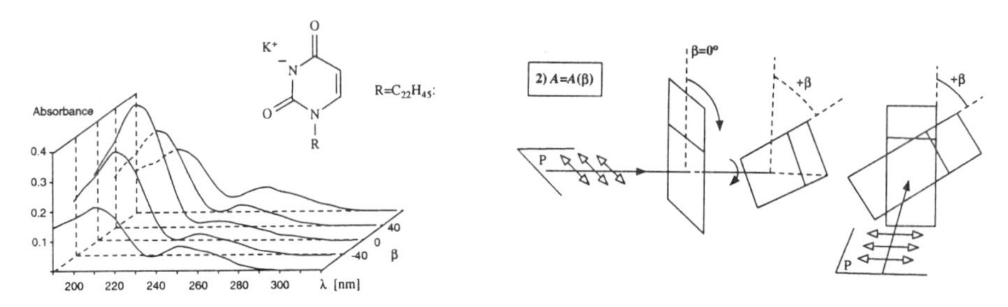

Figure 8.22 *Absorption spectra of a Langmuir–Bloggett film of uracil in polarized light. β is the angle between the electric vector of the beam and the film*

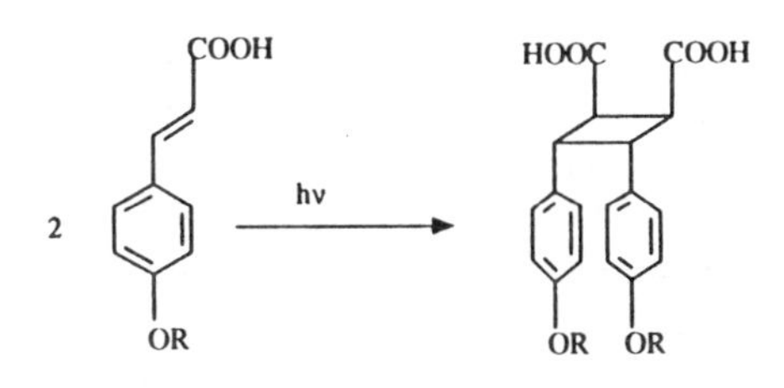

Figure 8.23 *A cycloaddition reaction in a Langmuir–Bloggett film*

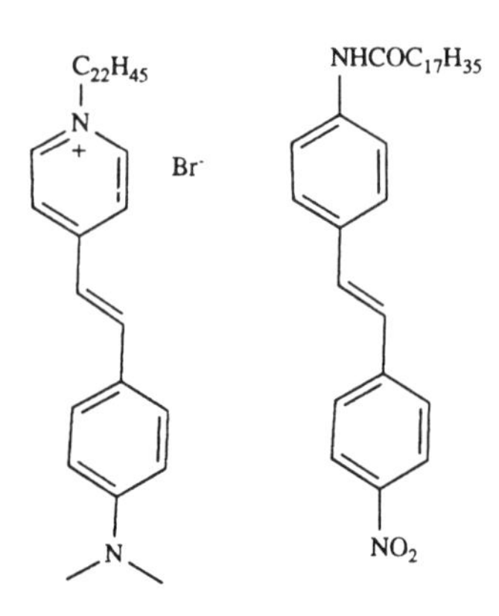

Figure 8.24 *Structures of molecules used for non-linear optical films*

these non-linear optical films. A common feature of all the molecules used in L–B films is the long aliphatic chain which must be attached to the chromophore. This may result sometimes in synthetic problems which are one of the challenges of this very promising field of research.

8.5 PHOTOCHEMISTRY IN MOLECULAR BEAMS

The most basic chemical processes can be studied in conditions of single collisions between two molecules or between a molecule and a photon. Such conditions are obtained in molecular beams in the gas phase at extremely low pressures, low enough to ensure single encounter conditions. A molecular beam is formed when gas molecules contained in a heated vessel are allowed to escape into a vacuum through a nozzle (Figure 8.25). In such conditions there are two limiting situations known as the 'effusive' beam and the 'supersonic' beam; the latter only will be considered here. In the supersonic beam the gas molecules are 'cooled down' by a few collisions with other molecules within the space of the nozzle, but they suffer no further collisions thereafter. Although the concept of thermodynamic temperature breaks down in this case, it is still possible to consider various 'temperatures' according to the Boltzmann distribution of rotational and vibrational energies

$$n_v/n_0 = g \exp(\Delta E/kT) \tag{8.5}$$

where ΔE is the energy difference between these states. The translational distribution plays no further role since the molecules in the beam move independently. Very low rotational and translational 'temperatures' can be obtained in supersonic beams, and here the conditions of single molecule reactions can be reached.

In a photochemical experiment the molecular beam crosses a laser beam which can be pulsed or continuous, and the photoproducts are analysed in a

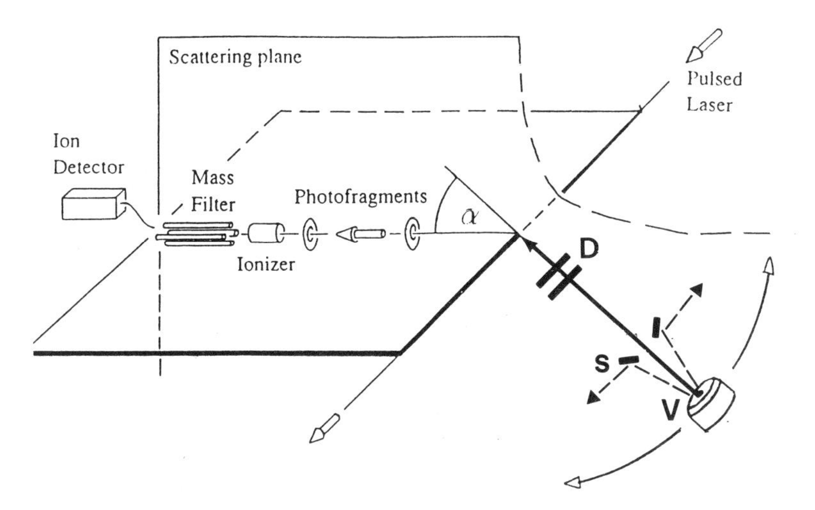

Figure 8.25 *Outline of a laser photolysis molecular beam experiment. V, heated vessel; S, skimmer; D, pair of rotating discs*

mass spectrometer to determine their speed and their angular distribution. It is also possible to observe the vibrational energy levels of the photoproducts and to gain insight into the most fundamental photoinduced dissociation reactions. Molecular beams therefore provide information about the photophysics and photochemistry of isolated molecules, or of molecules in very simple environments, in contact with a small number of 'solvent' molecules for example (clusters). This is a field of research of considerable interest at this time, and we shall look at several examples of dissociation and isomerization processes.

Figure 8.25 shows a diagram of a supersonic molecular beam/laser apparatus. The sample is a gas held within the heated vessel V, which has a narrow nozzle open to the observation space which is kept at a very low pressure by means of powerful pumps. The translational velocity of the molecules can be controlled in several ways, for example by passing the beam first through a skimmer which selects a narrow solid angle and then through two rotating discs D. Molecules which have the right speed pass through the holes of the discs, the other molecules being scattered and removed by the pumps. In the observation area a laser beam excites some of the molecules of the beam, and as these molecules dissociate the fragments fly off in different directions at different velocities.

The path of the laser beam and the detector (the ionizer in this case) define a plane perpendicular to the scattering plane in which the source of the molecular beam can be rotated to some scattering angle α. Neutral photofragments are detected through ionization by electron impact. Ions formed in this way can be mass selected in electric and/or magnetic fields.

When a pulsed molecular beam is used with a pulsed laser beam, the velocity of the photofragments can be obtained from the delay time between the collision of the beams and the appearance of the ionized photofragments. In this way a very complete picture of the reaction mechanism is drawn.

Molecular beam techniques have been used mostly with small molecules, and most of the information about the dynamics of elementary photochemical processes are restricted to such species. It is however also possible to bring relatively large molecules (*e.g.* aromatics) into molecular beams through 'seeding' in a carrier gas such as He.

8.5.1 Molecular Beam Photoinduced Reactions

Even the homolytic dissociation of a very simple molecule reveals an unexpected wealth of information in time-of-flight translational spectroscopy. The molecule SO_2 undergoes dissociation when excited with far UV light (193 nm)

$$SO_2 + h\nu \rightarrow SO + O \tag{8.6}$$

The energy balance can be written as follows.

$$E_{avl} = E_{int}(SO_2) + h\nu - D_0(SO\text{–}O) = E_{trans} + E_{rot} + E_{vib} + E_{el} \tag{8.7}$$

the available energy becoming translational (trans), rotational (rot), vibrational (vib) and sometimes electronic (el) energy of the fragments. When the SO radical is vibrationally excited there is of course less translational energy, so the velocity of the SO fragment is lower. Figure 8.26 shows the separation of SO radicals of different vibrational energies as a function of the time-of-flight from the irradiation area to the detector.

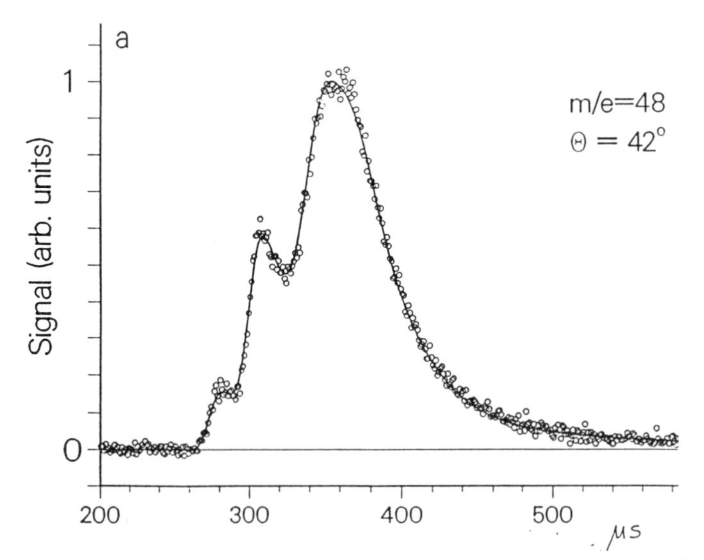

Figure 8.26 *Time-of-flight spectrum of the photofragments of the dissociation of SO_2 measured at mass 48; horizontal axis, time in μs*

The detailed mechanism of the dissociation of formaldehyde (H_2CO) has been investigated in molecular beam experiments. The reaction itself may appear quite simple

$$H_2CO \xrightarrow{h\nu} H_2 + CO \tag{8.8}$$

but there has been much discussion about a rather mysterious 'intermediate' state in this process, because the fragments have a considerable translational energy and the CO molecule appears to be formed after a substantial time delay. It now seems probable that this delay results in fact only from the rotational relaxation of the fragments. If the observation wavelength corresponds to the absorption of CO in its ground rotational state ($J = 0$) there will be some apparent delay if the CO fragments are formed mostly in higher rotational levels.

In formaldehyde it is clear that the dissociation is a concerted process of the breaking of two CH bonds and the making of a new H–H bond. In a larger molecule such as glyoxal, $H_2C_2O_2$, the situation is more complex and the time-of-flight spectra show that there are two distinct channels for the formation of the fragments CH_2O and CO

$$H_2C_2O_2 \rightarrow H_2CO + CO \tag{8.9}$$

$$\xrightarrow{h\nu} HCOH + CO \tag{8.10}$$

where HCOH is hydroxymethylene.

A third quite different reaction has also been observed in these molecular beam experiments: the direct formation of H_2 according to

$$H_2C_2O_2 \xrightarrow{h\nu} H_2 + 2CO \tag{8.11}$$

This must be a concerted process rather than a secondary dissociation of formaldehyde or hydroxymethylene from the previous reactions, because of the high energy barriers of these secondary dissociations.

8.5.1.1 Population Inversion of Vibrational States. The planar molecule CH_3ONO dissociates in the S_1 ($n-\pi^*$) excited state to radicals $\dot{C}H_3O$ and $\dot{N}O$. A detailed analysis of the fragments shows that these are formed within 150 fs (close to vibrational motion) with a remarkable selectivity in the vibrational distribution. The jet-cooled reactant CH_3ONO absorbs light in its zero vibrational level of the ground state S_0, but the $\dot{N}O$ radical is formed in vibrationally excited states so that population inversion is actually obtained. According to the Boltzmann distribution all energy states should be equally populated at infinite temperature, but if the vibrational level $v = 3$, for example, is populated in preference to the lower levels ($v = 0$, 1 or 2) the conditions of population inversion are reached and the system can be used in principle as a 'chemical' laser. In the mathematical formalism of the Boltzmann distribution the temperature would then be negative, but this has of course no physical significance.

8.5.1.2 Spectral Resolution with Large Molecules. Although most molecular beam studies deal with small molecules, there are some observations of reactions of fairly large species which do not undergo dissociation. The intramolecular proton transfer in 2,5-bis(2-benzoxazolyl)hydroquinone has been observed through fluorescence measurements shown in Figure 8.27.

One important feature of the spectroscopy in molecular beams is nicely illustrated in this example, namely the very high resolution of absorption and emission spectra. This results from the low effective temperatures which can be reached through supersonic expansion.

8.6 VIBRATIONAL PHOTOCHEMISTRY WITH INFRARED LIGHT

Photochemical reactions take place from electronically excited states of molecules, these states being reached by absorption of light in the UV, VIS, or near IR regions of the spectrum. In this respect, 'light' is defined as electromagnetic radiation, and this extends to the IR, microwave and radiowave regions. The absorption of far IR light promotes molecules to excited vibrational states, not the electronically excited states. In most cases absorption of light of such very low photon energy does not lead to chemical changes, but special effects are observed when very intense IR laser light is used to irradiate a chemical sample. The vibrational energy levels of molecules can have relatively long lifetimes, so that sequential multiphoton absorption becomes possible when the light intensity is sufficiently high (Figure 8.28). It takes some 10 to 20 photons to reach the dissociation limit of the ground electronic state.

At one stage this 'vibrational' photochemistry held the promise of 'mode selectivity', the possibility of tuning the excitation wavelength to excite one particular vibrational mode in a polyatomic molecule. It would then be possible to select one particular bond for dissociation, simply by changing the wavelength of the irradiation beam according to the vibrational frequency (mode) of that bond. In practice this mode selectivity has not been observed. No matter which specific vibration is excited by multiphoton IR absorption, the energy is very quickly distributed statistically over all the available vibrational modes, so that the weakest bonds dissociate faster.

Although mode selectivity has not been reached in practice, isotopic separation is a reality. The C–H and C–D bonds for example have quite distinct vibrational stretching frequencies, so that IR light of the right wavelength will be absorbed by one isotopic species only. The difference in irradiation wavelength depends on the relative mass difference of the isotopes, it is of course greatest for H/D but is still significant for $^{12}C/^{14}C$ and even atoms of higher mass numbers. The wavelength of the IR light can be tuned to one of the specific vibrations, and dissociation via multiphoton absorption can be used as a means of isotopic separation.

Figure 8.28 illustrates the principle of multiphoton IR dissociation with a simplified potential energy diagram. There are in principle two ways to reach the dissociation limit from the zeroth vibrational level of the ground state.

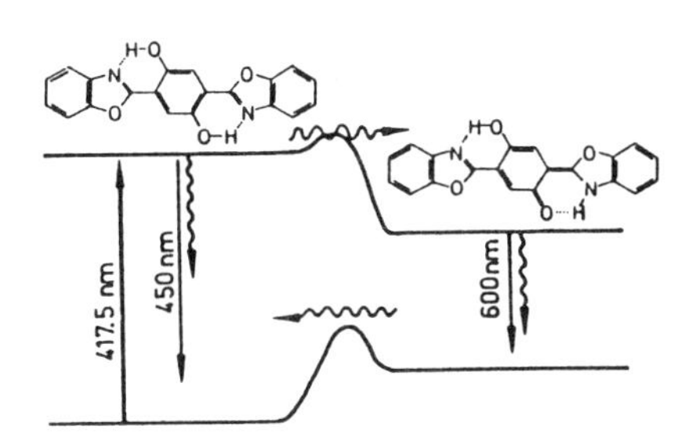

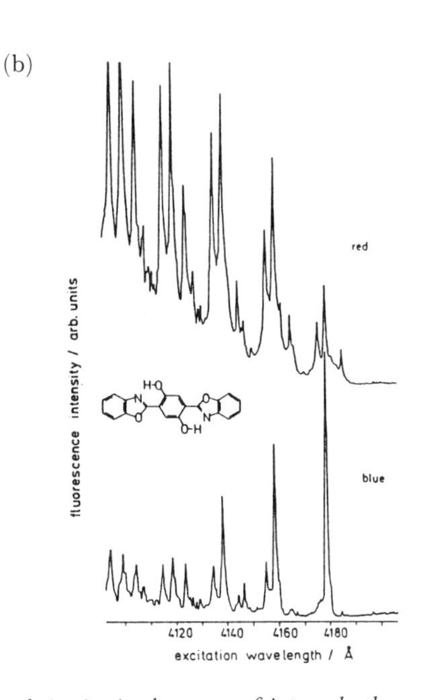

Figure 8.27 *Structures and energies of a hydroquinone derivative in the course of intramolecular proton transfer (a), and examples of fluorescence spectra of these species under jet-cooled conditions (b)*

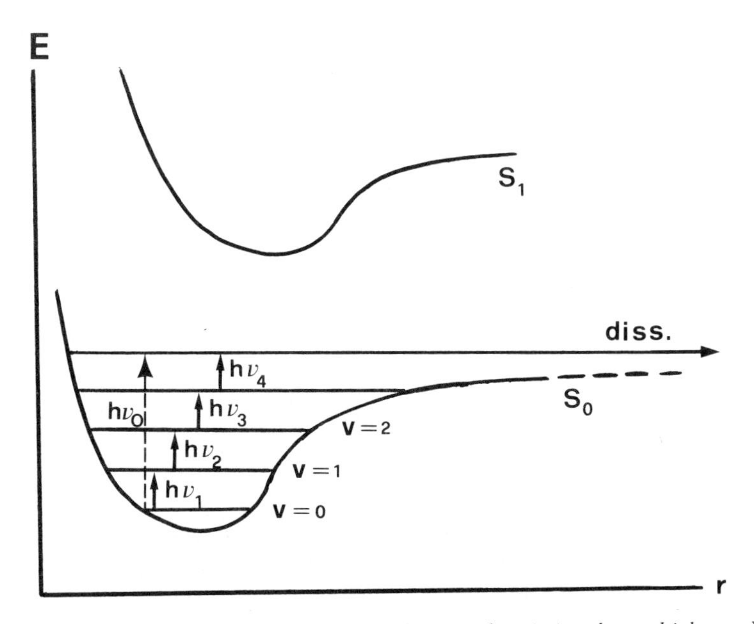

Figure 8.28 *Potential energy diagram of a ground state dissociation by multiphoton IR excitation. diss. = dissociation limit*

(1) Through the absorption of a single photon ($h\nu_0$) of near IR light, the photon energy being equal to the bond dissociation energy. The molecule is promoted to a very high vibrational state, and this process is known as 'overtone excitation'.
(2) Through the sequential absorption of several low-energy photons of far IR light. Figure 8.28 shows only four vibrational levels, but in reality there would be many more.

The choice of breakable bonds in such processes is rather limited, the relatively weak carbon–halogen bonds being of prime interest. The intensity of the excitation light must be very high; in practice powerful CO_2 lasers have been used for such reactions in many cases. When the tunability of the excitation light wavelength is essential, dye lasers provide the most suitable sources.

8.7 SPECTRAL HOLE BURNING

This somewhat dramatic name given to a special photochemical process should not be taken too literally. It does not involve any actual burning nor the making of any real holes. A better definition would be "spectro-selective photoreactivity", but the expression "hole burning" has become widely accepted.

The spectral hole is in fact the decrease in the absorbance of a sample in a narrow part of its absorption spectrum, a decrease which results from a photochemical reaction of molecules which absorb the irradiation light of a specific wavelength. This can take place mostly in solid samples in which the molecules are trapped in different environments, so that the overall absorption spectrum is a superposition of many site-selected spectra. Figure 8.29 gives an example of the change in the absorption spectrum described as 'hole burning': the spectral hole is located at around λ_0.

In solid samples these local decreases of absorbance can last indefinitely, but this is not true for liquids in which molecular motions would soon restore the equilibrium distribution.

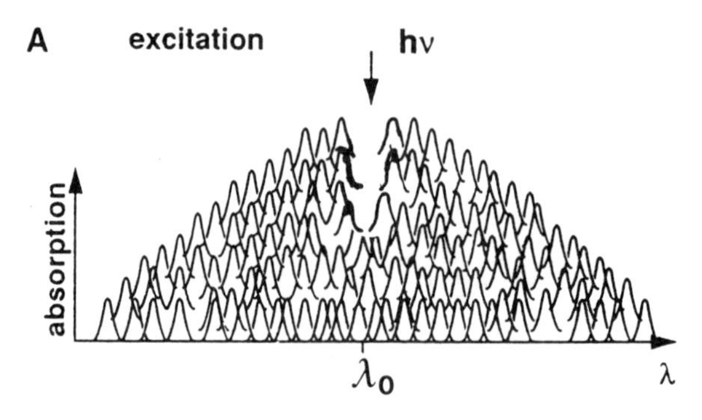

Figure 8.29 *Principle of spectral hole burning*

Spectral hole burning must be seen as a type of supramolecular photochemistry, because the interaction of the dye molecule (guest) with its environment (host) is of the greatest importance. The absorption spectrum of a guest molecule in a rigid host matrix at very low temperature (a few K) consists of a sharp zero-phonon line (ZPL) and a broad phonon wing. Only the former is useful for spectral hole burning since its wavelength is very well defined (Figure 8.30). The phonons are the vibrational excitations of the medium, so that the ZPL is somewhat similar to the 0–0 line of the absorption band of a molecule, higher excitations (0–v) being the phonon wing. The difference is that the number of vibrational modes of the medium surrounding a molecule is virtually infinite, therefore the phonon wing is a continuum.

The process of spectral hole burning relies on photochemical or photophysical changes in the selected molecules. An intramolecular hydrogen shift is an example of photochemical hole burning (Figure 8.31) while the rearrangement of hydrogen bonds in the surrounding matrix is its photophysical counterpart.

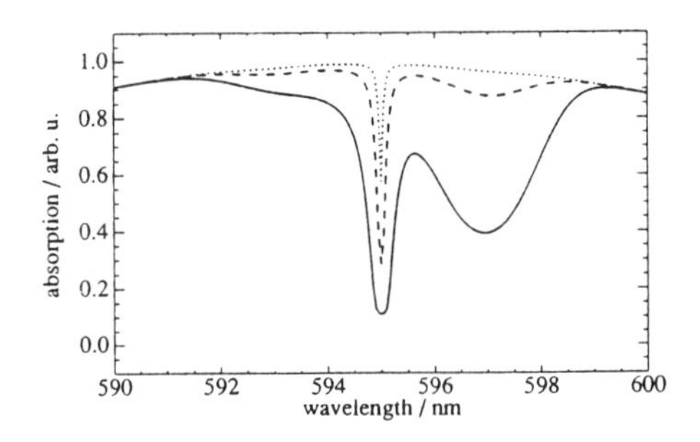

Figure 8.30 *Example of the absorption spectrum of a low-temperature solid*

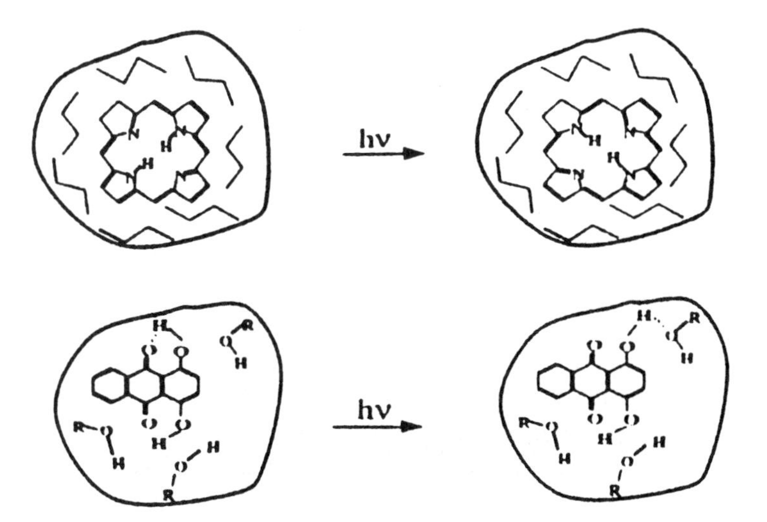

Figure 8.31 *Schemes of chemical and physical hole burning processes*

Different detection methods can be used for monitoring the spectral hole. Conventional absorption spectrophotometry has the disadvantage of rather low sensitivity, since it measures a small change in absorbance over a large background of transmitted light. Fluorescence spectroscopy is often preferable, being a zero background experiment. There are more sophisticated methods which rely on polarized light and holographic techniques which will not be considered here.

Interest in spectral hole burning comes from its potential application as high density computer memory. An external electric field can be used to shift the spectral holes so as to obtain an even larger number of detectable holes than with the excitation wavelength alone.

APPENDIX 2A

Experimental Evidence of the Photon Spin

When circularly polarized light falls on a quartz half-wave plate in the shape of a disc, a torque is developed. The transmitted light is of the opposite circular polarization. The half-wave plate disc can rotate as it hangs on a thin quartz fibre, and the magnitude of the photon spin can be deduced from the torsion angle α.

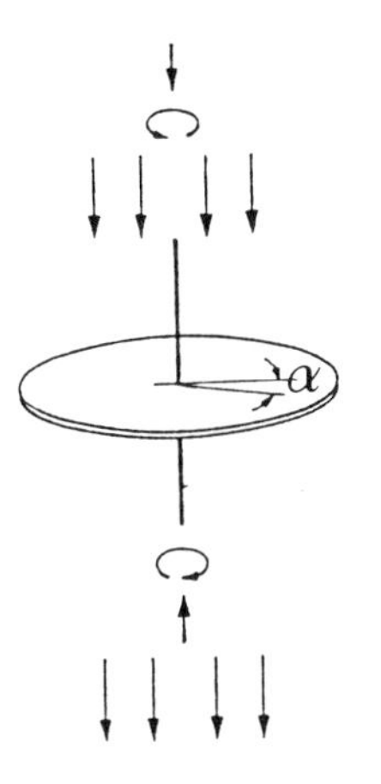

APPENDIX 3A

Relationships Between Emission Rate Constants and Absorption Coefficients

The natural radiative lifetime of a luminescence is

$$\tau_r = 1/k_f$$

k being the rate constant. The relationship with the extinction coefficient is

$$1/\tau_r = 2900 n^2 \bar{\nu}_0^2 \int \varepsilon \, d\bar{\nu}$$

in units of μm^{-1}, τ_r being given in s. This assumes however a mirror image relationship between the emission and absorption spectra. A useful rough estimate of τ_r can be obtained from ε_{max} only, according to

$$1/\tau_r \approx 10^4 \varepsilon_{max}$$

for a transition in the wavelength region of 200 to 500 nm.

APPENDIX 3B

Bandgap Energies of Some Semiconductors (see Figure below)

The potentials are referred to the normal hydrogen electrode (NHE). The energy levels for the oxidation and reduction of water at pH 7 are shown by horizontal lines. Energy scale in volts.

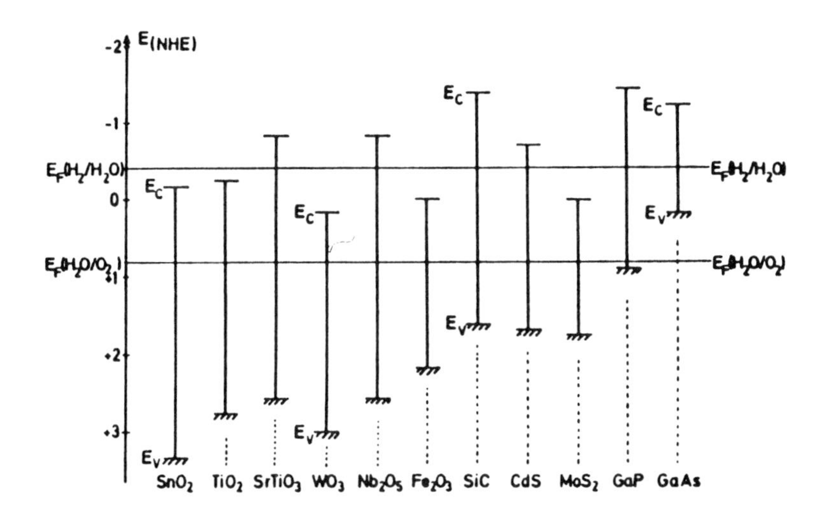

Work Functions W of Some Metals

Energies in eV. K 2.3; Na 2.75; Al 4.28; Cu 4.65; Pt 5.65.

Some Standard Electrode Potentials

These are referred to the NHE electrode (potential 0).
Saturated calomel electrode (SCE) 0.24 V
Silver chloride electrode (AgCl) 0.22 V

APPENDIX 7A

Polarizability and Hyperpolarizability

The polarizability is in general a tensor which relates the electric field vector to the induced dipole moment vector

$$\mu_i = \alpha E$$

It must therefore be defined by three numbers in space. Only in the case of an isotropic sphere can it be given by a single number which has the dimensions of a volume, L^3, given in m^3 or cm^3, or $Å^3$ at the molecular scale ($1\ cm^3 = 10^{24}\ Å^3$).

The induced dipole moment μ_i is not linear with the electric field at high field strengths and the total dipole moment is

$$m = \mu + \alpha E_0 + \tfrac{1}{2}\beta{:}E_0E_0 + \tfrac{1}{6}\gamma{\vdots}E_0E_0E_0 + \cdots$$

where β and γ are the first and second hyperpolarizabilities. These are third- and fourth-degree tensors, respectively.

APPENDIX 7B

Signal-to-noise Ratio

In practice any physical signal contains some 'noise' which is a random variation of the signal over the course of time. The signal-to-noise ratio S/N is defined as shown in the Figure below.

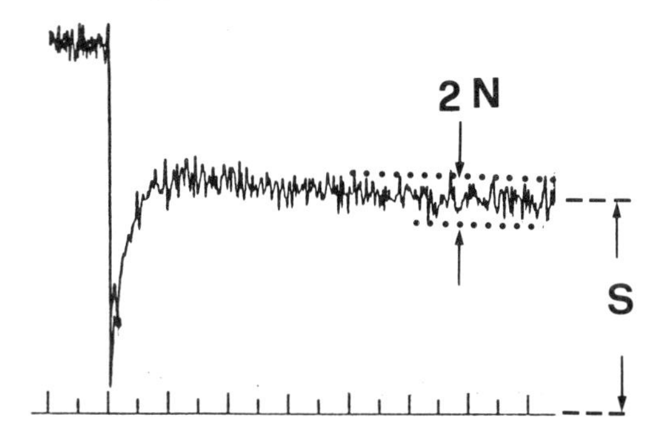

In kinetic experiments the S/N ratio can be improved in two ways.
(1) By the use of frequency filters (RC circuits) which cut off the high frequency noise from a low frequency signal. Care must be taken to avoid distortion of the signal. An RC circuit as shown below is a 'low pass filter' of time constant t = RC. This gives a rough value of the 'cut-off' frequency.

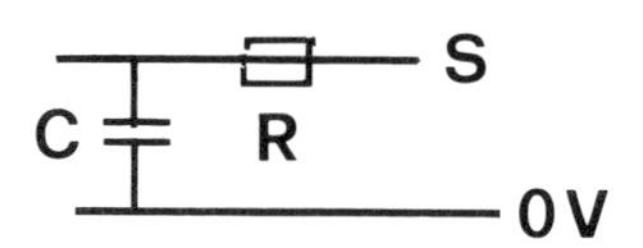

(2) By signal averaging in repetitive experiments. This technique is useful when the noise is truly random so that positive and negative deviations from the true signal are equally likely. The S/N ratio increases then with the square root of the number of experiments

$$S/N \propto \sqrt{n}$$

This technique does not present the potential danger of distortion of the signal mentioned above for frequency filters. An example of the improvement of the S/N ratio with averaging is shown in the Figure below.

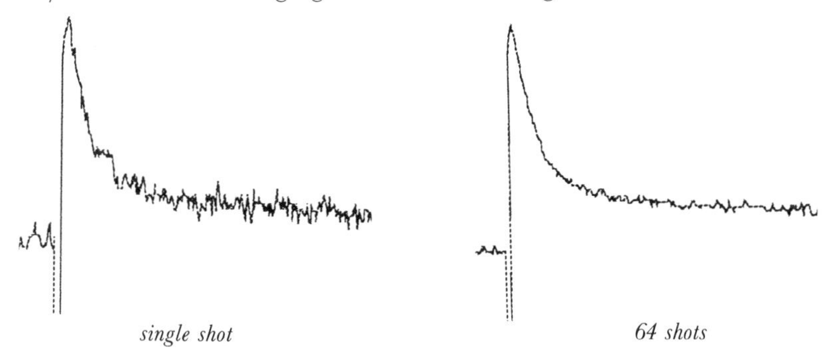

single shot *64 shots*

APPENDIX 7C

Procedures of Deoxygenation

In liquids, under normal conditions of temperature and pressure, the concentration of molecular oxygen $[O_2] \approx 10^{-4}$. This can be reduced either by bubbling an inert gas through the solution for a few minutes, or by the 'freeze-pump-thaw-shake' procedure which gives much better results. In the example of the preparation of a liquid sample for luminescence or flash photolysis experiments, the sample is held in a 'degassing' reservoir in which it is first frozen at 77 K through immersion in liquid nitrogen. The gases present in the cell–reservoir assembly are then pumped off on a vacuum line (see Figure below). The stopcock which links the assembly to the vacuum line is then closed and the degassing reservoir is warmed up to room temperature. The liquid is then thoroughly shaken to force the gases to bubble out, and then the whole procedure is repeated by freezing the sample again, *etc.*

Depending on the number of these cycles the final oxygen concentration is determined by the residual pressure of the pump. Quantitative analysis of traces of oxygen in liquids can be done by the measurements of long-lived luminescence lifetimes (*e.g.* pyrene fluorescence) and applying the Stern–Volmer equation.

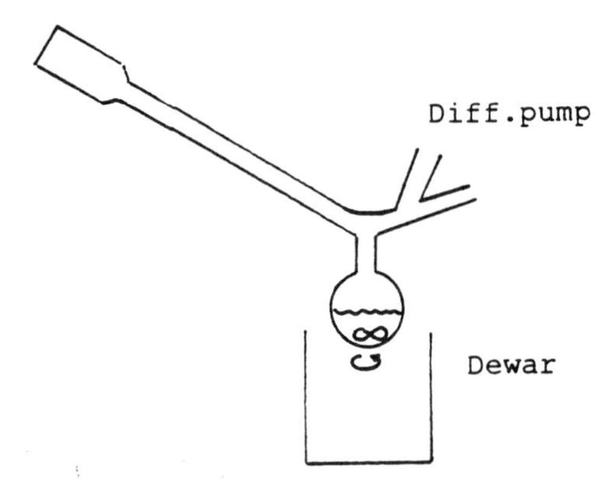

Further Reading

Chapter 1

J.N. Pitts Jr., F. Wilkinson, and G.S. Hammond, The 'Vocabulary' of Photochemistry, in *Adv. Photochem.*, 1963 **1**, 1.

J.W.T. Spinks and R.J. Woods, 'An Introduction to Radiation Chemistry', Wiley, New York, 1964.

Chapter 2

D.A. Brown, 'Quantum Chemistry', Penguin Library of Physical Sciences, London, 1972.

G.S. Monk, 'Light', Dover Publ. Inc., New York, 1963.

C.A. Coulson, B. O'Leary, and R.B. Mallion, 'Huckel Theory for Organic Chemists', Academic Press, London, 1978.

R.P. Feynman, R.B. Leighton, and M. Sands, 'The Feynman Lectures on Physics', Addison–Wesley Publ. Co., Reading, Massachusettes, 1965.

Chapter 3

'Radiationless Transitions', ed. S.H. Lin, Academic Press, New York, 1980.

J.T. Yardley, 'Introduction to Molecular Energy Transfer', Academic Press, New York, 1980.

J.N. Murrell, 'The Theory of the Electronic Spectra of Organic Molecules', Wiley, New York, 1963.

C.J.F. Böttcher, 'Theory of Electric Polarization', Elsevier, Amsterdam, 1973.

P. Suppan, *J. Photochem. Photobiol.*, 1990, **A50**, 293.

Z.I. Zink and K.-S. Shin, *Adv. Photochem.*, 1991, **16**, 119.

J.C.D. Brand and D.G. Williamson, *Adv. Phys. Org. Chem.*, 1963, **1**, 365.

Chapter 4

H. Okabe, 'Photochemistry of Small Molecules', Wiley Interscience, New York, 1978.

A. Gilbert and J. Baggott, 'Essentials of Moleculer Photochemistry', Blackwell, London, 1991.

N.J. Turro, 'Modern Molecular Photochemistry', Benjamin/Cummings, Menlo Park, 1978.

J.D. Coyle, 'Introduction to Organic Photochemistry, Wiley, Chichester, 1986.

'Photoinduced Electron Transfer', ed. M.A. Fox and M. Chanon, Elsevier, Amsterdam, 1988.

O.L. Chapman, *Adv. Photochem.*, 1963, **1**, 323.

M. Niclause, J. Lemaire, and M. Letort, *Adv. Photochem.*, 1966, **4**, 25.
K. Schaffner, *Adv. Photochem.*, 1966, **4**, 81.
C.R. Bock and E.A.K. von Gustorf, *Adv. Phtochem.*, 1977, **10**, 221.
A. Terenin and F. Vilessov, *Adv. Photochem.*, 1964, **2**, 385.
G. von Bünau and T. Wolff, *Adv. Photochem.*, 1988, **14**, 273.
J. Cornelisse, G.P. de Gunst, and E. Havinga, *Adv. Phys. Org. Chem.*, 1975, **11**, 225.
J.F. Ireland and P.A.H. Wyatt, *Adv. Phys. Org. Chem.*, 1976, **12**, 131.
G.B. Schuster and S.P. Schmidt, *Adv. Phys. Org. Chem.*, 1982, **18**, 187.
V. Balzani and V. Carassiti, 'Photochemistry of Coordination Compounds', Academic Press, London, 1970.
'Concepts of Inorganic Photochemistry', ed. A.W. Adamson and P.D. Fleischauer, Wiley-Interscience, New York, 1975.

Chapter 5

R.K. Clayton, 'Light and Living Matter', Krieger Publ. Co., New York, 1977.
K. Schaffner, S.E. Braslavsky, and A.R. Holwarth, *Adv. Photochem.*, 1990, **15**, 229.
T.J. Dougherty, *Adv. Photochem.*, 1992, **17**, 275.
N.Y. Dodonova, *J. Photochem. Photobiol.* 1993, **B18**, 111.
A.C. Wilbraham, *J. Chem. Educ.*, 1984, **61**, 540.

Chapter 6

T.H. James, 'The Theory of the Photographic Process', MacMillan Press, New York, 1966.
G.A. Delzenne, *Adv. Photochem.*, 1979, **11**, 1.
J.E. Byrd and J.M. Perona, *J. Chem. Educ.*, 1982, **59**, 335.
I. Irving Ziderman, *Chem. Br.*, 1986, **22**, 419.
J. Bailey and D.N. Rogers, *Chem. Br.*, 1985, **21**, 251.
R. Philips, 'Sources and Applications of Ultraviolet Radiation', Academic Press, London, 1983.
'Photochemical Energy Conversion', ed. J.R. Norris and D. Meisel, Elsevier, New York, 1989.
H. Levy, *Adv. Photochem.*, 1974, **9**, 369.
R. Hill and M.D. Archer, *J. Photochem. Photobiol.*, 1990, **51**, 45.
R. Dessauer and J.P. Paris, *Adv. Photochem.*, 1963, **1**, 275.
M.T. Spitler, *J. Chem. Educ.*, 1983, **60**, 330.
H. Niki and P.D. Maker, *Adv. Photochem.*, 1989, **15**, 69.

Chapter 7

C.A. Parker, 'Photoluminescence of Solutions', Elsevier, Amsterdam, 1968.
S.W. Beavan, J.S. Hargreaves, and D. Phillips, *Adv. Photochem.*, 1979, **11**, 207.
G. Porter and M.R. Topp, *Proc. R. Soc. London Ser. A.*, 1970, **315**, 163.

R.V. Bensasson, E.J. Land, and T.G. Truscott, 'Flash Photolysis and Pulse Radiolysis', Pergamon, London, 1983.
A.J. Kresge, *Acc. Chem. Res.*, 1990, **23**, 43.
S.E. Braslavsky and G.E. Heibel, *Chem. Rev.*, 1992, **92**, 1381.

Chapter 8

'Ultra-short Laser Pulses', ed. D.J. Bradley, G. Porter, and M.H. Key, The Royal Society, London, 1980.
H. Miyasaka, K. Morita, K. Kamada, and N. Mataga, *Chem. Phys. Lett.*, 1991, **178**, 504.
D.J. McGarvey, F. Wilkinson, D.R. Worrall, J. Hobley, and W. Shaik, *Chem. Phys. Lett.*, 1993, **202**, 528.
J.Z. Zhang, B.J. Schwartz, J.C. King, and C.B. Harris, *J. Am. Chem. Soc.*, 1992, **114**, 10921.
W. Frey and T. Elsaesser, *Chem. Phys. Lett.*, 1992, **189**, 565.
K.D.M. Harris, *Chem. Br.*, 1993, **29**, 132.
A.M. Brouwer, C. Eijckelhoff, R.J. Willemse, J.W. Verhoeven, W. Schuddeboom, and J.M. Warman, *J. Am. Chem. Soc.*, 1993, **115**, 2988.
H. Kuhn and D. Möbius, *Angew. Chem.*, 1971, **83**, 672.
M.J. Rosker, M. Dantus, and A.H. Zewail, *J. Chem. Phys.*, 1988, **89**, 6113 and 6128.
A.H. Zewail, *Science*, 1988, **242**, 1645.

Subject Index